IET CONTROL, ROBOTICS AND SENSORS SERIES 113

Design of Embedded Robust Control Systems Using MATLAB®/Simulink®

Other volumes in this series:

Design of Embedded Robust Control Systems Using MATLAB®/Simulink®

Petko H. Petkov, Tsonyo N. Slavov and Jordan K. Kralev

The Institution of Engineering and Technology

Published by The Institution of Engineering and Technology, London, United Kingdom

The Institution of Engineering and Technology is registered as a Charity in England & Wales (no. 211014) and Scotland (no. SC038698).

The Institution of Engineering and Technology
Michael Faraday House
Six Hills Way, Stevenage
Herts, SG1 2AY, United Kingdom

www.theiet.org

British Library Cataloguing in Publication Data
A catalogue record for this product is available from the British Library

ISBN 978-1-78561-330-2 (hardback)
ISBN 978-1-78561-331-9 (PDF)

Typeset in India by MPS Limited

To our teachers and students

Contents

Preface

The aim of the book

The aim of this book is to give the necessary knowledge about the implementation of MATLAB® and Simulink® in the development of embedded control systems. Together, MATLAB and Simulink present a sophisticated programing environment which may be used for the design as well as for the implementation of embedded control systems. In this book, the authors exploit the opportunity to generate automatically and embed control code from Simulink models which allows to develop quickly efficient and error free code. The automated code generation and the availability of powerful processors make possible the implementation of complex high-order controllers which achieve fast and high-performance closed-loop dynamics.

The book is oriented toward the application of modern Control Theory to the development of high-performance control laws which ensure good dynamics and robustness of the closed-loop system to plant uncertainties. The theoretical developments are reduced to the possible minimum the accent being put on the application issues. The basic results of Control Theory are given without proofs, and for more information, the reader is advised to consult the notes and references given at the end of the corresponding chapter. The presentation contains lots of nontrivial examples, which allow to illustrate the practical implementation of theoretical results. Most of the examples are taken from the area of motion control, but the book may also be used by designers in other areas.

The book covers mainly the design of linear controllers which are most frequently used in practice. This approach is justified by the principle of linearity of small increments which states that almost any natural process is linear in small amounts almost everywhere. Fortunately, as noted by Kostrikin and Manin [1], the small neighborhood in which this principle is valid is sufficiently large.

An important part of the book is the freely downloadable material which contains MATLAB and Simulink files for all examples presented in the corresponding chapters. The usage of this material can help in understanding the different issues arising in the analysis and design of embedded control systems.

Expected audience

The book is intended as a reference source for MSc and PhD students who study in the field of Control Engineering as well as for control engineers working in the industry. It can also be used as a reference for researchers in Control Engineering who are interested in the implementation of MATLAB and Simulink in the design of Control Systems. The first four chapters may also be used for a masters course on the design of embedded control systems.

The contents

The book consists of seven chapters and six appendices.

Chapter 1 presents a brief overview of the embedded control systems and the corresponding design process.

In Chapter 2, we describe several fundamental issues related to the development of plant model, like linearization, discretization, stochastic modeling, and identification. This chapter contains also a section on uncertainty modeling.

Chapter 3 is entirely devoted to the performance requirements and design limitations arising in embedded controller design. A significant part of this chapter are the sections on robust stability and robust performance analysis of uncertain systems.

In Chapter 4, we present in detail the design of five basic controllers used in the modern Control Theory: proportional-integral-derivative (PID) controllers, linear-quadratic-Gaussian (LQG) controllers, and linear-quadratic (LQ) regulators with $\mathcal{H}_\infty$ filters, $\mathcal{H}_\infty$, and μ controllers. For comparison purposes, all controllers are implemented on the same plant which represents the well-known cart–pendulum system. We consider the possible difficulties in the design of these controllers and give a comparison of the properties of corresponding closed-loop systems. These properties are illustrated by the hardware-in-the-loop (HIL) simulation of the closed-loop systems for the worst combination of plant parameters values.

In the last three chapters, we present three case studies which describe in detail the theoretical and practical issues arising in the design of three embedded control systems.

In Chapter 5, we consider the design of a low-cost control system for a two-tank plant. This chapter should be of interest to the readers who want to use low-cost processors in the design of embedded systems.

Chapter 6 is devoted to the robust control of a miniature helicopter. We consider the implementation of high-order controller which ensures robust performance of the closed-loop system in the presence of severe wind disturbances.

Finally, in Chapter 7, we present the design of embedded control system of a two-wheeled robot. In this case, we demonstrate experimentally the implementation of 30th-order controller which ensures robust stability and performance of the closed-loop system in the presence of plant uncertainty.

In Appendices A–D, we give some necessary facts from matrix analysis, linear system theory, stochastic processes, and identification of linear models, respectively. In Appendices E and F, we discuss important practical issues like connection between sensors and DSP and measurement of angular velocities by Hall encoders, respectively.

Acknowledgments

The authors are indebted to several people and institutions who helped them in the preparation of the book. We are particularly grateful to The MathWorks, Inc. for their continuous support, and to Professor Da Wei Gu from Leicester University, and Professor Nicolai Christov from Université Lille 1 for the numerous discussions and help. The assistance from IET editors and comments from the Reviewers are highly appreciated. We are also very grateful to Professor Tasho Tashev, Dean of the English Department of Engineering of the Technical University of Sofia, for his continuous support of our work in the recent years.

Using downloadable material

As a supporting online material for this book, we present seven folders with more than 250 .M- and .SLX-files intended for the design, analysis, and HIL simulation of embedded control systems which may found at

https://groups.google.com/d/forum/Book_PSK2018

In order to use the .M- and .SLX-files, the reader should have at his/her disposition MATLAB and Simulink version R2016a or higher, with Control System Toolbox™ and Robust Control Toolbox™. The programs described in Chapter 5 require the availability of Simulink Support Package for Arduino hardware and Arduino IDE. The programs presented in Chapters 6 and 7 require the installation of Code Composer Studio™ release 6.0.0, Control Suite version 3.3.9 and C2000 Code Generation Tools version 6.4.6.

Sofia, Bulgaria
December 2017

Petko Petkov
Tsonyo Slavov
Jordan Kralev

Chapter 1

Embedded control systems

In this chapter, we make a concise overview of embedded control systems and discuss some aspects of the corresponding hardware and software which is used in these systems. The embedded control systems are digital systems and their performance is affected by sampling and quantization errors. That is why, we present some basic elements of fixed-point and floating-point computations and describe the rounding errors associated with these computations. In case of fixed-point arithmetic, the emphasis is put on the scaling problem, which is the most important issue in using such arithmetic. We describe briefly the stages of embedded controller design, controller simulation, and implementation.

1.1 Introduction

According to a popular definition, every electrical or mechanical system that contains a controller, implemented on the base of digital processor, is called *embedded system*. The *embedded control systems* are systems in which are implemented algorithms for real-time control using feedback. The embedded control systems represent synthesis between modern digital technologies and control theory methods. It is necessary to distinguish between general purpose computational devices (computers) and the embedded system processors. The computers may execute a great number of programs with different purpose which are used to solve computational problems. On the other hand, the embedded system processor, which may be very powerful, performs only a special control program. Also, the embedded system controller may contain additional hardware which distinguishes it from the general purpose computer. Most of the contemporary embedded systems are implemented on the basis of microcontrollers—computational devices whose functional blocks (central processor, memory, input and output devices, and interface buses) are combined on a single chip. From hardware point of view, the microcontrollers represent a very-large-scale integration (VLSI) circuits.

To work successfully in real time, the embedded system should be developed so that the required computational cycle fits in the given time interval. For this aim, it is necessary to choose processor with appropriate computational efficiency, to develop fast control algorithm and to create interface schemes with minimum possible delay of signal transmission. On the second place, the embedded control system should possess stability in respect to external data. If, for instance, the data, necessary to obtain the result, do not arrive in time, then the system cannot produce the required result in time. In such a case, the system should not lock, but has to continue to give appropriate result in real time.

The process of developing embedded control systems has strongly multidisciplinary character, since it is required to perform a system integration of problems, associated with

- derivation of mathematical models of physical plants, sensors, communication hardware, and so on,
- development of methods for high performance control,
- embedding of control algorithms in different hardware and software platforms,
- carrying out communication with remote plants,
- solving problems associated with power supply.

The theoretical foundation of the embedded control systems is the theory of *hybrid systems*. The hybrid systems combine continuous processes described by differential or difference equations and discrete-event processes, described by finite automata. Such systems arise in a natural way in the control of continuous processes by the aid of digital devices and their investigation require a synthesis of control theory and computer science. The usage of complicated robust and adaptive control laws leads to the necessity of developing embedded system which work under the conditions of restricted processor accuracy and relatively small sampling interval. This represents a serious challenge both from theoretical and practical point of view.

1.2   Structure and elements of embedded control systems

1.2.1   Typical block diagram

From control theory point of view, the embedded control system for a continuous-time plant represents a closed-loop multivariable digital control system with a block diagram, shown in Figure 1.1. Very few plants encountered in practice are inherently digital, so we assume that generally the plant is continuous-time. Such systems are called *sampled-data systems*.

The aim of the control system is to ensure desired behavior of controlled plant outputs in accordance with the reference signals in presence of unknown disturbances and noises in the closed loop. In the general case, the plant has m analog control inputs produced by actuators and r analog outputs measured by the respective sensors. The measurements are corrupted by noises which, along with the plant disturbances, may significantly affect the closed-loop system behavior. The analog sensor

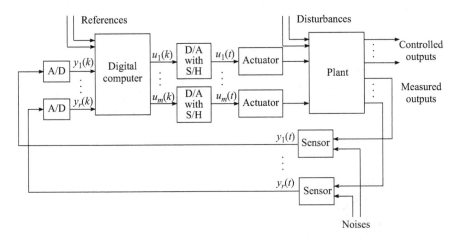

Figure 1.1   Block diagram of an embedded control system

signals are sampled with *sampling period* T_s by an impulse *sampler*, which produces digital representations $y_1(k)$, $y_2(k)$, ..., $y_r(k)$ of the measured signals, where $y_i(k)$ means the value of $y_i(t)$ for $t = kT_s$, $k = 1, 2, ...$. The sampler is a device, driven by the system clock, that is converting a continuous-time signal into a sequence of numbers. Normally, the sampler is combined into *analog-to-digital (A/D) converter*, which also quantizes the sequence of numbers into a finite precision number. (Note that the impulse sampler by itself does not have any physical meaning.) The digital measurement signals are used by the controller algorithms, embedded in the digital computer, to produce the digital control signals $u_1(k)$, $u_2(k)$, ..., $u_m(k)$. These signals are converted to the corresponding actuator inputs $u_1(t), u_2(t), ..., u_m(t)$ by using *digital-to-analog (D/A) converters*. The purpose of the D/A converters is to produce analog approximations of the digital signals using appropriate *reconstruction* algorithms. The analog signals $u_1(t), u_2(t), ..., u_m(t)$ are determined by *hold devices* during the sample period until the next sample arrives and the process of holding each of the samples is termed *sample and hold* (S/H). The analog signals $u_i(t)$ are used as actuator inputs to control the plant behavior. The work of the A/D and D/A converters is synchronized by the system clock with the work of the digital computer. Note that the block diagram may contain additional elements like *antialiasing filters* whose function is described in the next section. Also, it is possible that the outputs of the analog sensors are sampled at different periods and the system may have many controllers with different sampling periods. In the cases where the sensors have digital outputs or/and the actuators have digital inputs, the block diagram shown in Figure 1.1 may still be valid taking into account that the A/D and D/A conversion is performed inside the corresponding devices.

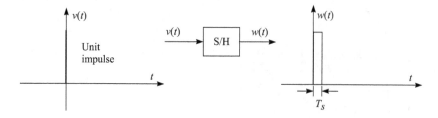

Figure 1.2   *Operation of the zero-order hold (ZOH)*

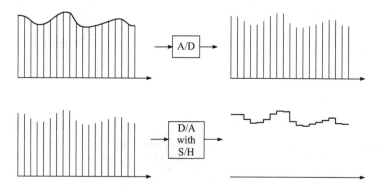

Figure 1.3   *Operation of the analog-to-digital converter and digital-to-analog converter with zero-order hold*

1.2.2 A/D and D/A conversion

In the simplest case, which is assumed in this book, the sample and hold process reduces to a *zero-order hold (ZOH)* whose operation is illustrated in Figure 1.2. The digital code of the signal at the sampling instant is converted into analog signal with magnitude corresponding to the value of the digital signal and duration, equal to the sampling period T_s. The frequency $\omega_s = 2\pi/T_s$ is called *sampling frequency*. The zero-order hold keeps constant the magnitude of the analog signal between two sampling instants.

The operation of A/D converter and D/A converter in case of using zero-order hold is illustrated in Figure 1.3. Commonly using a process of successive approximation, the A/D conversion (ADC) maps the analog input signal to a digital output. This digital value is composed of a set of binary values called *bits* (often represented by 0s and 1s). The set of bits represents a decimal or hexadecimal number that can be used by the microcontroller. The D/A converter transforms the digital code to signal samples and then converts the binary-coded digital signal to analog signal. It is seen that the D/A conversion with zero-order hold produces a staircase signal from the samples sequence.

The A/D converter has two functions:

1. Sampling of the analog signal: the continuous-time signal is replaced by a sequence of values equally spaced in the time. These values correspond to the amplitude of the continuous-time signal at sampling instants.
2. Quantization: the signal amplitude is approximated by a finite precision number coded with a binary sequence. Typically, the A/D converter has 8–24 bits resolution giving 2^8–2^{24} levels of quantization.

A drawback of the zero-order hold is that the output of the hold device is discontinuous. The discontinuities can excite poorly damped mechanical modes of the physical process and also cause wear in the actuators of the system. That is why in some cases, a more sophisticated hold device is used by allowing the continuous-time signal to be a higher order polynomial between the sampling points. Using a *first-order hold* the signal between the sampling points is obtained by a linear interpolation which leads to better reconstruction of the sampled signal.

1.2.3 Sensors

Sensor is a device that when exposed to a physical phenomenon (displacement, force, temperature, pressure, etc.) produces a proportional output signal (electrical, mechanical, magnetic, etc.). The term *transducer* is often used synonymously with sensors. However, ideally, a sensor is a device that responds to a change in the physical phenomenon. On the other hand, a transducer is a device that converts one form of energy into another form of energy. The new generation sensors involve smart material sensors, microsensors, and nanosensors.

Sensors can be classified as *passive* or *active*. In passive sensors, the power required to produce the output is provided by the sensed physical phenomenon itself (such as a thermometer), whereas the active sensors require external power source (such as a gyroscope). Furthermore, sensors are classified as analog or digital on the basis of the type of output signal. Analog sensors produce continuous signals that are proportional to the sensed parameter and typically require ADC before feeding to the digital controller. Digital sensors on the other hand produce digital outputs that can be directly interfaced with the digital controller. Often, the digital outputs are produced by adding an A/D converter to the sensing unit. If many sensors are required, it is more economical to choose simple analog sensors and interface them to the digital controller equipped with a multichannel A/D converter.

A number of static and dynamic factors must be considered in selecting a suitable sensor to measure the desired physical parameter. The following list involves the typical factors [2, Chapter 17]:

Range	Difference between the maximum and minimum value of the sensed parameter
Resolution	The smallest change the sensor can differentiate
Accuracy	Difference between the measured value and the true value
Precision	Ability to reproduce repeatedly with a given accuracy
Sensitivity	Ratio of change in output to a unit change of the input
Zero offset	A nonzero value output for no input
Linearity	Percentage of deviation from the best-fit linear calibration curve
Zero drift	The departure of output from zero value over a period of time for no input
Response time	The time lag between the input and output
Bandwidth	Frequency at which the output magnitude drops by 3 dB
Resonance	The frequency at which the output magnitude peak occurs
Operating temperature	The range in which the sensor performs as specified
Deadband	The range of input for which there is no output
Signal-to-noise ratio	Ratio between the magnitudes of the signal and the noise at the output

Models of sensors, appropriate for usage in embedded control system design, are described in Section 2.7.

1.2.4   Actuators

The purpose of actuators is to control a physical device or affect the physical environment. The three commonly used actuators are solenoids, motors, and servos. *Solenoids* are devices containing a movable iron core that is activated by a current flow. The movement of this core can then control some form of hydraulic or pneumatic flow. The next type of actuator is the *electric motors*. There are three main types: direct current (DC), alternating current (AC), and stepper motors. DC motors may be controlled by a fixed DC voltage or by pulse width modulation (PWM). In a PWM signal, such as shown in Figure 1.4, a voltage is alternately turned on and off while changing (modulating) the width of the on-time signal, or duty cycle. AC motors are generally cheaper than DC motors, but require variable frequency drive to control the rotational speed. Stepper motors move by rotating a certain number of degrees in response to an input pulse. *Servos* are DC motors with encapsulated electronics, built in gear and feedback for PWM control. The servos should perform fast changes in the position, velocity and acceleration. Most servo can rotate up to 90° or 180°, but some may perform a full revolution. The servos cannot rotate permanently in one direction, i.e., they cannot be used to actuate wheels, but they are precise in positioning and are convenient in control of mobile robots, drones, and so on.

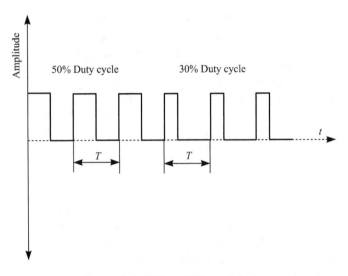

Figure 1.4   Pulse width modulation

1.2.5   Processors

For the purpose of embedded control systems, one of the following processor technology is frequently used.

Programmable logical controller (PLC)—The programmable logical controller is a specialized industrial controller. PLC is a device working in a real time: the inputs from switches and sensors are processed on the basis of a logical program and the output controller states change to steer machine or process. PLC may work under heavy operating conditions (dust, electrical interferences, vibration, and shock). These controllers can be used to implement sufficiently complicated control laws.

Microcontroller unit (MCU)—In essence, this is a small computer which constitutes of processor, memory, and periphery on a single chip. The components of most MCU are: processor, buses (address bus, data bus, and control bus), interruption controller, DMA (direct memory access) controller, ROM memory, RAM memory, timers, inputs, and outputs. Usually, MCU have ADC, digital outputs, digital inputs, PWM outputs. Due to the comparatively low microcontroller prices, they are widely used in mass production.

Some of the popular microcontrollers are from the PIC (programmable intelligent computer or peripheral interface controller) family. These are general purpose microcontrollers with affordable price which have applications in robotics, servo-controllers, and so on. Other microprocessors from this family include Parallax SX and the series Holtek HT48FxxE. Widely used are also microcontrollers from ARM (advanced RISC machines) family, which are based on the 32-bit architecture of RISC processors. This family keeps up the market part of about 75 percent of all 32-bit processors and almost 90 percent of all embedded processors. Examples of ARM-processors are Intel X-Scale, the family Philips LPC2000, Atmel AT91SAM7, ST Microelectronics STR710, and the series Freescale MCIMX27.

Digital signal processor (DSP)—It is designed for specialized applications like matrix operations, real-time filtration, sound and image processing, and so on. In essence, the DSPs are microcontrollers (they have ROM memory, RAM memory, serial and parallel interfaces, DMA controller, timers, controllers for interrupt processing and digital and analog periphery in some cases). For real-time control, the 32-bit controllers of Texas Instruments series C-2000: Delfino, Piccolo, InstaSPIN and F28M3x are used. The highest speed have the microcontrollers from Delfino family, which may be used to implement complex control laws. These microcontrollers have coprocessors, which are used to perform floating point computations.

Field programmable gate arrays (FPGA)—Integrated circuits, whose structure usually represents a two-dimensional array of logical blocks, buses for interconnection between them, auxiliary memory blocks and functional blocks (for instance, multipliers). The functions of these circuits are post determined by programing (electrical configuration). FPGA can perform parallel computations (MCU and DSP cannot). In practice, some of the fastest DSPs are built on FPGA chips. Also, there are ready processor kernels, which are programed on FPGA chips, on which it is possible to set very high clock frequency, i.e., their performance increases. The desired functionality of FPGA can be configured after the device is produced, installed in a product and even in some cases after the product is supplied to the user. This makes the FPGA a device, which is fundamentally different from the other devices on integrated circuits.

More details about the architecture and operation of microcontrollers and FPGA are given in Section 1.8.

1.2.6 Software

Practical implementation of control algorithm is not a trivial problem. Target hardware platform which executes the control calculation enriches with numerous dynamical effects the original plant. Such effects are time sampling, quantization, sample time variation, information transport delays, real numbers formatting, and rounding. These effects have to be accounted in plant modeling if not as controllable dynamics at least as uncertainty. Nowadays, there are many software components with open or closed source code, which accelerate programing process. Therefore, software design issues are left mainly in system configuration and compatibility between software and hardware components.

There are several ways for embedded programing. Formal languages are ubiquitous tool for programing. Expressions from such languages are derived from certain formal grammar and have tree like structure. To accelerate further system development, there are many visual languages too.

Ultimately, the goal of control engineer is to program the formula of his control algorithm into the hardware platform in order to start its autonomous execution. This goal requires securing of the following software components:

- *Periodic task execution*—Usually, a hardware timer is programed to generate an interrupt signal periodically. Then timer interrupt routine calls step function of control algorithm to update its internal state and output according to elapsed time.

- *Arithmetic support*—Target hardware arithmetic capabilities have to be considered when developing control calculation. For example, if the target does not have native support for floating-point numbers, some software libraries for such support can be included.
- *Input data drivers*—Heterogeneous nature of sensors requires several layers of driver programs to process their signals to enable control algorithm to access them. For digital sensors, the programmer should install software components for communication busses and related protocols. Analog sensors require some ADC configuration software.
- *Output data drivers*—In software, the control signal is a number which have to be transferred to an actuator device in order to take its effect. Usually, this is related to driver installation for D/C or PWM peripherals. Sometimes, the control signal is transmitted over digital communication to a smart actuator device.
- *External communication*—Since control system development is iterative process designer needs a continuous feedback about internal state of control algorithm and sensor measurements. To achieve this, the programmer should install some high-speed communication (USB, RS232, Ethernet) software component to transmit data to an operator workstation.

1.2.6.1 Operational systems

The real-time operational systems (RTOS) are designed to control real-time applications. This means control under specified time restrictions, i.e., strictly specified time limits for control system reaction. The RTOS classification, depending on the importance of the timely response, defines two types RTOS—with hard real time and with soft real time. The difference between these two types lies in that in the hard real-time the delay of control reaction will be fatal for the whole system. In the other case—control with soft real time—the delay after the specified period will lead in the worst case to system performance degradation, but effect is not irreparable. In the case of single-task systems, the usage of RTOS is not indispensable. The operational system is necessary when the embedded control system should perform several complex tasks or to connect it to other devices. The most frequently used RTOS are the Linux-based systems RTLinux and UTLinux, QNX, VxWorks, FreeRTOS, and others. Embedded RTOS provide limited user interface compared to conventional OS.

Technically the OS is a collection of interacting modules invoked by a central program (the kernel). All information in the kernel is represented as objects encapsulating data and allowed operations on that data. Most common OS objects are tasks, mutexes, semaphores, timers, events, message queues, mailboxes, and files. Usually there is a separation in OS between kernel and user program scope. User programs (or applications) have limited access to kernel objects in an effort to guarantee stability of the system. User programs use dedicated API (application programing interface) to access hardware platform devices which are highly abstracted. For example, the file object represents a communication device and sending data through that device is equivalent to write that data to the corresponding file object.

1.2.6.2   Protocols

Communication in digital systems synchronize the data across the system—between processor, memory, sensors, and other peripherals. Information channel can be any physical media connecting communication parties. Communication process can be of various complexity so it is organized as a hierarchy of protocols. *Protocol* is a set of rules and recipes for data processing in order to achieve some quality of service (QoS).

The communication between the processor, sensors, and actuators is realized by the aid of different protocols like RS232 (Recommended Standard 232), I²C (two-wire serial bus), CAN (controller area network) and SPI (serial peripheral interface). For wireless connection between the embedded system and other devices in local network, are used protocols like WLAN (wireless local area network), Bluetooth, ZigBee, and others.

1.3   Sampling and aliasing

MATLAB® file used in this section

File	Description
sampling_aliasing	Aliasing illustration

Is any information lost when sampling a continuous-time signal? The answer of this question is given by the Kotelnikov–Shannon *sampling theorem* which states that if the signal contains no frequencies above ω_0, then the continuous-time signal can be uniquely reconstructed from a periodically sampled sequence provided the sampling frequency ω_s is higher than $2\omega_0$. The frequency $\omega_N = \omega_s/2 = \pi/T_s$ plays an important role in the analysis of sampled systems and is called the *Nyquist frequency*. A signal whose frequency is above the Nyquist frequency cannot be reconstructed after the sampling.

Example 1.1. Aliasing
Consider the signal

$$y(t) = 4 \sin (2\pi t) + \sin (20\pi t + \pi/6).$$

If the sampling period T_s is chosen equal to 0.1 s then

$$y(kT_s) = 4 \sin (0.2k\pi) + \sin (2k\pi + \pi/6)$$
$$= 4 \sin (0.2k\pi) + 0.5$$

In this way, as a result of the sampling, the high-frequency component $\sin (20\pi t + \pi/6)$ has been shifted to the constant 0.5, i.e., the high frequency component appears as a signal of low frequency (in the given case zero). This phenomenon is called *aliasing* or *frequency folding*.

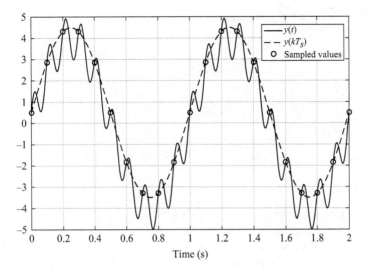

Figure 1.5 Aliasing effect for low sampling rate

The result of sampling is shown in Figure 1.5. It is seen that the sampled high frequency component appears as a constant upward shift of the low frequency component. In the given case, the signal contains frequency $\omega_0 = 20\pi$ rad/s and since the sampling frequency is $\omega_s = 2\pi/0.1 = \omega_0$, the condition imposed by the sampling theorem is not fulfilled ($\omega_0 > \omega_N = 10\pi$).   □

Due to the aliasing effect, it is necessary to choose the sampling frequency such that $\omega_N > \omega_{max}$, where ω_{max} is the maximum frequency of the signal which should be sampled. If the required sampling frequency is very high and cannot be implemented, then one has to remove all frequencies above the Nyquist frequency before sampling the signal. This may be done by using *antialiasing filters* which are analog filters whose bandwidth must be such that the attenuation above the Nyquist frequency is sufficiently high. It is stated in [3] that Bessel filters of orders 2–6 are in practice sufficient to eliminate most of the influence of higher frequencies. A second-order Bessel filter with bandwidth ω_B has the transfer function

$$\frac{\omega^2}{(s/\omega_B)^2 + 2\zeta\omega(s/\omega_B) + \omega^2}$$

with $\omega = 1.27$ and $\zeta = 0.87$. The Bessel filters has the property that they can be well approximated by a time delay. This is an advantage in the design of the controller since the dynamics of the antialiasing filter has to be included in the design of the sampled-data controller.

A more elaborated technique to avoid aliasing, suitable to the case of very low frequency sampling, is described in [4, Chapter 2].

1.4   Fixed-point arithmetic

MATLAB file used in this section

File	Description
`gyro_slop_bias`	Compute the slop and bias for a 14-bit gyro signal

Binary numbers are represented as either fixed-point or floating-point data types.

1.4.1   Fixed-point numbers

An n-bit binary word can represent integers between 0 and $2^n - 1$. Conversely, all the integers in this range can be represented by an n-bit binary word. This interpretation of binary words is called *unsigned integer* representation, because each word corresponds to a positive (or unsigned) integer. The drawback of the unsigned integers is that they can be used only in the case of positive integers.

A more general data type is the *fixed-point* representation. A fixed-point data type is characterized by the word length in bits, the position of the binary point, and whether it is signed or unsigned. The position of the binary point is the means by which fixed-point values are scaled and interpreted.

A binary representation of a generalized fixed-point number (either signed or unsigned) is shown in Figure 1.6 where the bit b_i is the ith binary digit. The fixed-point number is represented as

$$(\underbrace{b_{n-1} \ldots b_m}_{integer \ part} \bullet \underbrace{b_{m-1} \ldots b_0}_{fractional \ part})_2$$

where the number n is the word length in bits, $k = n - m$ is the *integer* part length and m is the *fractional* part length. The subscript 2 indicates the radix used. The integer part of the number, $b_{n-1}b_{n-2} \ldots b_m$ is separated from the fractional part $b_{m-1}b_{m-2} \ldots b_0$ by the *binary point* or *radix point*. The bit b_{n-1} is termed the *most significant*, or highest, bit (MSB) and b_0 is termed the *least significant*, or lowest, bit (LSB). The fixed-point number has the decimal value

$$N = \sum_{j=0}^{n-1} b_j 2^{j-m} = b_{n-1}2^{n-1} + b_{n-2}2^{n-2} + \cdots + b_{m+1}2^1 + b_m 2^0$$

$$+ b_{m-1}2^{-1} + \cdots + b_1 2^{-m+1} + b_0 2^{-m} \tag{1.1}$$

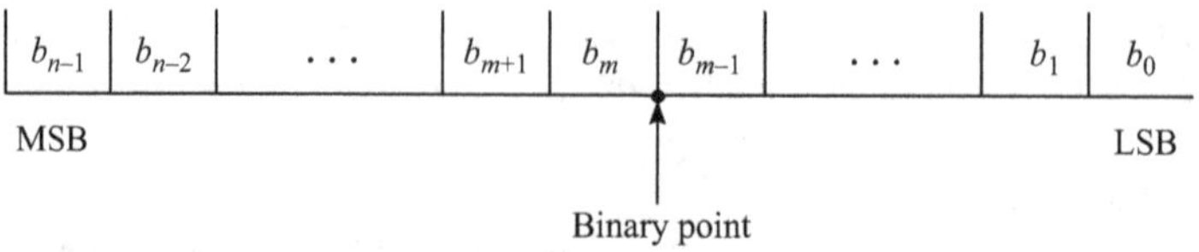

Figure 1.6   *Fixed-point number*

As an example, consider the fixed-point number 1011.011 which has word length $n = 7$ and fractional part length $m = 3$. It has the decimal value

$$1011.011 = 1(2^3) + 0(2^2) + 1(2^1) + 1(2^0) + 0(2^{-1}) + 1(2^{-2}) + 1(2^{-3})$$
$$= 11.375.$$

Particular case of the fixed-point numbers are the integer numbers for which $m = 0$.

Fixed-point data types can be either signed or unsigned. Signed binary fixed-point numbers are typically represented in one of the following ways:

- Signed-magnitude.
- One's complement.
- Two's complement.

In signed-magnitude representation, the sign and magnitude are specified separately. The first digit is the sign digit and the remaining $(n - 1)$ represent the magnitude. In the binary case, the sign bit is usually selected to be 0 for positive numbers and 1 for negative ones.

The ones' complement of a binary number is defined as the value obtained by inverting all the bits in the binary representation of the number (swapping 0s for 1s and vice versa). For instance, the ones' complement of 10111 is 01000. The ones' complement can represent integers in the range $-(2^{n-1} - 1)$ to $+(2^{n-1} - 1)$.

The two's complement of an n-bit number is another way to interpret a binary number. In two's complement, positive numbers always start with a 0 and negative numbers always start with a 1. If the leading bit of a two's complement number is 0, the value is obtained by calculating the standard binary value of the number. If the leading bit of a two's complement number is 1, the value is obtained by assuming that the leftmost bit is negative, and then calculating the binary value of the number. For example,

$$001 = (0 + 0 + 2^0) = (0 + 0 + 1) = 1$$
$$010 = (0 + 2^1 + 0) = (0 + 2 + 0) = 2$$
$$011 = (0 + 2^1 + 2^0) = (0 + 2 + 1) = 3$$
$$100 = ((-2^2) + 0 + 0) = (-4 + 0 + 0) = -4$$
$$101 = ((-2^2) + 0 + (2^0)) = (-4 + 0 + 1) = -3$$
$$110 = ((-2^2) + (2^1) + 0) = (-4 + 2 + 0) = -2$$
$$111 = ((-2^2) + (2^1) + (2^0)) = (-4 + 2 + 1) = -1$$

It is easy to check that in two's complement representation, an n-bit word represents integers from -2^{n-1} to $2^{n-1} - 1$.

The range of representable numbers for two's complement fixed-point numbers is illustrated in Figure 1.7.

Two's complement is the most common representation of signed fixed-point numbers and is the only representation used in MATLAB.

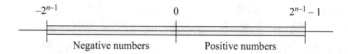

$$-2^{n-1} \qquad\qquad 0 \qquad\qquad 2^{n-1}-1$$

Negative numbers   Positive numbers

Figure 1.7   Range of representable numbers for two's complement numbers

Fixed-point numbers can be created in MATLAB by the aid of Fixed-Point Designer™ using the function `fi`. This function produces a fixed-point number with default signedness and default word and fraction lengths. The default value of the word length is 16. For example, the fixed-point representation of $-1/3$ is

```
a = fi(-1/3)

a =

  -0.333328247070313

          DataTypeMode: Fixed-point: binary point scaling
            Signedness: Signed
            WordLength: 16
        FractionLength: 16
```

It is possible to specify the signedness (1 for signed, 0 for unsigned) and the word and fraction length as follows:

```
fi(-1/3,1,15,12)

ans =

  -0.333251953125000

          DataTypeMode: Fixed-point: binary point scaling
            Signedness: Signed
            WordLength: 15
        FractionLength: 12
```

Several other properties of the `fi` object are described in [5].

The binary representation of the fixed-point number a may be displayed by the command

```
a.bin

ans =

111101010101011
```

The generation of unsigned fixed-point numeric object is done by the function `ufi`.

A particular case of the fixed-point numbers are the integers. In MATLAB, the integers may be represented as unsigned or signed 8, 16, 32, or 64 bit variables using

Table 1.1   MATLAB integer representation

Number of bits	Unsigned integers	Signed integers
8	uint8	int8
16	uint16	int16
32	uint32	int32
64	uint64	int64

the corresponding functions, shown in Table 1.1. If, for instance, $x = -81.3$, its 16 bit integer representation is obtained as

```
ix = int16(x)

ix =

    -81
```

The minimum and maximum representable numbers are determined by the functions intmin and intmax. For instance, the commands intmin('uint32') and intmax('uint32') produce the minimum and maximum 32-bit unsigned integers, respectively, representable in the workspace of MATLAB.

1.4.2   Scaling

Since fixed-point numbers and results of arithmetic operations are stored in registers with a fixed length, there is a finite number of distinct values which can be represented within the computer arithmetic unit. Let N_{min} and N_{max} denote the smallest and the largest representable fixed-point numbers. The interval $[N_{min}, N_{max}]$ is called *range* of the representable numbers. Any arithmetic operation that attempts to produce a result larger than N_{max} or smaller than N_{min} will give rise to an erroneous result. In such cases the arithmetic unit will produce a message which is called an *overflow* in the first case and *underflow* in the second case.

In programing with the fixed-point notation, special care must be taken to avoid overflow and underflow problems while maintaining a suitable precision. For this reason the fixed-point numbers must be *scaled*. It is possible to perform scaling by changing the position of the binary point of a fixed-point number or by implementing an arbitrary linear scaling. Both options are described briefly below.

A fixed-point number can be represented by the general *slope and bias encoding scheme*

$$\text{real-world value} = (\text{slope} \times \text{integer}) + \text{bias},$$

where the slope can be expressed as

$$\text{slope} = \text{slope adjustment factor} \times 2^{\text{fixed exponent}}.$$

The integer is the raw binary number, in which the binary point is assumed to be at the far right of the word.

The slope and bias together represent the scaling of the fixed-point number. In a number with zero bias, only the slope affects the scaling. A fixed-point number that is only scaled by binary point position is equivalent to a number in [Slope Bias] representation that has a bias equal to zero and a slope adjustment factor equal to one. This is referred to as *binary point-only scaling* or *power-of-two scaling*:

$$\text{real-world value} = 2^{\text{fixed exponent}} \times \text{integer}$$

or

$$\text{real-world value} = 2^{-\text{fraction length}} \times \text{integer}.$$

Fixed-Point Designer supports both binary point-only scaling and [Slope Bias] scaling.

1.4.2.1   Binary-point-only scaling

Binary-point-only or power-of-two scaling involves moving the binary point within the fixed-point word. The advantage of this scaling mode is to minimize the number of processor arithmetic operations.

With binary-point-only scaling, the components of the general slope and bias formula have the following values:

- $F = 1$
- $S = F2^E = 2^E$
- $B = 0$

The scaling of a quantized real-world number is defined by the slope S, which is restricted to a power of two. The negative of the power-of-two exponent is the fraction length (the number of bits to the right of the binary point). For Binary-Point-Only scaling, the fixed-point data types can be specified as

```
signed types fixdt(1, WordLength, FractionLength)
unsigned types fixdt(0, WordLength, FractionLength)
```

As mentioned previously, integers are a special case of fixed-point data types. Integers have a trivial scaling with slope 1 and bias 0, or equivalently with fraction length 0. The integers are specified as

```
signed integer fixdt(1, WordLength, 0)
unsigned integer fixdt(0, WordLength, 0)
```

1.4.2.2   Slope and bias scaling

When one scales by slope and bias, the slope S and bias B of the quantized real-world number can take on any value. The slope must be a positive number. Using slope and bias, the fixed-point data types are specified as

```
fixdt(Signed, WordLength, Slope, Bias)
```

1.4.2.3   Unspecified scaling

The fixed-point data types with an unspecified scaling are set as

```
fixdt(Signed, WordLength)
```

Simulink® signals, parameters, and states must never have unspecified scaling. When scaling is unspecified, one has to use some other mechanism such as automatic best precision scaling to determine the scaling that the Simulink software uses.

Example 1.2. Slope and bias scaling of a gyro sensor output

A microelectromechanical system (MEMS) gyroscope provides 14-bit fixed-point measurements of angular velocity. It is necessary to scale the fixed-point signal taking into account that the angular velocity may vary in the range $[-180, 180]$ deg/s.

First, enter the endpoints, signedness, and word length.

```
lower_bound = -180;
upper_bound = 180;
is_signed = true;
word_length = 14;
```

To find the range of a `fi` object with a specified word length and signedness, one may use the `range` function from Fixed-Point Designer.

```
[Q_min, Q_max] = range(fi([],is_signed, word_length, 0));
```

To determine the slope and bias, it is necessary to solve the system of equations, written in MATLAB notation as

```
lower_bound = slope * Q_min + bias
upper_bound = slope * Q_max + bias
```

These equations may be rewritten in matrix/vector form as

$$
\begin{bmatrix} \texttt{lower\_bound} \\ \texttt{upper\_bound} \end{bmatrix} = \begin{bmatrix} \texttt{Q\_min 1} \\ \texttt{Q\_max 1} \end{bmatrix} \begin{bmatrix} \texttt{slope} \\ \texttt{bias} \end{bmatrix}
$$

The vector containing the resulting slope and bias is computed by the following command lines:

```
A = double([Q_min 1; Q_max 1]);
b = double([lower_bound; upper_bound]);
x = A\b;
```

The slope, or precision, is

```
slope = x(1)

slope =

   0.021973997436367
```

and the bias is

```
bias = x(2)

bias =

    0.0109869987181835
```

It is convenient to create a `numerictype` object with slope and bias scaling which will be used in the generation of `fi` object.

```
T = numerictype(is_signed,word_length,slop,bias)

T =

          DataTypeMode: Fixed-point: slope and bias scaling
           Signedness: Signed
           WordLength: 14
                Slope: 0.021973997436366965
                 Bias: 0.010986998718183483
```

Now, it is easy to create a `fi` object with `numerictype` T.

```
a = fi(-160,T)

a =

  -160.003662332906

          DataTypeMode: Fixed-point: slope and bias scaling
           Signedness: Signed
           WordLength: 14
                Slope: 0.021973997436366965
                 Bias: 0.010986998718183483
```

Finally, it is appropriate to verify that the created `fi` object has the correct specifications by finding the range of a.

```
range(a)

ans =

   -180    180

          DataTypeMode: Fixed-point: slope and bias scaling
           Signedness: Signed
           WordLength: 14
                Slope: 0.021973997436366965
                 Bias: 0.010986998718183483
```

1.4.3 Range and precision

The range of representable numbers for a two's complement fixed-point number of word length n, scaling S and bias B is shown in Figure 1.8.

Because a fixed-point data type represents numbers within a finite range, overflows and underflows can occur if the result of an operation is larger or smaller than

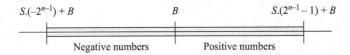

Figure 1.8   Range of representable numbers for scaled two's complement numbers

the numbers in that range. As shown by Example 1.2, the overflows and underflows can be avoided by appropriate scaling of the corresponding variables.

The precision of a fixed-point number is the difference between successive values representable by its data type and scaling, which is equal to the value of its least significant bit. The value 2^{-m} of the least significant bit, and therefore the precision of the number, is determined by the number m of fractional bits. A fixed-point value can be represented to within half of the precision of its data type and scaling. For example, a fixed-point representation with eight bits to the right of the binary point has a precision of 2^{-8} or 0.00390625, which is the value of its least significant bit. Any number within the range of this data type and scaling can be represented to within $(2^{-8})/2$ or 0.001953125, which is half the precision.

Fixed-Point Designer software currently supports the following rounding methods.

- `Ceiling` rounds to the closest representable number in the direction of positive infinity.
- `Convergent` rounds to the closest representable number. In the case of a tie, convergent rounds to the nearest even number. This is the least biased rounding method provided by the toolbox.
- `fix rounds` to the closest representable number in the direction of zero.
- `Floor`, which is equivalent to two's complement truncation, rounds to the closest representable number in the direction of negative infinity.
- `Nearest` rounds to the closest representable number. In the case of a tie, nearest rounds to the closest representable number in the direction of positive infinity. This rounding method is the default for `fi` object creation and `fi` arithmetic.
- `Round` rounds to the closest representable number. In the case of a tie, the round method rounds:
  - positive numbers to the closest representable number in the direction of positive infinity.
  - negative numbers to the closest representable number in the direction of negative infinity.

Recommendations about the choice of rounding methods for fixed-point numbers are given in [5].

1.4.4   Fixed-point arithmetic operations

1.4.4.1   Addition and subtraction

In the addition of two fixed-point numbers, it may be necessary to have an additional (carry) bit to correctly represent the result. For this reason, when adding two n-bit numbers (with the same scaling), the resulting value has an extra bit compared to the two operands used. For example, consider the addition of the numbers 0.3749 and 0.5681 which are represented in fixed-point arithmetic with word length 12 and fraction part length 8. The result has word length 13 and fraction part length 8.

```
a = fi(0.3749,0,12,8)

a =

   0.375000000000000

          DataTypeMode: Fixed-point: binary point scaling
           Signedness: Unsigned
           WordLength: 12
        FractionLength: 8

b=fi(0.5681,0,12,8)

b =

   0.566406250000000

          DataTypeMode: Fixed-point: binary point scaling
           Signedness: Unsigned
           WordLength: 12
        FractionLength: 8

c = a + b

c =

   0.941406250000000

          DataTypeMode: Fixed-point: binary point scaling
           Signedness: Unsigned
           WordLength: 13
        FractionLength: 8

a.bin

ans =

000001100000

b.bin

ans =

000010010001

c.bin

ans =

0000011110001
```

If one adds or subtracts two numbers with different precision, the radix point first needs to be aligned to perform the operation. The result is that there is a difference of more than one bit between the result of the operation and the operands. For instance,

```
a = fi(0.3634,0,12,8);
b = fi(5.2987,0,16,12);
c = a + b
```

```
c =

    5.661865234375000

        DataTypeMode: Fixed-point: binary point scaling
         Signedness: Unsigned
         WordLength: 17
     FractionLength: 12
```

Fixed-point subtraction is equivalent to adding while using the two's complement value for any negative values. To compute the negative of a binary number using two's complement, one may perform the following operations:

1. Take the one's complement, or "flip the bits."
2. Add a 2^{-m} using binary arithmetic, where m is the fraction length.
3. Discard any bits carried beyond the original word length.

For instance, the negative of 01101 (13) is 10011 ($-2^4 + 2^1 + 2^0 = -13$).

1.4.4.2 Multiplication

In the general case, a full precision product of two fixed-point numbers requires a word length equal to the sum of the word lengths of the operands. In the following example, the word length of the product c is equal to the word length of a plus the word length of b. The fraction length of c is also equal to the fraction length of a plus the fraction length of b.

```
a = fi(4.2961,1,20)
```

```
a =

    4.296096801757813

        DataTypeMode: Fixed-point: binary point scaling
         Signedness: Signed
         WordLength: 20
     FractionLength: 16
```

```
b = fi(2.167,1,18)
```

```
b =

    2.166992187500000
```

```
          DataTypeMode: Fixed-point: binary point scaling
            Signedness: Signed
            WordLength: 18
        FractionLength: 15
c = a*b

c =

   9.309608206152916

          DataTypeMode: Fixed-point: binary point scaling
            Signedness: Signed
            WordLength: 38
        FractionLength: 31
```

1.5   Floating-point arithmetic

A major disadvantage of the fixed-point numbers is that their range is much less than the range of floating-point values with equivalent word sizes. Also, the scaling of floating-point numbers is done automatically which facilitates the usage of floating-point arithmetic.

1.5.1   Floating-point numbers

The floating-point number system F is characterized by the base b, the *precision p* and the *exponent range* e_{min}, e_{max}. Here, b and p are positive integers, e_{min} is a negative integer and e_{max} is a positive integer. In this system, each *p-digit base b floating-point number* is represented in the *normalized form*

$$x = \pm 0.d_1 d_2 \ldots d_p \times b^e$$
$$= \pm \left(\frac{d_1}{b} + \frac{d_2}{b^2} + \cdots + \frac{d_p}{b^p} \right) \times b^e$$
$$= \pm f \times b^e,$$

where

$$1 \leq d_1 < b,$$

$$0 \leq d_i < b \quad i = 2, 3, \ldots, p$$

and $e_{min} \leq e \leq e_{max}$. The integer e is called the *exponent* and the number f—the *fractional part* or *mantissa*.

The most frequently used number bases are 2, 8, 10, and 16.

The smallest positive number, represented in the floating-point system, is

$$m = 2^{e_{min}-1}.$$

s	e (8 bits)	f (23 bits)

0   1   8 9   31

Figure 1.9   Single precision 32-bit word

s	e (11 bits)	f (52 bits)

0   1   11 12   63

Figure 1.10   Double precision 64-bit word

Table 1.2   IEEE arithmetic parameters

Precision	p	E_{min}	E_{max}	*bias*	ε
Single	24	−126	127	+127	$2^{-24} \approx 5.96 \times 10^{-8}$
Double	53	−1,022	1,023	+1,023	$2^{-53} \approx 1.11 \times 10^{-16}$

and the largest one is

$$M = b^{e_{max}}(1 - b^{-p}).$$

On some computers two floating-point systems are used which are called *single precision* and *double precision*. These system are characterized by different values of p, e_{min} and e_{max}.

1.5.2   IEEE arithmetic

The binary floating-point arithmetic standard 754-2008 [6], or subset of it, is commonly called "IEEE arithmetic." Virtually all modern processors implement IEEE arithmetic. In this standard, the single precision arithmetic is characterized by $b = 2$ and $p = 24$. The corresponding 32-bit word is organized as shown in Figure 1.9.

Two 32-bit words for double precision arithmetic are organized as shown in Figure 1.10.

A floating-point number is represented in IEEE arithmetic as

$$(-1)^s 2^E (b_0.b_1 b_2 \dots b_{p-1}),$$

where $s = 0$ or 1 determines the number sign, E is any integer between E_{min} and E_{max}, $b_i = 0$ or 1 for $i = 1, \dots, p - 1$, $e = E + bias$, where *bias* is used to avoid having a bit, corresponding to the exponent sign. Note that the bit b_0 is not used explicitly, since the floating-point number is normalized, i.e., b_0 is always equal to 1.

The parameters of the single precision and double precision arithmetic are shown in Table 1.2.

1	01111101	01010101010101010101011

0 1   8 9   31

Figure 1.11   Binary presentation of −1/3 in single precision

Example 1.3. Representation of −1/3 in IEEE arithmetic

The number −1/3 is represented in the single precision IEEE arithmetic as shown in Figure 1.11. This is checked in MATLAB by using the command lines

```
format hex, single(-1/3)

ans =

  beaaaaab
```

and converting the hexadecimal number *beaaaaab* to binary number.   ❏

A special bit pattern called NaN ("Not a Number") with $e = 255$ and $f \neq 0$ is generated by operations like $0/0$, $0 \times \infty$, ∞/∞, and $(+\infty) - (+\infty)$. Arithmetic operations involving a NaN return NaN as an answer.

The infinity symbol is represented by $f = 0$ and the same exponent field as a NaN, the sign bit distinguishing between $\pm\infty$.

Zero is represented by $e = 0$ and $f = 0$, with the sign bit providing distinct representation for $+0$ and -0 which can be useful in some cases.

In MATLAB, the permanent variables realmin and realmax represent the smallest positive and the largest positive normalized floating point numbers, respectively.

Let x be any real number which satisfies

$$m \leq x \leq M$$

The *rounded value* $fl(x)$ of x is defined as the floating-point number which is nearest to x and the transformation $x \rightarrow fl(x)$ is called *rounding*. There are several ways to break ties when x is equidistant from two floating-point numbers, including taking $fl(x)$ to be the number of larger magnitude (round away from zero) or the one with an even last digit d_p (round to even). The default rounding mode in IEEE arithmetic is to round to the nearest representable number, with rounding to even (zero least significant digit), in the case of a tie. Other supported modes are rounding to plus or minus infinity and rounding to zero. The last mode is called *truncation* or *chopping*. If $fl(x) > M$ one says that $fl(x)$ *overflows* and if $0 < fl(x) < m$ than it *underflows*.

It is possible to show that every real number x lying in the range of F can be approximated by an element of F with a relative error no larger than

$$\varepsilon = \frac{1}{2}b^{-p}$$

in the case of rounding to even and

$$\varepsilon = b^{-p}$$

in the case of truncation. The quantity ε is called *unit roundoff*. In case of IEEE arithmetic, the values of ε for single precision and double precision computations are given in the Table 1.1. Notice that $\varepsilon \gg m$.

In MATLAB, eps without arguments produces the quantity b^{-p} which is two times larger than the unit roundoff ε corresponding to correct rounding.

If x lies in in the range of F one has that

$$fl(x) = x(1 + \delta), \quad |\delta| \leq \varepsilon. \tag{1.2}$$

1.5.3 *Floating-point arithmetic operations*

Assume that x, y are floating-point numbers. According to the IEEE standard, all arithmetic operations involving x and y are to be performed as if they were first calculated to infinite precision and then rounded according to one of the four modes. This lies to the following model of the basic arithmetic operations.

$$fl(x \text{ op } y) = (x \text{ op } y)(1 + \delta), \quad |\delta| \leq \varepsilon, \text{ op} = +, -, *, /. \tag{1.3}$$

The model says that the computed value of x op y is "as good as" the rounded exact answer, in the sense that the relative error bound is the same in both cases [7]. Likewise, one may assume that

$$fl(\sqrt{x}) = \sqrt{x}(1 + \delta), \quad |\delta| \leq \varepsilon. \tag{1.4}$$

The model (1.3), (1.4) is basic in the error analysis of floating-point computations.

1.6 Quantization effects

MATLAB file used in this section

File	Description
sampling_adc	Step responses for different A/D converters

The conversion of analog quantities into binary signals in digital devices is associated with the introduction of quantization errors which may affect the stability and performance of the corresponding control systems.

1.6.1 *Truncation and roundoff*

As shown in the previous sections, the representation of numbers in fixed-point or floating point arithmetic may lead to errors. Denoting the number before quantization as x, the error introduced by quantization is

$$e_q = x_q - x.$$

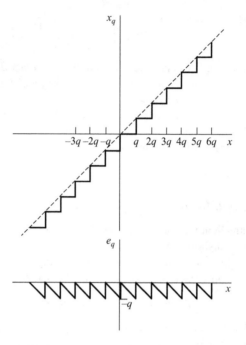

Figure 1.12   Truncation errors in fixed-point arithmetic

where x_q is the quantized value of x. The range of the quantization error depends on the type of arithmetic (fixed-point or floating point), the type of rounding (trunca-tion, rounding to the nearest representable number, etc.). In fixed-point arithmetic, truncation or roundoff errors are independent of the magnitude of the unquantized number x.

In Figure 1.12, we show the static characteristic (x_q as a function of x) of a digital device and the behavior of the truncation error in the case of fixed-point arithmetic with two's complement representation. The truncation error e_q in this case satisfies the inequality

$$-2^{-m} < e_q \leq 0$$

for all x, where m is the fractional part length of x_q.

In case of fixed-point arithmetic with rounding to the nearest representable number, the static characteristic of the device and the behavior of the rounding error are shown in Figure 1.13. In this case one has that

$$-\frac{1}{2}2^{-m} < e_q \leq \frac{1}{2}2^{-m}.$$

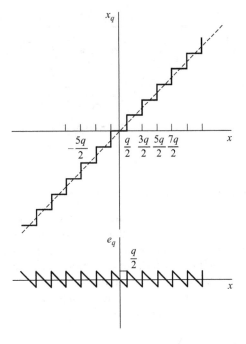

Figure 1.13   Round-off errors in fixed-point arithmetic

In floating point arithmetic, the truncation and roundoff errors depend on the magnitude of the unquantized number. According to (1.2) one has that for $x \neq 0$ the relative quantization error is given by

$$\frac{x_q - x}{x} = \delta, \quad |\delta| \leq \epsilon, \tag{1.5}$$

where ϵ depends on the rounding mode. In the case of truncation $\epsilon = 2^{-p}$ and in the case of rounding to the nearest representable number $\epsilon = \frac{1}{2}2^{-p}$ where p is 24 or 53 for IEEE single or double precision, respectively.

Apart of the number representation, the rounding error may have significant effect on the arithmetic operations performed during the computation of control actions in an embedded control system. At first glance, the relative error (1.5) in the arithmetic operations is small, especially in the case of double precision arithmetic. In some computations, however, such errors may lead to large errors in the final result, due to *catastrophic floating point cancelations* . The theoretical analysis of this effect in case of complicated high-order controllers is difficult. The most reliable way to analyze the effect of rounding errors is to simulate the closed-loop system taking into account the precision of different signals and control action computations.

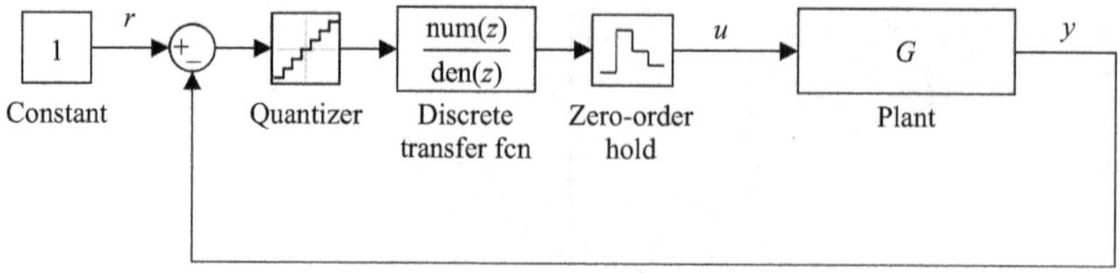

Figure 1.14   Simulink block diagram of discrete-time system

1.6.2   Quantization errors in A/D conversion

The conversion of analog quantities into binary signals in the digital sensors is equivalent to the introduction of quantization noise which is an example of discrete-time white noise. If the full scale measurement range of analog quantity is S and if the A/D converter has p bits resolution, then the quantization noise may be represented as a zero mean white random process having a uniform probability density from $-q/2$ to $q/2$ where $q = S/2^p$ is the quantum size. It is possible to show [8, Chapter 10] that this quantization noise has strength equal to

$$\sigma_v^2 = \frac{q^2}{12}.$$

For example, a shaft p-bit encoder with full measurement range $S = 2\pi$ has a quantization noise whose strength is given by

$$\sigma_v^2 = \left(\frac{2\pi}{2^p}\right)^2 \frac{1}{12}.$$

The quantization error of A/D converters may deteriorate the closed-loop system performance as demonstrated by the following example.

Example 1.4.   Influence of the ADC precision on system performance
Consider a discrete-time system whose Simulink block diagram is shown in Figure 1.14. The system involves a continuous-time second-order plant described by the transfer function

$$G = \frac{K_o}{T_o^2 s^2 + 2\xi T_o s + 1}$$

with parameters $K_o = 20$, $T_o = 1.0$, and $\xi = 0.1$, discrete-time controller with transfer function

$$K_d = 33.333 \frac{z - 0.9854}{z - 0.5134}$$

and an ADC converter, modeled as a Quantizer block in Simulink. The sampling time of the ADC is $T_s = 0.02$ s.

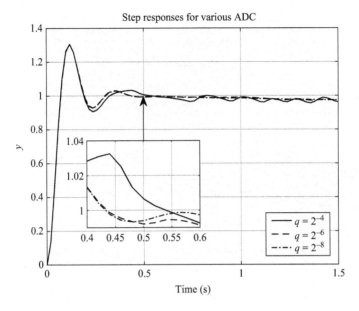

Figure 1.15   Unit step responses for different ADC

In Figure 1.15, we present the unit step responses of the closed-loop system for different ADC with resolution 4 bits, 6 bits, and 8 bits, respectively. It is seen that the step response for 4 bits ADC becomes oscillatory, due to the large dead zone of the ADC.   □

1.7   Design stages

MATLAB file used in this section	
File	**Description**
sampling_period	Step responses for various sampling intervals

The design of embedded control systems with MATLAB and Simulink is described in simplified form as follows.

The process of controller design, controller code generation, and embedding of the code in the digital computer is shown in Figure 1.16. The discrete-time controller structure and parameters are determined in MATLAB based on plant, sensor and actuator models incorporating disturbances and noises. For this aim the designer uses general purpose toolboxes such as Control System Toolbox™, Robust Control Toolbox™, System Identification Toolbox™, Optimization Toolbox™,

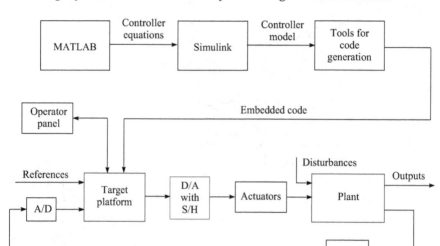

Figure 1.16   Design of embedded control system with MATLAB

Signal Processing Toolbox™ or special purpose toolboxes like Aerospace Blockset™ and Robotics System Toolbox™. The controller description is obtained in the form of discrete-time state-space equations or transfer functions. A controller model, based on this description, is built in Simulink using appropriate blocks. Using the tools for code generation (Simulink® Coder™ and Embedded Coder® or HDL Coder™) a C and C++ code is generated and embedded in the target computer.

Specifically, the design process may involve the following main stages.

- Plant modeling
  The plant model is obtained by theoretical modeling based on first (physical) principles or experimental modeling (identification) using measured input and output variables. Generally, the plant model may contain parts, described by nonlinear algebraic or differential equations. For the purpose of controller design, these equations are linearized and discretized. Some plants have parameters that are not known exactly or may vary in some intervals. Instead of using a single plant model, this may lead to the necessity of using a family of plant descriptions. The determination of appropriate plant model is a difficult stage of system design which may require a large volume of theoretical and experimental work. Note that the building of good plant model represents an iterative process aimed to fulfill the closed-loop system performance requirements.
  The plant modeling is considered in more details in Sections 1.2.1 and 1.2.5.
- Controller design
  The modern control theory offers a rich variety of methods intended for controller design. These include linear and nonlinear controllers designed by using different

techniques. In this book, we focus on the linear controllers designed by using the optimal and robust control theory. The purpose of these controllers are to ensure precise tracking of reference commands in the presence of deterministic and stochastic disturbances and plant parameter variations. Depending on the plant and disturbances model order, the controller order may have to be reduced in order to implement the controller algorithm in real time. Note that the contemporary microcontrollers make possible to use complicated high-order control laws even for sufficiently small sampling periods. For instance, linear controllers up to 50th order can be implemented without difficulties in case of sampling periods larger than 0.001 s (1 kHz sampling frequency) on microcontrollers with clock frequency 150 MHz.

- Software-in-the-loop simulation (SIL)
  In SIL testing, the system hardware is represented entirely by software models in Simulink. Initially, the software models may work in full (double) precision, but at a later stage, they can take into account the different precision of sensors, actuators, and controller. The operational software is automatically generated from Simulink models and tested in a non-real-time simulated environment for different references, disturbances and noises.
- Rapid control prototyping
  The aim of rapid control prototyping is a real-time controller simulation (emulation) with hardware (e.g., off-the-shelf signal processor) other than the final series production hardware may be performed. It is used to test software control algorithms in a real-time environment before implementing code on an embedded processor. The plant, the actuators, and sensors can then be real. The rapid control prototyping may reduce the models and algorithms to meet the requirements of cheaper mass production hardware and help in defining the specifications for final hardware and software.
- Processor-in-the-loop (PIL) simulation
  When the real-time embedded processor is available, the operational software is tested in the real-time embedded processor, using simulated hardware. This PIL testing validates software functions and performance. In this phase of testing, estimates of processor throughput, memory, and timing are also obtained.
- Hardware-in-the-loop (HIL) simulation
  After the PIL test, the operational software and prototyped hardware are ready for HIL testing, which is used to verify the integrated functional and operational performance. Based on the test data, a fine tuning of the regulators and filters is done and the closed-loop system performance is verified to meet requirements.

The advantages of the HIL simulation are [2, Chapter 2].

- Design and testing of the control hardware and software without operating a real process ("moving the process field into the laboratory")

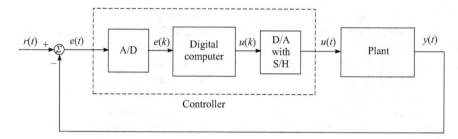

Figure 1.17   Digital implementation of "analog" type controller

- Testing of the control hardware and software under extreme environmental conditions in the laboratory (e.g., high/low temperature, high accelerations and mechanical shocks, aggressive media, and electromagnetic compatibility)
- Testing of the effects of faults and failures of actuators, sensors, and computers on the overall system
- Operating and testing of dangerous operating conditions
- Reproducible experiments, frequently repeatable
- Easy operation with different man—machine interfaces
- Saving of cost and development time

HIL simulation of different controllers is described in Chapter 4.

1.7.1   Controller design

As it is well known (see for instance [4]), the design of discrete-time controllers can be carried out in two distinctly different ways. The first approach is illustrated in Figure 1.17. According to this approach, the controller is designed in continuous-time and then one derives a discrete version of it using some discretization method. The advantage of this approach is that the performance specifications are done in the continuous-time domain and the design is fulfilled by using the well-developed methods for continuous-time system design. In this case, the cascade A/D—Digital computer—D/A behaves like an analog controller provided the sampling frequency is sufficiently high. With the increasing of the sampling period the behavior of the discrete-time closed-loop system deviates from the behavior of the continuous-time system which leads to the deterioration of closed-loop performance.

Example 1.5. Influence of the sampling period on system performance
Consider again the discrete-time system whose Simulink block diagram is shown in Figure 1.14. The plant transfer function is

$$G = \frac{K_o}{T_o^2 s^2 + 2\xi T_o s + 1}$$

with parameters $K_o = 20$, $T_o = 1.5$, and $\xi = 0.1$.

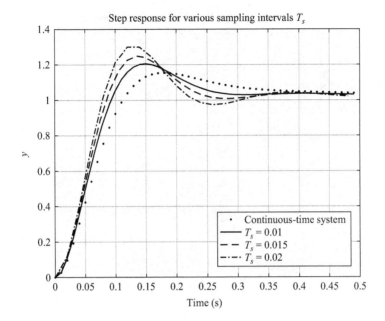

Figure 1.18   Unit step responses for different sampling periods T_s

A continuous-time controller with transfer function

$$K = c\frac{T_1 s + 1}{T_2 s + 1},$$

where $c = 4$, $T_1 = 0.5$, $T_2 = 0.03$, ensures an acceptable performance of the closed-loop system. This controller is discretized for three different sampling periods $T_s = 0.01$, $T_s = 0.015$, and $T_s = 0.02$ s. The corresponding closed-loop unit step-responses and control actions are shown in Figures 1.18 and 1.19, respectively. It is seen from Figure 1.18 that the overshoot of the system response increases from 16 percent (the case of continuous-time controller) to 30 percent (discrete-time controller for $T_s = 0.02$). This shows that the approach using continuous-time controller has limited capabilities especially for large sampling periods.   ❏

The alternative approach to the design of discrete-time systems is illustrated in Figure 1.20. With this approach, one discretizes the continuous-time plant model and performs the controller design in discrete time. The performance specifications are usually formulated in continuous-time due to the clear physical interpretation in this case. After plant discretization, a discrete-time controller is computed which may be implemented directly for simulation or real-time control. The usage of such approach avoids the performance degradation and the computational errors which might be introduced in converting a continuous-time controller to discrete-time. For better results, in this book we try to adopt the second approach.

Different controller design methods are represented and compared in Chapter 4.

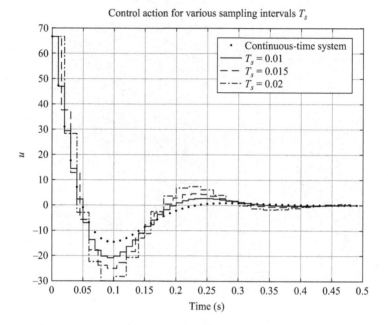

Figure 1.19   Control actions for different sampling periods T_s

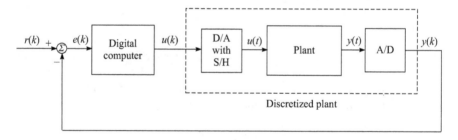

Figure 1.20   Digital system with dedicated discrete-time controller

1.7.2   Closed-loop system simulation

The controller design is usually followed by simulation of the closed-loop system and assessment of the system performance. In practice, it is not easy to find appropriate controller after the first attempt, so that the procedure consisting of controller design and simulation is repeated iteratively until an acceptable solution is found.

Initially, the simulation is done by using a simplified linearized model as the one shown in Figure 1.21. This model reflects the plant behavior in a single equilibrium state and may not take into account the effect of disturbances and noises. In case of time-invariant systems this allows to find relatively quickly the structure and

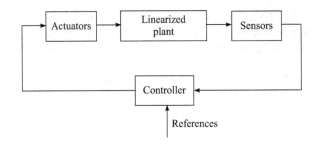

Figure 1.21   Simulation of a linear system

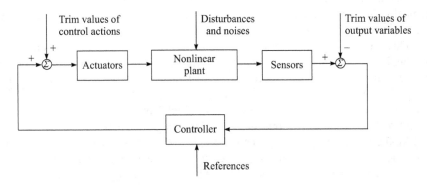

Figure 1.22   Simulation of a nonlinear system

the approximate controller parameters using simple simulation methods, like `step`, `lsim` and `dlsim` from MATLAB. At this stage of the simulation, it is also appropriate to use an uncertainty plant model in order to verify the closed-loop robustness. For this aim one may use the uncertainty models presented in Section 1.2.5.

After an appropriate controller is found, a more complicated system model can be used, as shown in Figure 1.22. It involves nonlinear plant model, specific equilibrium point determined by the trim values of control and output variables, as well as disturbances and noises. This model is made more and more involved, reflecting the different precision used by sensors, actuators, and controllers, and including more nonlinear effects (nonlinear static characteristics, friction, etc.). The price to be paid for such refinements is the increasing computational time necessary to simulate the system. In this case it is necessary to pay some attention to the choice of appropriate method for integration of the plant differential equations, choice of integration step and studying the system behavior between sampling instances.

In case of uncertain parameters it is appropriate to use the Monte-Carlo simulation technique.

The Monte-Carlo simulation is a heuristic method which consists of random sampling the uncertain model parameters and simulation of the closed-loop system for various fixed values of these parameters in the uncertainty range. The averaged model responses then give an impression about the closed-loop system behavior in real-world conditions. The Monte-Carlo technique is especially convenient in the simulation of uncertain nonlinear system when it is difficult to use analytical methods to assess the closed-loop performance. Simulink makes possible to use easily the Monte Carlo method in case of uncertain models.

Examples of Monte-Carlo simulation are given in Chapter 4.

1.7.3   Embedded code generation

After a suitable controller is determined, it is possible to go to the next design step generating control code from the Simulink control model.

Using the MATLAB and Simulink capabilities, the designer can generate automatically the control code which increases efficiency, improve performance, and promote the innovation of control algorithms.

One of the widely used technologies for automatic generation of code intended for loading in microcontrollers and digital signal processors is based on the programing tools Simulink Coder™ and Embedded Coder, which are included in the programing system MATLAB.

The Simulink Coder (with former name Real-Time Workshop) generates and executes C and C++ codes from Simulink block diagrams, diagrams of Stateflow®, and MATLAB functions. The generated output code can be used in applications in real time and non-real time including accelerated simulation, rapid prototyping, and HIL simulation. The generated code can be tuned and debugged using Simulink or can be executed out of MATLAB and Simulink.

Embedded Coder generates compact and fast C and C++ code for using on embedded processors and microcontrollers for mass production. Embedded Coder gives additional opportunities for configuration MATLAB Coder and Simulink Coder and optimization of the generated code, files, and data. These optimizations improve the code efficiency and facilitate the integration with previous code, data types, and calibration parameters, used in the embedding. The Embedded Coder supports SIL and PIL simulations.

To generate diagrams for embedding in FPGA, one uses the program systems Simulink and HDL Coder. Initially, a diagram model is created in Simulink, after that the diagram is translated in the hardware description language (HDL) VHDL and then it is passed on the custom software products, submitted by the device producer. Then, after design, visualization, and wiring, a configuration file is generated, which is loaded in the FPGA device. This process is automatized, so that the designed works only in Simulink environment and makes fine tuning in the rest environments.

The embedded code generation is considered in more details in Section 1.9.

1.8   Hardware configuration

Generally, the interface between control algorithm and target machinery provided by embedded system can be decomposed into hardware and software configuration. The *hardware configuration* supplies an execution environment of interconnected electronic devices. This environment should be capable of fast and deterministic actions ordered in predefined but conditional sequence which is called program. It is planned by the system designer and written inside digital memory. Another hardware function is continuous processing of information packed in discrete units of data. This section will briefly discuss and demonstrates how a hardware configuration is developed.

The basic building block of digital electronics is the transistor working as a two-state switch. Switching behavior of these building elements puts the design with them in domain of Boolean algebra and discrete mathematics. Nowadays transistors for computing are scaled down to nanometer units to allow faster switching between states and larger spatial density. Integral circuits such as microcontroller or FPGA are made of millions of transistors printed on a piece of silicon base (called a *chip*). Here comes the problem of complexity in design—a large number of simple elements have to be interconnected with exponentially increasing number of connections to achieve some useful function. Since humans can work with limited number of elements in a time the best way of building a complex system is through modular approach. A module introduces a level of isolation of its internal mechanism and provides a convenient external user interface. When module design is completed there is no need to consider its internal working during the next design phase as long as its interface stays compatible. The complexity in integral circuit design is managed through introduction of levels of increasing granularity. Design is hierarchically organized such that each module from a given level is composed of modules from the level below. Resultant tree of modules supports all device functions and target system design has to account at least for highest levels of that architecture.

1.8.1 Microprocessing architectures

Microarchitecture concerns interconnections inside an integrated circuit. Since the distances between semiconductor switches are several nanometers the distances between modules of these building blocks are in the micrometer range. There are several key computing architectures which are common in embedded control. A typical microcontroller architecture is presented on Figure 1.23. It is organized around two main buses—memory bus and peripheral bus. Each bus is composed of large number of logical signals carrying addressing information—typically 32 lines for address, 32 lines for data, clocking signal for synchronization, request and response signals for line arbitration.

The central element is central processing unit (CPU). There are mainly two kind of architectures of CPU nowadays—complex instruction set computer and reduced instruction set computer (RISC). For embedded application more common are RISC architectures like ARM, MIPS, SPARC, PowerPC, and other custom company specific. CPU executes instructions from the program with the help of its internal control logic and storage resources (registers)—Figure 1.24. Each instruction is a data unit which encodes an operation upon one or several operands. An instruction decoder inside CPU determines operation and operands. Next, the operands are distributed

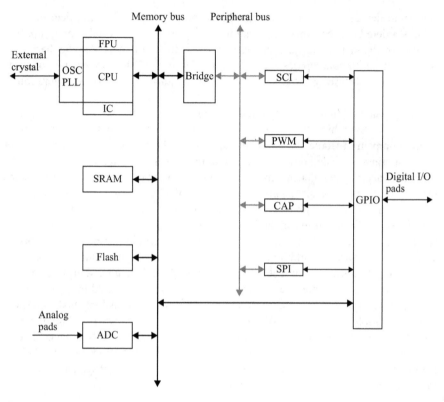

Figure 1.23   Typical microcontroller architecture. CPU, central processing unit; FPU, floating-point unit; IC, interrupt controller; PLL, phase-locked loop; OSC, oscillator; SRAM, static random access memory; ADC, analog to digital convertor; SCI, serial communication interface; PWM, pulse width modulation; CAP, capture module; SPI, synchronous peripheral interface; GPIO, general purpose input/output

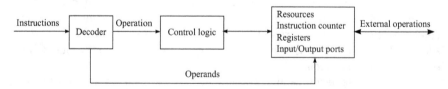

Figure 1.24   CPU principal operation

to internal CPU resources and the operation selects particular logical function to be applied with them.

The three key CPU resources are instruction counter which point to the address of next instruction that is read from SRAM or Flash memory, a set of general or special

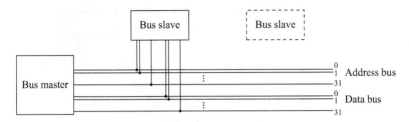

Figure 1.25   Typical bus architecture (arbitration signals not shown)

purpose registers for computation, addressing and flow control, an input/output communication module supporting memory bus control.

There are several hardware units closely supporting CPU operation—FPU, OSC and PLL, IC. Floating-point unit extends CPU mathematical capabilities with floating-point operations which are important in scientific calculation and in many higher order control systems. Microcontrollers are synchronous devices, all signal events (logical level changes) inside them are synchronized with clock signal edges. Clock signals in the microcontroller derive from a CPU clock signal with the help of PLL feedback system. Master CPU clock is generated by an OSC module with the help of external crystal with fixed resonant frequency. Real-time systems must be responsible to external events represented as logical signal level changes. The response time τ is a critical parameter and should be several times smaller than control system sample time T_s. Interrupt controller (IC) registers external to CPU events, holds current program execution and loads interrupt service program (ISR). After completion of ISR the CPU returns to main program execution.

Devices which require high rate of CPU communication (10–100 MHz) like memory devices or A/D convertors are located on a CPU memory bus. Other devices which operate at slower rate (10–100 kHz) are located on a secondary peripheral bus. Devices on memory bus can access devices on peripheral bus through a bridge device. It translates addresses, synchronizes clock rate and buffers data. Peripheral devices are usually low-speed communication devices (SCI, SPI, I2C), signal generators (PWM, DAC, sensors) or event captures (CAP). Each device has a function specific architecture supporting its function. Most of the microcontroller pads are defined as general purpose (GPIO) and work as a routing station. The program can configure each GPIO pad as input, output, or peripheral signal.

Several devices share a common information media called bus to achieve data integrity between them (Figure 1.25). In microcontrollers, usually, the CPU is bus master and the rest devices are bus slaves. All information inside slave devices is represented as a set of 32-bit registers (memory cells) and each register has its unique 32-bit address for the bus. A register works as a sequence of binary states which are related to slave device functions. For example, to start an ADC the program switches a particular state inside ADC control register. Since CPU is the bus master device only it can generate addresses to access internal device registers. All registers of

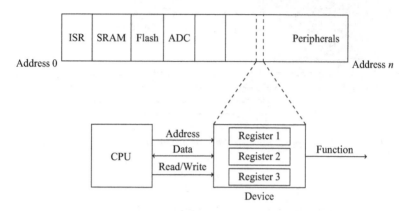

Figure 1.26   Memory map and peripheral device control

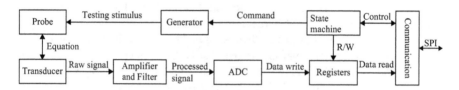

Figure 1.27   Typical smart sensor architecture

a microcontroller are described by the memory map which is always presented in documentation.

In the beginning of address space is an interrupt vector table. Each interrupt vector contains the address of ISR which is started when a particular event occur. After the interrupt table there are memory bus devices mapped—SRAM, Flash, and ADC (Figure 1.26). At the end of address space are located registers of peripheral devices. These last addresses are translated to peripheral bus addresses by the Bridge Device. When developing a device driver for a particular device there should be considered its internal architecture. Here, we will present briefly some of architectures common in such peripherals.

Much of modern sensors are scaled down to integrated circuits using microme-chanical structures like springs, levers, or oscillating masses, called MEMS. Figure 1.27 presents the internal architecture of such a MEMS device. The probe subsystem reveals information about physical variable (like velocity, temperature, or pressure) through specific geometric, material, or resonance phenomena. This information reflects on some component of electrical circuit (Transducer). Produced raw electrical signal have to be amplified and filtered to expose further the useful information about measured physical variable. ADC converts the conditioned signal to digital format which is recorded in sensor memory location (Registers). As noted,

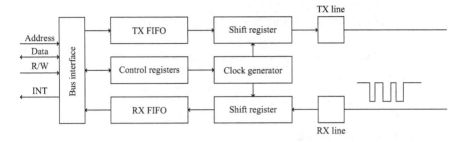

Figure 1.28   Typical serial communication device architecture

registers transfer their data to CPU through peripheral bus or some serial interface like synchronous peripheral interface (SPI) . Dedicated state machine controls the measurement process by generating testing stimuli for excitation of mechanical structure of probe. State machine also manage communication with the host.

Serial communication transmits data over single electrical line with binary states (Figure 1.28). Since only 1 bit can be presented on the line in a moment, the data must be represented as a sequence of bits distributed over available time slots. This is in contrast to bus data transmission where multiple parallel lines carry all the bits of data word. In asynchronous communication, the clock signal isn't transmitted to the receiver and in synchronous communication it is. Generated clock signal defines the boundaries of available time slots for data transmission. Dedicated shift registers convert data from parallel to serial form in tact with the clock signal. Due to differences in bus and serial line rates of operation FIFO (first-in–first-out) buffers store temporally abundant data. These buffers are with limited size and when they are full or empty the CPU is signaled through the interrupt line.

1.8.2   Hardware description language

The main design tool for microstructures from the previous subsection is HDL. Most popular HDLs are Verilog and VHDL. They are similar in syntax and here we consider only the VHDL. It is a derivative from Ada which is the ancestor of Pascal. In contrast to common programing languages which support only sequential control flow HDL support also concurrent operations. HDL represents internal state of an electronic module as *signals* or *variables*. The signal is physical wire connecting two points. Signal level is usually voltage in digital circuits taking one of three states (high, low, high-impedance). Signal can have only one driver point which controls its level possibly as a function of other signals or variables. Variable is an internal state which is sequentially processed before to drive a signal. Ultimately in implementation all variables are mapped to signals. A skeleton of VHDL program is shown in Listing 1.1. The process block defines processing of input signals from its sensitivity list to output signals using some internal variables. Hence VHDL has operations either for variables or signals.

Listing 1.1   General structure of VHDL program

```
1   LIBRARY IEEE; USE IEEE.std_logic_1164.ALL; USE IEEE.numeric_std.ALL;
2
3   ENTITY test IS
4     PORT( clk: IN std_logic ; Value : IN std_logic_vector (3 DOWNTO 0);
5         ...; IValue: OUT std_logic_vector(3 DOWNTO 0));
6   END test;
7
8   ARCHITECTURE rtl OF test IS
9   signal s1: std_logic ; ...
10  BEGIN
11    process (clk) variable delay_out: std_logic := '0'; ... begin
12      if clk'event and clk = '1' then delay_out := set; ... end if;
13          ...
14            IValue <= std_logic_vector ( integrator_out );
15    end process;
16        ...
17        process (clk) begin ... end process;
18  END rtl;
```

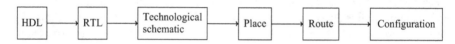

Figure 1.29   HDL translation process

After HDL description, the design is translated to register transfer level (RTL) schematic which looks like logical circuit (Figure 1.29). Every such circuit is decomposed to combinatorial (static) and sequential (dynamic) subsystems. Combinatory logic defines all the values of a multivariable logical function with some basic elements (OR, AND, NOT gates, or XOR gates). Sequential logic is made of simple memory elements (triggers) which can store a current state of a signal. RTL schematic is then translated to silicon technology specific primitives which are implementable on the chip. Two optimization procedures define final configuration of the microarchitecture—for primitive placement and for routing of interconnections. Configuration file can be implemented on ASIC (application specific integrated circuit) or on FPGA.

The three possible programing styles with VHDL are behavioral (Figure 1.32), structural (Figure 1.30), or dataflow (Figure 1.31). Same function can be achieved using behavioral or structural approach but these ways of programing are two alternatives and can't be mixed. *Data Flow* models integrate behavioral and structural programing style through employment of basic principles from control system theory—causality, concurrency, and dynamics.

Data is numerical information related to the model which is either a signal if it varies with time or a parameter if it being time invariant. In data flow programing, one

```
process (clk)
    variable delay_out        : std_logic;
    variable integrator_out : unsigned(3 DOWNTO 0);
    variable switch_out       : unsigned(3 DOWNTO 0);
    variable sum_out          : unsigned(3 DOWNTO 0);
begin
  if clk'event and clk = '1' then
      Delay_out := set;
      integrator_out := sum_out;
  end if;
  sum_out := switch_out + unsigned(Value);
  if (set and (not delay_out)) = '0' then
    switch_out := integrator_out;
  else
    switch_out := to_unsigned(16#2#, 4);
  end if;
  IValue <= std_logic_vector(integrator_out);
end process;
```

Figure 1.30   Behavioral program

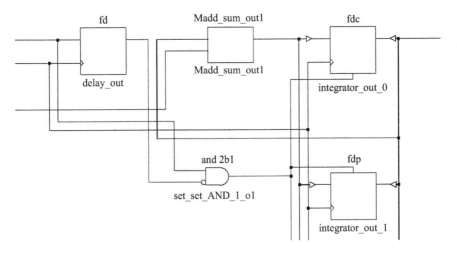

Figure 1.31   Structure Program (D triggers 2 and 3 of "integrator_out_" not visible)

could express sequence of operations like in behavioral language because of causality relation between blocks. At the same time, he/she could decompose the system into independent modules like in hardware schematic because of concurrency relation. All blocks share same mathematical description (in Simulink as hybrid dynamical system) which could be easily ported to hardware or software program.

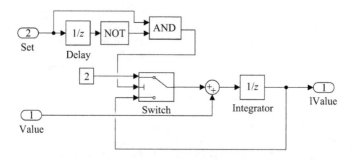

Figure 1.32   Data flow program

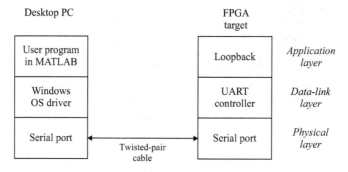

Figure 1.33   OSI model conformance of serial communication

Clear representation of directed system activity is important advantage of these models because one can easily understand how intended function emerges from the collection of simple blocks. When a diagram is mapped into a hardware, the input data of each block affects its output data after some *hardware dependent time delay*. Therefore, one can follow the shift of current information through diagram in timely manner which creates the impression of a flow.

1.8.3 Module level development

This subsection presents how to design a simple asynchronous serial interface controller, compatible with RS232 standard and implement it on Spartan-3E FPGA, through automatic HDL code generation and hardware synthesis. Simulink HDL coder is a tool that can produce VHDL or Verilog description from Simulink diagrams. Generated code is compatible with standard synthesis tools and environments offered by FPGA vendors, for example, Xilinx Inc.

The interconnection between communicating devices must have layered architecture (Figure 1.33). For embedded applications typical communication system is

Start bit	Data 0	Data 1	Data 2	Data 3	Data 4	Data 5	Data 6	Data 7	Stop bit

Figure 1.34   Schematic representation of typical serial data frame

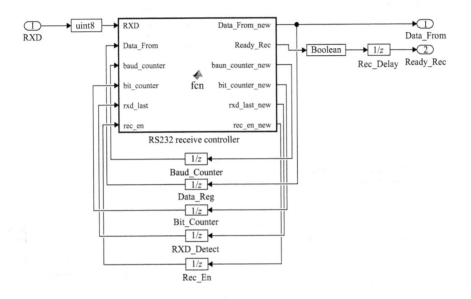

Figure 1.35   Simulink diagram of asynchronous serial receiver

composed of application layer (AL), data-link layer, and physical layer. For transmission, the controller converts a byte to a sequence of bits, then forms a message frame by addition of protocol information (Figure 1.34) and transfers the bits one by one. Reception is a reverse process. Asynchronism results from undefined timing between consecutive messages because there is no shared timing system between devices. Each message is composed of one start bit, 5–8 data bits, 1–2 stop bits and optional parity bit. Parameters for the presented here UART controller are: 115,200 bps baud rate, 8 data bits, 1 stop bit, no parity.

To model a serial controller in Simulink, we need to represent it in state-space form. The controller is composed of two subsystems—receiver module (Figure 1.35) and transmitter module. For modeling of digital devices, it is convenient to represent corresponding Simulink model in discrete time. Simulation configuration is set to fixed step discrete solver with sample time $T_s = 1$ clock period.

Unit delay block stores, the current state of each module. It acts as a D-Trigger. Embedded MATLAB Function block represents system transfer function. This block allows describing system behavior in the algorithmic language of

Table 1.3   States of receiver

Name	Size (bit)	Description
Data_From	8	Data register
baud_counter	9	Baud rate counter
bit_counter	4	Number of received bits
rxd_last	1	Last state of RXD line
rec_en	2	Stage of reception
dcnt	5	Delay counter

Listing 1.2   Pseudocode for receiver subsystem

```
1    if ( start  bit  detected ) then
2      initialize   baudrate  counter, delay counter and bit counter;
3           stage = delay;
4    elseif  (stage = delay)  then
5           increment  delay counter;
6           if (delay  counter = N_D) then stage = reception ; end
7    elseif  (stage = reception ) then
8      increment  baudrate  counter;
9              if ( baudrate  counter = N_115200) then
10     reinitialize   baudrate  counter;
11         Data_Reg = (Data_Reg shl 1)  bitor  RXD;
12         increment  bit counter;
13         if ( bit  counter = 8) then stage = idle ;  rise  ready  signal ;   reinitialize   bit counter;
14   end, end, end
```

MATLAB. The block executes each time step only once, according to static relationship. Input port and Output port blocks in Simulink model correspond to in port and out port declarations of the generated entity definition in VHDL. Each Simulink signal is associated with specific data type for data it carries. Data type conversion blocks are used to change the data type of a signal.

Receiver module has one input state of RXD signal line. It has two outputs. First is the content of data register, containing last received byte. Second is a signal line which is asserted when reception is completed, to inform higher processing layers. Receiver module has six states (Table 1.3). Three of them are for counters. Others are for storage of received byte, for detection of start bit (rxd_last), and for stage of reception. Reception is a three-stage process. First, a start bit is detected. Second, a delay counter is waited to finish. Third, a sequence of bits is shifted into data register.

At the third stage, a baud rate counter is used to synchronize the proper time of bit capture from RXD line. A bit counter stores the number of received bits. A delay counter is used to generate small delay after start bit detection to assure proper bit capture. A pseudo code for receiver module is given on Listing 1.2. This algorithm is coded in MATLAB language in the Embedded MATLAB function block.

Transmitter module has two input ports. First is data byte to be transmitted and second is signal line to trigger transmission event. The module controls the state of

Table 1.4   States of transmitter

Name	Size (bit)	Description
Data_To	8	Data register
baud_counter	9	Baud rate counter
bit_counter	4	Number of sent bits
txd_last	1	Last state of TXD line
send_en	2	Stage of transmission

Listing 1.3   Pseudocode for transmitter subsystem

```
1   if (send request detected) then
2      initialize baudrate counter and bit counter;
3      Data_Reg = Data_To_Send; TXD line is low (start bit); stage = transmit;
4   elseif (stage = transmit) then
5      increment baudrate counter;
6      if (baudrate counter = N_115200) then
7            TXD line = Data_Reg (bit 0); Data_Reg = Data_Reg shr 1;
8                increment bit counter; reinitialize baudrate counter;
9                    if (bit counter = 8) then stage = stop bit; TXD line is high (stop bit); end
10     end
11  elseif (stage = stop bit) then
12     increment baudrate counter; if (baudrate counter = N_115200+N_D) then
13           rise ready signal; end
14  end
```

TXD output line. When a byte is sent a ready signal is generated. Simulink model of Transmitter module is analogous to that of receiver module (Figure 1.35). The difference is on signal and algorithmic level. Transmitter module has five states (Table 1.4). Two of them are for counters. Others are for storage of transmitted byte, for storage of TXD line state and for stage of transmission. Transmission is a three-stage process. First is an initialization stage of counters and internal shift register. Second state is for generation of a start bit. And in the third stage the sequence of bits is shifted from data register to the TXD line (Listing 1.3).

Generation of HDL code from Simulink model is possible only if certain subset of supported blocks is used. Embedded MATLAB Function, Unit Delay, Data Type Conversion are all supported. For every atomic block in the model there is corresponding section in VHDL file. This allows us to review whether the generated code is appropriate or is as expected to be. In our case, we have one main file and two subsystems corresponding to Receiver and Transmitter module. Generated VHDL code is about 700 lines.

Unit Delay block is represented in VHDL as a process block which is sensitive to clock signal. When clock rise event is detected then the input signal is assigned to the state signal. Embedded MATLAB Function block is represented in VHDL as a process block with a sensitivity list composed of input ports of Simulink block. Sensitivity list of process block contains temporary variables used in MATLAB Function and

Table 1.5   Resource utilization

Parameter name	Used/Available
Number of slice flip flops	56/9,312
Number of 4 input LUTs	231/9,312
Number of occupied slices	123/4,656
Number of bounded IOBs	23/232
IOB flip flops	2

auxiliary variables needed to represent arithmetic and logic operations in HDL. The body of process block is translation of algorithm defined in MATLAB Function.

From generated VHDL code, one builds a Xilinx ISE project file. It contains the files corresponding to Simulink model and additional constraint file. UCF Location Constraints including I/O pin assignment and the I/O standard used must be specified in Xilinx ISE project according to target board documentation. The Xilinx Synthesizer generates representation of VHDL code in technological basis of target FPGA device. We can see from the reports that Unit Delay blocks are represented by D-Trigger components. The algorithm in an Embedded MATLAB Function block is represented by combinatorial circuit. It contains elementary logic gates, summator, and comparator components. This description can be easily optimized and embedded in FPGA device by Xilinx tool chain. Some numerical characteristics of device resource utilization are given in Table 1.5.

Design of the UART controller is validated on several levels corresponding to different stages of design process which are: Simulation in Simulink, Behavior simulation of VHDL code, Experiments. Figures 1.36 and 1.37 represent simulation results in Simulink environment for Transmitter module. The purpose of simulation is to transmit byte 73 (letter "I") two consecutive times. To have correct results for time scale we set the sample time as $T_s = 20$ ns.

Experiments are executed by connecting desktop PC to Spartan 3E Starter Kit board by RS232 twisted pair cable. From MATLAB environment, we can use serial command to open serial communication port. Then by `fread` and `fwrite` commands we can receive or send data. We send a sequence of data bytes to FPGA and wait to receive the same sequence back. A diagram of experimental setting is in Figure 1.33.

1.8.4   System level development

According to modular approach, system development is a composition of modules interacting through as set of compatible interfaces. Overall system functionality results from internal module functionality and module interactions. Hence, system level development rather concerns selection of modules and their configuration than keeping track of algorithmic control and temporal synchronization which are more related to module level development. However, it is not an easy task to choose a system configuration with higher performance and lowest cost especially when designing

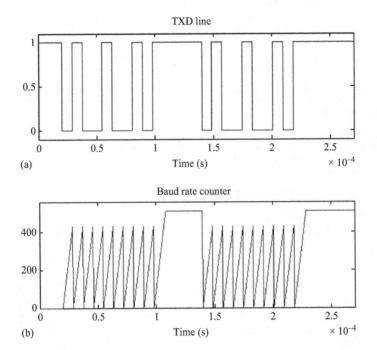

Figure 1.36   Simulation results: (a) logical state of TXD line and (b) state of baud rate counter

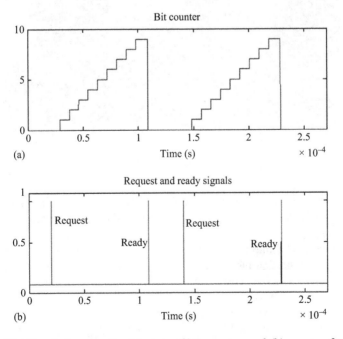

Figure 1.37   Simulation results: (a) state of bit counter and (b) events of start and end of transmission

Listing 1.4   System components in VHDL syntax

```
1   architecture Behavioral of ab_contr is
2
3   component gen_tick is
4           port (clk : in std_logic ; clk1 : out std_logic ; clk1s : out std_logic );
5   end component;
6
7   component counter_test is
8   port (clk : in std_logic ; count : out std_logic_vector (7 downto 0);
9                   reset : in std_logic ; en :in std_logic );
10  end component;
11      ...
12  signal clk100,clk100_90,stop, start , start_reg : std_logic ;
13  begin
14  clock1 :  gen_tick port map (clk => clk, clk1 => clk100, clk1s => clk100_90);
15  conter1 :  counter_test port map (clk => clk100, count => stage, reset => stop, en=> start_reg );
16      ...
17  end Behavioral ;
```

Figure 1.38   Top view of Spartan 6 FPGA SP601 evaluation board

ASIC which will be produced in large amount afterwards. Specifying, subsystems in VHDL are achieved with *component* declaration and interconnection between them with *port map* operator (Listing 1.4).

The modules building complex integrated circuits are called *cores*. This name is given because module printed image on the chip looks like a concentrated structure with highest density. There are open source cores and proprietary IP cores (Figure 1.38). Xilinx Platform Studio (XPS) offers a database of cores for system on chip design. Here we give as an example a *Microblaze* microprocessor system for Spartan 6 FPGA of Xilinx on SP601 evaluation board (Figure 1.38).

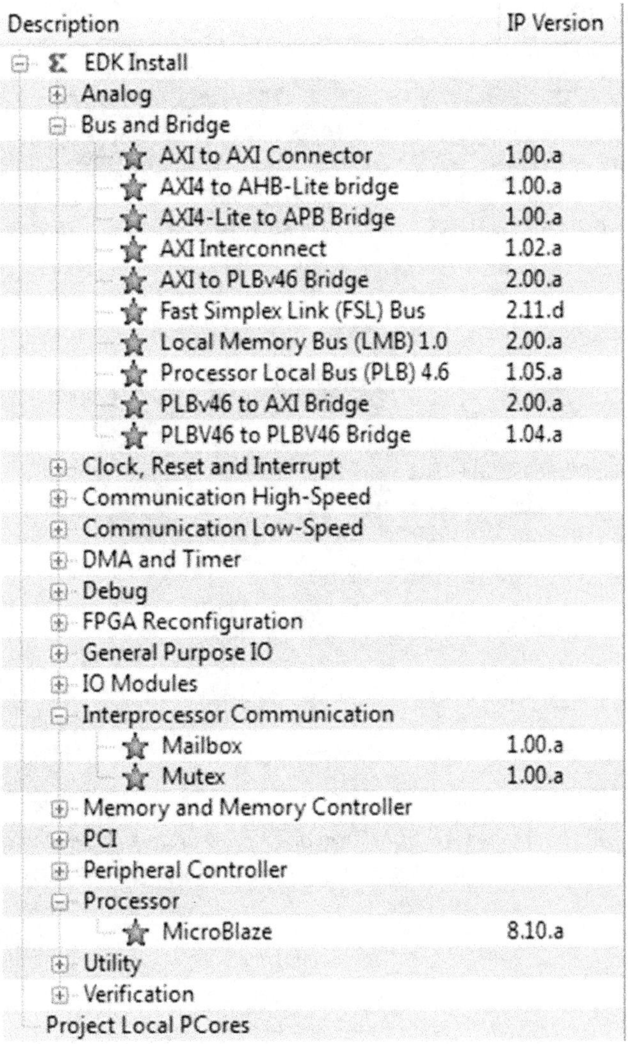

Description	IP Version
⊟ ∑ EDK Install	
⊞ Analog	
⊟ Bus and Bridge	
☆ AXI to AXI Connector	1.00.a
☆ AXI4 to AHB-Lite bridge	1.00.a
☆ AXI4-Lite to APB Bridge	1.00.a
☆ AXI Interconnect	1.02.a
☆ AXI to PLBv46 Bridge	2.00.a
☆ Fast Simplex Link (FSL) Bus	2.11.d
☆ Local Memory Bus (LMB) 1.0	2.00.a
☆ Processor Local Bus (PLB) 4.6	1.05.a
☆ PLBv46 to AXI Bridge	2.00.a
☆ PLBV46 to PLBV46 Bridge	1.04.a
⊞ Clock, Reset and Interrupt	
⊞ Communication High-Speed	
⊞ Communication Low-Speed	
⊞ DMA and Timer	
⊞ Debug	
⊞ FPGA Reconfiguration	
⊞ General Purpose IO	
⊞ IO Modules	
⊟ Interprocessor Communication	
☆ Mailbox	1.00.a
☆ Mutex	1.00.a
⊞ Memory and Memory Controller	
⊞ PCI	
⊞ Peripheral Controller	
⊟ Processor	
☆ MicroBlaze	8.10.a
⊞ Utility	
⊞ Verification	
Project Local PCores	

Figure 1.39   Xilinx IP (intellectual property) cores

Cores are grouped in categories according to their functions—processor, bus, arithmetic, memory, crypto, input/output, and so on. Since the micro system is organized around buses, these are several supported types in Xilinx like PLB (Processor Local Bus), AXI (Advanced eXtensible Interface), DCR (Device Control Register Bus) and Xilinx P2P (Peer to Peer) (Figure 1.39).

The clock_generator is the source of synchronization signal of the microcontroller, dlmb is data local bus, ilmb is instruction local memory bus. Microblaze is the only PLB master. There is a separate PLB controller not shown in the

SLAVES OF mb_plb

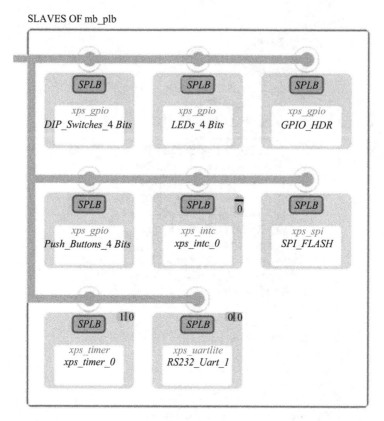

Figure 1.40   Architecture view of Microblaze FPGA system. Device with rectangular port is master and with circular port is slave

Instance	Base Name	Base Address	High Address	Size	Bus Interface(s)	Bus Name
microblaze_0's Address Map						
dlmb_cntlr	C_BASEADDR	0x00000000	0x0000FFFF	64K	SLMB	dlmb
ilmb_cntlr	C_BASEADDR	0x00000000	0x0000FFFF	64K	SLMB	ilmb
Push_Buttons_4Bits	C_BASEADDR	0x81400000	0x8140FFFF	64K	SPLB	mb_plb
LEDs_4Bits	C_BASEADDR	0x81420000	0x8142FFFF	64K	SPLB	mb_plb
GPIO_HDR	C_BASEADDR	0x81440000	0x8144FFFF	64K	SPLB	mb_plb
DIP_Switches_4Bits	C_BASEADDR	0x81460000	0x8146FFFF	64K	SPLB	mb_plb
SPI_FLASH	C_BASEADDR	0x83400000	0x8340FFFF	64K	SPLB	mb_plb
xps_timer_0	C_BASEADDR	0x83C00000	0x83C0FFFF	64K	SPLB	mb_plb
RS232_Uart_1	C_BASEADDR	0x84000000	0x8400FFFF	64K	SPLB	mb_plb
mdm_0	C_BASEADDR	0x84400000	0x8440FFFF	64K	SPLB	mb_plb
xps_intc_0	C_BASEADDR	0x90000000	0x9000001F	32	SPLB	mb_plb

Figure 1.41   Address mapping of bus devices

Figure 1.40 which is an interface between the processor and PLB slaves which are digital input/output controller (xps_gpio), IC (xps_intc), serial peripheral interface controller (xps_spi), universal asynchronous receiver transmitter (xps_uartlite) and programmable timer (xps_timer) (Figure 1.41).

Listing 1.5   PLB peripheral user logic

```
1   entity  user_logic  is
2     generic
3     (C_SLV_DWIDTH : integer := 32; C_NUM_REG : integer := 1);
4     port
5     ( user_specific_ports ;   plb_specific_ports );
6   end entity  user_logic ;
7
8   architecture  IMP of  user_logic  is
9     signal  slv_reg0              :  std_logic_vector (0 to C_SLV_DWIDTH−1);
10     ...
11     signal  slv_ip2bus_data    :  std_logic_vector (0 to C_SLV_DWIDTH−1);
12   begin
13     slv_reg_write_sel   <= Bus2IP_WrCE(0 to 0);  slv_reg_read_sel    <= Bus2IP_RdCE(0 to 0);
14
15     SLAVE_REG_WRITE_PROC : process( Bus2IP_Clk ) is
16     begin
17       if  Bus2IP_Clk'event and Bus2IP_Clk = '1'  then
18         ...
19         slv_reg0 (0  to  7)  <= Bus2IP_Data(0 to 7);
20       end if ;
21     end process  SLAVE_REG_WRITE_PROC;
22
23     SLAVE_REG_READ_PROC : process( slv_reg_read_sel, slv_reg0 ) is
24     begin
25         ...
26       slv_ip2bus_data   <= slv_reg0;
27     end case;
28     end process  SLAVE_REG_READ_PROC;
29
30     IP2Bus_Data  <= slv_ip2bus_data  when slv_read_ack = '1'  else  (others => '0');
31     IP2Bus_WrAck <= slv_write_ack; IP2Bus_RdAck <= slv_read_ack;
32   end IMP;
```

There is also an additional debug bus where dedicated debug controller is the master and microblaze is single slave device. This configuration allows debug controller to pause (halt) the program execution at a particular point (breakpoint) and examine the state of address space (memory and peripherals) as well as internal CPU state (registers). Usually, this is necessary when a problematic deviation from normal system function is detected and debugging is the processes of identifying the cause. Usually debugging module is required only for system development process and not in production so this module is removed in production design which saves some space on the chip.

User-defined peripherals are supported when properly connected to Microblaze PLB. XPS has a *Create Peripheral Wizard* for creating necessary wrapping VHDL modules. A typical PLB slave peripheral is composed of two subsystems—bus attachment and user logic. Bus attachment is responsible to respond to PLB master commands and address decoding for user logic registers. User logic implements some peripheral-specific function and provides it to the system through a dedicated set of registers. Listing 1.5 shows VHDL skeleton of user logic entity with a single register. PLB slave interface can select 1 of 4,096 possible internal registers. Bus attachment subsystem provides Bus2IP_WrCE and Bus2IP_RdCE signals for register selection. In the example, the single register is internally represented by the signal slv_reg0. User logic entity contains two process blocks for controlling reading and writing operations to the internal registers. Write process block takes the data from Bus2IP_Data input and puts it in a selected data register synchronously with Bus2IP_Clk signal. Read process block activates when register is selected by the slv_reg_read_sel signal.

1.9   Software configuration

The *software configuration* is a kind of virtual stuff emerging from inherent properties of hardware configuration and from pursuit to increase its quality of service. Usually software looks like a diagram or language which is essentially a conceptual representation, not a real thing, but always translatable to some material or measurable effect. Since embedded control system design requires more interaction with software than hardware the section details the steps of development of some basic software framework for diverse embedded applications.

The main driver of textual programing is human ability to interpret symbolically presented information. However visual field perception is another ability possibly even more powerful that textual comprehension. So there is a natural jump in computer software to the level of visual programing languages where one manipulates spatial arrangements of graphical and textual symbols. There are many visual programing languages—EICASLAB, Flowcode, LabVIEW, Ladder logic, Microsoft VPL, OpenDX, OpenWire, Blender, Simulink, GNU Radio, PLUS+1 GUIDE, WebML, and so on. Visual languages are strongly domain specific and hide lots of hardware specific detail (with few exceptions like xUML). They are targeted mostly to application design professionals rather than computer programmers. Mapping to hardware of visual program is trough process called *elaboration* meaning that some additional information is included to complete the design (Figure 1.42). And so one small graphical diagram can generate a thousands of textual code. In the past, there was some skepticism that elaboration can lead to reliable and optimized code as had been the case with compilation from C to assembler a few decades earlier. The development of validation techniques like PIL for component level and HIL for system level as well as introduction of code certification practices can give strong confidence in the automatically generated code from visual programs.

1.9.1   Board support package

First pieces of software which populate a particular microprocessor architecture are low-level device control programs such as clock signal activation, communication interface support, peripheral device initialization, boot loading, kernel image loading, etc. This primary tasks are strongly hardware dependent. Structure of program algorithm should closely match underlying hardware configuration in order to achieve target board support. Target architecture is made of various pieces of hardware which require to be accessed by the program in a correct way to be able to fulfill their function. Such a programmatic access to system devices supports more complex

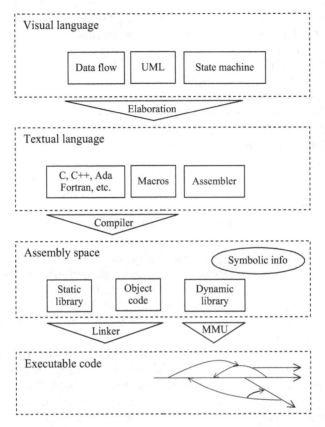

Figure 1.42   Code generation hierarchy. UML, unified modeling language; MMU, memory management unit

Listing 1.6   Low-level device control

```
1  interrupt  void  TIMER_ISR(void) {
2          unsigned  long  enc = EQep1Regs.QPOSCNT;
3          SciaRegs.SCITXBUF = comm_cnt; SciaRegs.SCITXBUF = enc & 0xFF;
4          SciaRegs.SCITXBUF = (enc >> 8) & 0xFF; SciaRegs.SCITXBUF = (enc >> 16) & 0xFF;
5          SciaRegs.SCITXBUF = (enc >> 24) & 0xFF; SciaRegs.SCITXBUF = comm_cnt*10;
6          comm_cnt++;
7          if  (comm_cnt == 16) comm_cnt = 1;
8          PieCtrlRegs .PIEACK.bit.ACK1 = 1; }
```

software execution. In temporal domain supporting code execution should be fast and reliable.

Listing 1.6 gives an example for low-level timer ISR of *TMS320F28335* micro-controller. A special *interrupt* keyword instructs the compiler to insert appropriate processor context switching code which store processor register state into program stack and after ISR restores state back. This timer ISR sends some data over asynchronous serial line (SCI) by writing bytes to the transmitter FIFO buffer which is accessed by an internal register SCITXBUF. The program puts sequentially bytes in the FIFO which SCI module sends over the serial line when ready. In listing 1.6 variable enc is 32 bit long quadrature encoder measurement so it is composed of 4 bytes. Each byte is taken from enc variable through corresponding bit shift bit mask. BSP for the microcontroller includes some helpful definitions like peripheral address space representation as C structures, unions, and bitfields. C-bitfields allow easy modification of a single bit of a register which is often the case.

BSP can be either presented as collection of low-level *stand-alone* driver or as an operating system *kernel*. If designed embedded system has to execute just a few tasks then usually stand-alone BSP option is preferred due to smaller size of compiled code and simplicity of application development. However when embedded system supports a large number of tasks (more than 10 for instance) then operating system kernel can considerably reduce development time.

Stand-alone BSP is simply a collection of software modules supporting user program execution through very basic hardware abstraction software. For example, a software module encapsulates some of microcontroller peripherals controlled by specific set of registers and access protocols. Usage of such kind of lightweight software packages can considerably reduce code size and sometimes also development time of embedded applications.

Operating system kernel BSP provide capability for multitasking execution on one or several processor cores. Hence the kernel have to regulate the access to shared resources by these tasks. The kernel is composed of several kinds of objects: tasks, mutex, semaphores, timers, mailboxes, events, and so on. The key algorithmic component in kernel is task scheduler which distributes processor execution time (runtime) among created tasks considering their priorities. There are couple of scheduling algorithms most commonly used—round robin and priority based. In real-time applications with limited processor resources good scheduling depend on number of tasks.

1.9.2   Application programing interface

Interface works like an imaginary border where application can access provided functionality by the underlying system. In programing, an interface is a set of callable functions and data structures which act as an isolation between implementation and usage. There are many instances of programing interfaces for purposes like device control (GDI, OpenGL, device drivers) or inter process communication (TAPI, file access). In order to use particular API the user have to be familiar with its data structures, functions, and concept of operation. For example, Table 1.6 shows how typical file access is organized.

1.9.3   Code generation

Traditionally simulation models arise as a means for understanding logical relations in the system under study or design. Often the model can replace the actual system

Table 1.6   API organization of file access

Data structures	Functions	Concept of operation
File handle, file name, access mode (read/write)	Open or close file	File is linear byte indexed space like a tape, identified by handle number
Buffer, operation size	Write or read data	Buffers of data are sequentially written or read
File position	Seek	There is a current file position changed by data transfer

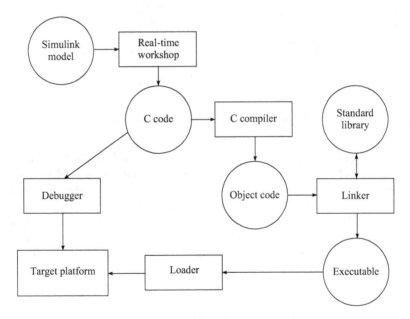

Figure 1.43   General concept of software design flow

to make conclusions about its behavior under extreme conditions. In the recent years, simulation models are used also as an executable specification becoming a functional part of the design process. Therefore, the model can produce not only conclusions about some extreme scenario but also can directly generate some part of an actual system behavior.

The general concept of software design flow is shown in Figure 1.43 and the code generation in Simulink is represented in Figure 1.44.

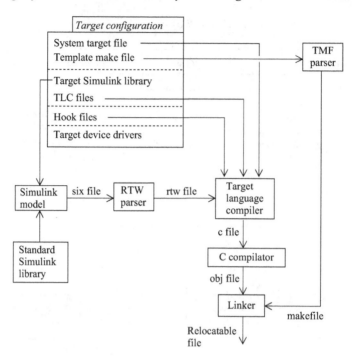

Figure 1.44   Code generation in Simulink

Simulink Coder generates C and C++ code from Simulink diagrams, Stateflow charts, and MATLAB functions. The generated source code can be used for real-time and non-real-time applications such as simulation acceleration, rapid prototyping, HIL simulation, embedded algorithm design, and so on.

MathWorks® documentation discuss in details the translation of a Simulink model into suitable description for target hardware platform [9]. This process generates a sequence of intermediate representations of the primary Simulink functional view. A key transition is the one from mathematical expressions to C formal language. Consequently generated code is integrated with additional device drive libraries and compiled to stand alone executable. Alternatively, one can integrate generated code with application interface calls and create an application for real-time operating system. The process finishes with compilation and loading into target microcontroller. These transformations of the original functional descriptions are required because Simulink model is more abstract form of representation than target hardware resources can handle. Code generation is controlled through template-like configuration files.

In the following points are summarized some general properties of above components:

- *Target language compiler (TLC)*. TLC is an intermediate script language representing the transition from Simulink graphical blocks to the target code. Usually each Simulink block has a corresponding TLC file. When S-function block is used the designer has to specify also its TLC description in order to inline the block into target software. TLC parser can read generated *rtw* file from the Simulink model. The parser also can access block input and output signals which are necessary to properly reproduce block function in C language.

- *System target file (STF)*. This is the main TLC file which is read when Build Command is executed. It calls the rest of block specific TLC files. The STF also defines some target specific interface for user interaction through Model Configuration Settings. For example one can turn on or off a particular software future in the generated code.

- *Template make file (TMF)*. In a standard compiler toolchain one can control the whole process of compilation and linking with a huge set of options through a dedicated compiler and linker switches. The various switches for a particular project are integrated in a single makefile which specify how the whole project is build. The makefile is fed into a make tool which in turn invokes other programs from the toolchain. TMF is a template for makefile generation for a particular Simulink model. It is like a makefile with a lot of placeholders for model specific names and user settings.

- *Hook files*. The build process is guided by the *make_rtw* program. It is the entry point for code generation. The hook files allow the user to interact with code generation process for example to setup some include and source file directories.

- *rtw file*. This file contains a representation of the Simulink model as a hierarchical record structure including all block interconnections and parameters. This is the form which TLC parser can read in order to produce C programs.

- *Target Simulink library*. Each target defines its specific blocks for device drivers or other platform dependent resources.

Programing of embedded control systems is commonly based on C programing language. However, some RTL functionality is still programmed in lower level language, i.e., target assembler. The compilation and linking programs are essential for generation of executable code for a target hardware platform.

First, the compilation program transforms C program to a corresponding object code which is program representation with hardware dependent instructions (Figure 1.43). The object file contains also a symbolic table for declared identifiers by the program. The symbolic table allows address independent program representation which is necessary to generate relocatable and modular programs. Second, the linking program combines all modular object file and other static libraries into a single relocatable file. In this form, the program is a set of memory sections which can be mapped to the target physical address space by the loader program (Figure 1.45).

Common practice is programing in native target language (assembler) to be as little as possible because of portability issues. However, knowledge of target assembler is required in certain situations—hardware initialization, memory usage or execution speed optimization, debugging on instruction basis.

The interface between platform specific software and Simulink model is based on custom developed blocks (S-function driver blocks). They support separate representation for simulation and implementation. Usually during simulation, these

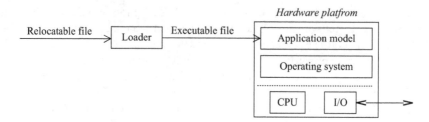

Figure 1.45   *Simulink Coder TMF infrastructure*

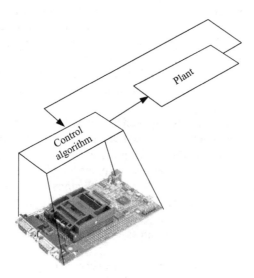

Figure 1.46   *HIL simulation concept*

driver blocks have only structure presence in diagram sourcing and sinking dummy signals. However during code generation, driver blocks TLC script can access connected signals and other block parameters with the aim to produce required C program for the block (Figure 1.46).

1.9.4 Code validation

There exists some helpful validation techniques which are used frequently with automatic code generation. The purpose of code validation is to experiment with automatically generated program in hardware or in cycle-accurate simulator. The problem is that in visual programing, the same task can be achieved in many different manners which are equivalent on visual level but not on a textual or machine level. Therefore, if one doesn't understands the internal mechanism of code generation process, there is a high risk to produce a nonoptimal or nonreliable solution. For this reason, code validation have an important place before actual experimental work with the plant to begin.

Many of the problems in a system under design can be found trough precise numerical simulation which accounts for target hardware specifics. In cycle-accurate simulation, timing of events is precise but there is no need the simulation to use exact hardware models.

Replacing a physical system like a vehicle, plane, or robot with a real-time simulation of a virtual system drastically reduces the cost of testing control software and hardware. The corresponding methodology is called HIL simulation. HIL designer decides which subsystems to be implemented as hardware and which as simulations. Hardware implementation in the early project phases could bring some considerable risk or unpredictable costs. Simulations are safe to execute and easy to handle but to be useful require extensive experimental knowledge and consequent verification and validation procedures. In the field of control usually the plant is simulated the controller is in hardware so HIL simulation decisions are straightforward. Still appropriate interface points and interconnections should be established.

Usually hardware implementation of the controller doesn't cause large variations from host-simulated controller except for some critical conditions caused by limited platform resources. However under HIL simulation generated control signal u_{hil} is more or less different from non-HIL simulated u_{sim} so

$$u_{hil} = u_{sim} + \tilde{u}_{hw} \tag{1.6}$$

where $\tilde{u}_{hw}$ reflects several hardware-dependent effects which are summarized in the following points:

- Target hardware may use different finite precision arithmetic (fixed or floating-point) than host model for representation of numbers and arithmetic operations. This causes rounding errors subject to numerical analysis.
- Communication between host and the target can experience uncertain transport delays caused by loss of data during transmission related to line collision conditions, electromagnetism noise sources or protocol negotiations.
- Reaction of real-time system to external events is temporally deterministic but still variable in some interval. This reflects as slight deviations in defined by designer sample times which can be represented as impulse-like input disturbance acting upon continuous plant model.

1.10 Notes and references

Important aspects of embedded control system design are considered in Hristu-Varsakelis and Levine [10], Marwedel [11], Popovici *et al.* [12], Westcot [13], Basten *et al.* [14], Ledin [15], Lozzano [16]. A good overview of the design process is presented in Forrai [17].

A variety of sensors and actuators is described in detail in Fraden [18], Anjanappa *et al.* [2, Chapter 17], Bräunl [19, Chapters 3 and 4], and Isermann [20, Chapters 9 and 10].

The theory of digital control systems is presented in depth in Åström and Wittenmark [21]; Landau and Zito [4]; Franklin, Powell, and Workman [8]; Isermann [22,23]; Fadali and Visioli [24]; and Houpis and Lamont [25]. For good introductions to this subject, see Santina and Stubberud [26, Chapters 10, 12, and 13] and Wittenmark, Åström and Årzén [3]. The book of Pelgrom [27] is entirely devoted to ADC.

The computer arithmetic and floating point computations are described in Higham [7, Chapter 2], Moler [28], Muller *et al.* [29], Koren [30], Laub [31], and Overton [32].

The techniques of control prototyping and HIL simulation are discussed in Iserman [20, Chapter 12], [33], [2, Chapter 2] and Forrai [17, Chapter 8]. More on HIL and PIL simulations may be found in the user guide of Embedded Coder [34, Chapter 36].

Different aspects of hardware and software configuration of embedded systems are presented in depth in Hardin [35], Marwedel and Goossens [36], Mogensen [37], Noergaard [38], and Pedroni [39]. The development of embedded systems with FPGA is considered in Dubey [40], Kilts [41], and Sass and Schmidt [42].

Chapter 2

System modeling

This chapter is devoted to the mathematical description of the basic elements and processes pertaining to the embedded control systems. The models obtained as a result of this description are important for the design of controllers which have to ensure the necessary performance and robustness of the closed-loop system. The main point of the chapter is the derivation of adequate continuous-time and discrete-time models of the plant, sensors, and actuators. For this aim, we implement various analytic and numeric tools available in control theory and control engineering practice. These tools include modeling, linearization, and discretization of dynamic plants, system identification, modeling of uncertain systems, and stochastic modeling. We demonstrate the usage of different MATLAB® functions and Simulink® blocks intended to build accurate and reliable models of embedded system components.

2.1 Plant modeling

An important stage in building the system model is the control plant modeling. In the general case, the plant is a nonlinear multiple-input–multiple-output (multivariable) high-order dynamic system whose properties to a large extent determine the performance which may be achieved by the closed-loop system. In some cases, the plant involves processes described by partial differential equations, like gas and liquid flows, combustion processes, and so on, which for the design purposes should be approximated by low-order models described by ordinary differential equations. The plant modeling is thus a trade-off between the accuracy of dynamics approximation and the complexity of model obtained.

There is a variety of techniques used in the modeling of control plants which lead to different mathematical models. In the relatively simple cases, it is possible to derive *analytic models* implementing the laws (or principles) of physics, chemistry, biology, economics, and so on. Such models are convenient to use but frequently do not embrace the whole complexity of the processes under control. That is why, these models are added with blocks whose input–output behavior is determined experimentally or by using some identification method applied on the experimental data. In such case, one says that a *numeric model* is derived to distinguish it from the analytic models.

The mathematical description of plant dynamics can be obtained in the form of differential and algebraic equations, transient and frequency responses, tabular data, and so on. For the purpose of system analysis and design, we need a plant model in the form of linear state-space model or transfer function matrix. Such description is convenient to obtain by using the capabilities of MATLAB and Simulink which are suitable to determine high-order models.

Usually, the plant modeling is a difficult process which is performed in an iterative way. The systematic presentation of this process is out of the scope of this book and we refer the reader to the literature given at the end of this chapter, where the modeling techniques are described in sufficient details. In this section, we shall assume that the basic mathematical description of the plant in the form of (non)linear differential and algebraic equations is already available and it is necessary to find a linear plant model in the form of state-space equations or transfer function matrix. This leads to the necessity of linearizing the plant dynamics in the vicinity of an equilibrium state and subsequent discretization of the plant model for given sampling frequency used in the real time control system.

Example 2.1. Cart–pendulum system

Several embedded control problems considered in this book will be illustrated by the cart–pendulum system, used as a laboratory set-up in many universities. The system is shown schematically in Figure 2.1. It consists of an inverted pendulum with mass m (assumed concentrated at the tip) which is balanced in upright position by moving the cart on a rail in appropriate direction. The control problem is to stabilize the pendulum around the vertical axis keeping at the same time the cart at reference position. As we will see latter on, this system has some peculiarities which makes its stabilization a difficult control problem.

To find the mathematical description of the cart–pendulum system it is appropriate to use the Newton–Euler dynamics. For this aim, we introduce as state variables the cart position p and cart velocity $\dot{p}$, the pendulum angle θ and pendulum angular rate $\dot{\theta}$. The force applied at the base of the system is denoted by F, and it is aligned

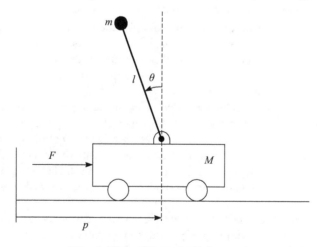

Figure 2.1   Cart–pendulum system

with p in the horizontal direction. Further on, we shall assume that this force is produced by the aid of a DC motor with pulse-width modulated (PWM) input signal u so that $F = k_F u$, where k_F is a constant parameter. As system outputs, we consider the cart position and pendulum angle. The system is described by two second-order nonlinear differential equations in the form

$$
\begin{bmatrix} M+m & -ml\cos\theta \\ -ml\cos\theta & I+ml^2 \end{bmatrix} \begin{bmatrix} \ddot{p} \\ \ddot{\theta} \end{bmatrix} + \begin{bmatrix} f_c \dot{p} + ml\sin\theta\dot{\theta}^2 \\ f_p \dot{\theta} - mgl\sin\theta \end{bmatrix} = \begin{bmatrix} k_F u \\ 0 \end{bmatrix}
\tag{2.1}
$$

where M is the cart mass, I is the moment of inertia of the system to be balanced, l is the distance from the base to the center of mass of the balanced body, f_c and f_p are coefficients of viscous friction related to the motion of cart and pendulum, respectively, and g is the acceleration of gravity.

In Table 2.1, we give the nominal parameters of a real cart–pendulum system which are used in the subsequent analysis and design of the system.

Let us introduce the state vector as $x = [p, \ \theta, \ \dot{p}, \ \dot{\theta}]^T$ and the output vector as $y = [p \ \theta]^T$. Defining the total mass as $M_t = M + m$ and the total moment of inertia as $I_t = I + ml^2$, (2.1) is represented as

$$
\frac{d}{dt}\begin{bmatrix} p \\ \theta \\ \dot{p} \\ \dot{\theta} \end{bmatrix} = \begin{bmatrix} \dot{p} \\ \dot{\theta} \\ \dfrac{-mlI_t s_\theta \dot{\theta}^2 + m^2 l^2 g s_\theta c_\theta - I_t f_c \dot{p} - f_p m l c_\theta \dot{\theta} + I_t k_F u}{M_t I_t - m^2 l^2 c_\theta} \\ \dfrac{-m^2 l^2 s_\theta c_\theta \dot{\theta}^2 + M_t m g l s_\theta - f_c m l c_\theta \dot{p} - f_p M_t \dot{\theta} + m l c_\theta k_F u}{M_t I_t - m^2 l^2 c_\theta} \end{bmatrix},
\tag{2.2}
$$

$$
y = \begin{bmatrix} p \\ \theta \end{bmatrix}
$$

where s_θ stands for $\sin(\theta)$ and c_θ stands for $\cos(\theta)$.

The nonlinear cart–pendulum model (2.2) is implemented by the Simulink model `pendulum_cart.slx`.

Table 2.1 Cart–pendulum nominal model parameters

Parameter	Description	Value	Units
m	Equivalent pendulum mass	0.104	kg
M	Equivalent cart mass	0.768	kg
l	Distance from the base to the system center of mass	0.174	m
I	Pendulum moment of inertia	2.83×10^{-3}	kg m^2
f_c	Dynamic cart friction coefficient	0.5	N s/m
f_p	Rotational friction coefficient	6.65×10^{-5}	N ms/rad
k_F	Control force to PWM signal ratio	9.4	N

Equation (2.2) can be written in the general form

$$\frac{d}{dt}x(t) = f(x, u),$$
$$y(t) = h(x, u)$$

(2.3)

where $x(t)$ is the *state*, $u(t)$ is the *control*, and $y(t)$ is the *output*.

Equation (2.3) represents the nonlinear plant model. This model is used to obtain a linearized plant model and is implemented in the nonlinear simulation of the closed-loop system. ☐

2.2   Linearization

MATLAB files used in this section

File	Description
`linearize_pendulum_cart`	Linearization of the cart–pendulum system
`transfer_functions`	Transfer functions of the cart–pendulum system
`pendulum_cart.slx`	Simulink model of the cart–pendulum system

To linearize the nonlinear plant (2.3), it is possible to implement different techniques. In this section, we shall describe three methods which will be provisionally named as *analytic, symbolic, and numeric linearization*.

2.2.1   Analytic linearization

The analytic linearization aims to obtain linear plant equations by using the analytic plant model (2.3). In this case, the plant dynamics is approximated by linear differential and algebraic equations in the vicinity of an equilibrium point x_e, u_e characterized by the condition $\dot{x} = f(x, u) = 0$. This is done by expanding the nonlinear functions f and h in Taylor series in respect to the vectors $\Delta x = x - x_e$, $\Delta u = u - u_e$ and $\Delta y = y - y_e$. Since the deviations from the equilibrium state are assumed small, the quadratic and higher order terms in Δx, Δu, and Δy can be neglected. As a result one obtains the linearized state equation

$$\frac{d}{dt}\Delta x(t) = A\Delta x + B\Delta u,$$
$$\Delta y(t) = C\Delta x + D\Delta u$$

(2.4)

where A, B, C, D are matrices with corresponding dimensions. These matrices are obtained as partial derivatives (Jacobian matrices) of the vector-functions f and h in respect to the arguments x and u, evaluated at the equilibrium (trim) values x_e, u_e as

$$A = \left.\frac{\partial f}{\partial x}\right|_{\substack{x=x_e \\ u=u_e}}, \quad B = \left.\frac{\partial f}{\partial u}\right|_{\substack{x=x_e \\ u=u_e}}, \quad C = \left.\frac{\partial h}{\partial x}\right|_{\substack{x=x_e \\ u=u_e}}, \quad D = \left.\frac{\partial h}{\partial u}\right|_{\substack{x=x_e \\ u=u_e}}.$$

The trim values are obtained as a solution of the nonlinear algebraic equation $f(x_e, u_e) = 0$ and may be nonunique. In the case of time-invariant models, the matrices A, B, C, D have constant elements. Further on the symbol Δ will be dropped out to simplify notation and instead of (2.4), we shall use the equation

$$\frac{d}{dt}x(t) = Ax + Bu,$$
$$y(t) = Cx + Du.$$

(2.5)

However, the reader should always bear in mind that x, y, u involved in this equation are actually the deviations of the corresponding variables from their trim values.

Example 2.2. Analytic linearization of cart–pendulum system

Consider the linearization of the model (2.3) for small deviations of the pendulum angle around zero. The equilibrium state in the given case corresponds to the straight-up position $x_e = [0, \ 0, \ 0, \ 0]^T$. (Note that this equilibrium state is unstable.) Taking into account that for small θ it is fulfilled that $\sin(\theta) \approx \theta$, $\cos(\theta) \approx 1$ and $\dot{\theta}^2 \approx 0$ we obtain the linearized model

$$\frac{d}{dt}x(t) = Ax(t) + Bu(t),$$
$$y(t) = Cx(t)$$

where

$$x(t) = [p \ \theta \ \dot{p} \ \dot{\theta}]^T, \quad y(t) = [p \ \theta]^T.$$

and

$$A = \begin{bmatrix} 0 & 0 & 1 & 0 \\ 0 & 0 & 0 & 1 \\ 0 & m^2 l^2 g/\eta & -f_c I_t/\eta & -f_p lm/\eta \\ 0 & M_t mgl/\eta & -f_c ml/\eta & -f_p M_t/\eta \end{bmatrix},$$

$$B = k_F \begin{bmatrix} 0 \\ 0 \\ I_t/\eta \\ lm/\eta \end{bmatrix}, \quad C = \begin{bmatrix} 1 & 0 & 0 & 0 \\ 0 & 1 & 0 & 0 \end{bmatrix},$$

$$\eta = M_t I_t - m^2 l^2, \quad I_t = I + ml^2.$$

Note that in the given case

$$D = \begin{bmatrix} 0 \\ 0 \end{bmatrix}.$$

The presence of a zero column of matrix A prompts that there is an integrator in the plant.

For the nominal parameters of the cart–pendulum system, one obtains the matrices (up to four decimal digits)

$$
A = \begin{bmatrix}
0 & 0 & 1 & 0 \\
0 & 0 & 0 & 1 \\
0 & 6.5748 \times 10^{-1} & -6.1182 \times 10^{-1} & -2.4629 \times 10^{-4} \\
0 & 3.1682 \times 10^{1} & -1.8518 \times 10^{0} & -1.1868 \times 10^{-2}
\end{bmatrix},
$$

$$
B = \begin{bmatrix}
0 \\
0 \\
1.5736 \times 10^{1} \\
4.7629 \times 10^{1}
\end{bmatrix}, \quad
C = \begin{bmatrix}
1 & 0 & 0 & 0 \\
0 & 1 & 0 & 0
\end{bmatrix}.
$$

The plant is completely controllable and completely observable. Its poles, found by the function `pole`, are

$$
p_1 = 0, \quad p_2 = 5.6054, \quad p_3 = -5.6561, \quad p_4 = -0.5730
$$

which confirms that there is a zero pole corresponding to an integrator.   ☐

The analytic linearization has the advantage that the linear model (2.5) can be obtained so that the elements of matrices A, B, C, D depend explicitly on plant parameters. This is important in deriving uncertain plant models in which the parameters vary around their nominal values. The drawback of this linearization method is that its implementation by hand is difficult in case of high-order systems and may be associated with errors.

2.2.2 Symbolic linearization

The symbolic linearization is a kind of analytic linearization performed by using the Symbolic Math Toolbox™ of MATLAB. The functions of this toolbox allow to manipulate analytically complicated expressions arising in the derivation of nonlinear plant models and to evaluate the Jacobian matrices participating in the linearized models. This makes possible to obtain error free expressions for the linearized models of high-order systems.

Example 2.3. Symbolic linearization of cart–pendulum system
To find the matrices A, B, C, D of the linearized cart–pendulum system, one should first enter in symbolic form the nonlinear model (2.1). This can be done by the command lines

```
syms p p_dot theta theta_dot u
syms m g l I_t M_t f_c f_p k_F
%
M_q = [M_t      -m*l*cos(theta)
       -m*l*cos(theta) I_t]
C_q = [f_c*p_dot+m*l*sin(theta)*theta_dot^2-k_F*u
       f_p*theta_dot-m*g*l*sin(theta)]
```

The nonlinear expressions for the variables $\dot{p}$, $\dot{\theta}$ are found easily by using the single command

```
f = -inv(M_q)*C_q
```

which inverts symbolically the matrix M_q. Then, the Jacobian matrices are computed by the lines

```
d_p = p_dot;
d_theta = theta_dot;
d_p_dot = f(1);
d_theta_dot = f(2);
%
F_x = jacobian([d_p; d_theta; d_p_dot; d_theta_dot], ...
               [p theta p_dot theta_dot])
F_u = jacobian([d_p; d_theta; d_p_dot; d_theta_dot], [u])
H_x = jacobian([p; theta], [p theta p_dot theta_dot])
H_u = jacobian([p; theta], [u])
```

Setting the trim conditions as

```
p_trim = 0;
theta_trim = 0;
p_dot_trim = 0;
theta_dot_trim = 0;
u_trim = 0;
```

the symbolic expressions for the matrices of the linearized model are found by the lines

```
F_x = subs(F_x,{p,theta,p_dot,theta_dot,u}, ...
               {p_trim,theta_trim,p_dot_trim,theta_dot_trim,u_trim})
F_u = subs(F_u,{p,theta,p_dot,theta_dot,u}, ...
               {p_trim,theta_trim,p_dot_trim,theta_dot_trim,u_trim})
H_x = subs(H_x,{p,theta,p_dot,theta_dot,u}, ...
               {p_trim,theta_trim,p_dot_trim,theta_dot_trim,u_trim})
H_u = subs(H_u,{p,theta,p_dot,theta_dot,u}, ...
               {p_trim,theta_trim,p_dot_trim,theta_dot_trim,u_trim})
```

As a result, we obtain expressions for the elements of matrices A, B, C, D as explicit functions of the plant parameters. Numerical values of these matrices may be found by substituting the plant parameters by their numerical values in the derived expressions for F_x, F_y, H_x, H_u. ❏

The using of symbolic computations is also convenient in determining the plant transfer functions. To find the transfer functions between the plant input u and the outputs p and θ, one may solve symbolically the linear system of equations

$$Mf = q$$

in respect to f where

$$M = \begin{bmatrix} \eta s^2 + f_c l_t s & f_p m l s - m^2 l^2 g \\ f_c m l s & \eta s^2 + f_p M_t s - M_t m g l \end{bmatrix}, \quad q = \begin{bmatrix} I_t k_F \\ m l k_F \end{bmatrix}$$

As a result, one obtains the transfer functions

$$G_{pu}(s) = k_F \frac{I_t s^2 + f_p s - mgl}{(M_t I_t - m^2 l^2)s^4 + (M_t f_p + I_t f_c)s^3 + (f_c f_p - M_t mgl)s^2 - f_c mgls}$$

and

$$G_{\theta u}(s) = k_F \frac{mls}{(M_t I_t - m^2 l^2)s^3 + (M_t f_p + I_t f_c)s^2 + (f_c f_p - M_t mgl)s - f_c mgl}$$

Note that due to cancelation between zero and pole in the nominator and denominator of the transfer function $G_{\theta u}$, the order of this function is equal to three, while the order of G_{pu} is four. This cancelation does not affect the system controllability and observability because it takes place only in the transfer function $G_{\theta u}$. Using the nominal plant parameters, one obtains that

$$G_{pu}(s) = \frac{15.74s^2 + 0.175s - 467.2}{s^4 + 0.6237s^3 - 31.68s^2 - 18.17s},$$

$$G_{\theta u}(s) = \frac{47.63s}{s^3 + 0.6237s^2 - 31.68s - 18.17}.$$

The transfer function G_{pu} has zeros at

$$z_1 = -5.45463, \quad z_2 = 5.4435$$

and $G_{\theta u}$ has one zero at 0. The plant is nonminimum phase.

The block diagram of cart–pendulum system is represented in Figure 2.2. The system is described by the equation

$$y(s) = G(s)u(s)$$

where

$$G(s) = \begin{bmatrix} G_{pu}(s) \\ G_{\theta u}(s) \end{bmatrix}.$$

Note that $G(s)$ is a 2×1 (rectangular) transfer function matrix so that the frequency response $G(j\omega)$ has only one singular value. This means that the closed-loop system may track only one reference and in the given case this is the cart position. The pendulum can only be stabilized around the vertical position.

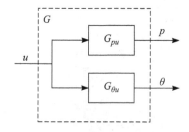

Figure 2.2　Block diagram of cart–pendulum system

The symbolic linearization is recommended for using in all cases when an analytic nonlinear plant model is available.

2.2.3 *Numeric linearization*

A linearized model of the nonlinear plant (2.2) can be determined by using a Simulink model of the plant and implementing the functions `trim` and `linmod`. The purpose of the function `trim` is to find the state x_e and the control u_e corresponding to the equilibrium point of the system. It also gives information about the value of the state derivative at equilibrium point which should be theoretically equal to zero. `trim` has several options allowing to satisfy input, output, and state constraints. Using the information obtained by `trim`, the function `linmod` determines the matrices of the plant model linearized at equilibrium point. Note that the linearized model may have a state vector which differs from the state vector of the analytic model. This will affect the matrices of the linearized model but will not affect the input–output properties of the model.

Example 2.4. Numeric linearization of cart–pendulum system
The numeric linearization of cart–pendulum system utilizes the nonlinear plant model `pendulum_cart.slx` and is done by the command lines

```
x0 = [0 0 0 0]';
u0 = 0;
[x,u,y,dx] = trim('pendulum_cart',x0,u0,[],[],[],[])
[a,b,c,d] = linmod('pendulum_cart',x,u);
G_num = ss(a,b,c,d);
```

The state-space model of the linearized plant is obtained in the variable `G_num`. The elements of matrices *a*, *b*, *c*, *d* obtained by this linearization method coincide with the corresponding elements of the matrices *A*, *B*, *C*, *D* found by analytic linearization at least to then decimal digits and the vector *dx* is exactly zero vector. ❑

It is interesting to compare the three linearized models of the cart–pendulum system derived in Examples 2.2–2.4.

In Figures 2.3 and 2.4, we compare the Bode plots of analytic, symbolic, and numeric models for the nominal parameters with outputs cart position *p* and pendulum angle θ, respectively. The three models practically coincide.

2.3 Discretization

MATLAB files used in this section

File	Description
discrete_freq_response	Discrete-time frequency responses
discretization_pendulum_cart	Comparison of different discretization methods
discrete_transfer_functions	Discrete-time transfer functions of cart–pendulum system

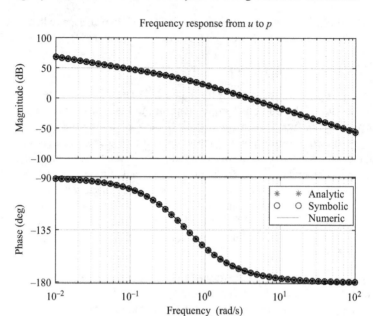

Figure 2.3   Bode plot of the linearized models with output p

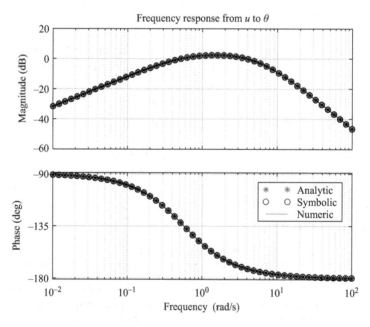

Figure 2.4   Bode plot of the linearized models with output θ

2.3.1 *Discrete-time models*

A discrete-time control system may be described by different mathematical models. In the simple case of a linear discrete-time system with scalar input $u(k)$ and scalar output $y(k)$, it is possible to use input–output relation in the form of nth order difference equation

$$y(k+n) + a_1 y(k+n-1) + \cdots + a_n y(k) = b_0 u(k+m) + \cdots + b_m u(k) \quad (2.6)$$

where $n \geq m$.

In the analysis of the system (2.6), it is convenient to use the so-called *forward shift operator q* defined by

$$qf(k) = f(k+1)$$

Multiplying a time-sequence by the operator q implies that the sequence is shifted one step ahead. Multiplying with q^i implies a shift of i steps forward. In the same way, multiplying with q^{-i} means a backward shift of i steps. The operator q^{-1} is called the *backward shift operator*.

In terms of the forward shift operator, the model (2.6) becomes

$$q^n y(k) + a_1 q^{n-1} y(k) + \cdots + a_n y(k) = b_0 q^m u(k) + \cdots + b_m u(k) \quad (2.7)$$

or

$$y(k) = \frac{B_q(q)}{A_q(q)} u(k) \quad (2.8)$$

where the polynomials $A_q(q)$ and $B_q(q)$ are defined as

$$A_q(q) = q^n + a_1 q^{n-1} + \cdots + a_n,$$
$$B_q(q) = b_0 q^m + b_1 q^{m-1} + \cdots + b_m$$

Taking Z-transform on each side of (2.7) and assuming zero initial conditions, one obtains that

$$A_q(z)Y(z) = B_q(z)U(z) \quad (2.9)$$

where $Y(z)$, $U(z)$ are the Z-transforms of the sequences $y(k)$ and $z(k)$, respectively, and

$$A_q(z) = z^n + a_1 z^{n-1} + \cdots + a_n,$$
$$B_q(q) = b_0 z^m + b_1 z^{m-1} + \cdots + b_m$$

Equation (2.9) can be rewritten as

$$Y(z) = G(z)U(z) \quad (2.10)$$

where

$$G(z) = \frac{B_q(z)}{A_q(z)} \quad (2.11)$$

is called the *discrete transfer function* of the system (2.9). As in the continuous-time case, the transfer function uniquely determines the input–output behavior at the discrete sampling times for zero initial conditions. The roots of $A_q(z) = 0$ are the *poles* and the roots of $B_q(z) = 0$ are the *zeros* of the system. Stability of the system then requires that all the poles should be strictly inside the unit circle.

In several cases instead of the model (2.6), it is preferable to use a state-space description in the form

$$x(k+1) = Ax(k) + Bu(k), \qquad (2.12)$$

$$y(k) = Cx(k) + Du(k) \qquad (2.13)$$

where $x(k)$, $u(k)$, and $y(k)$ are the state, input, and output sequences, respectively, and A, B, C, D are matrices with appropriate dimensions.

Using the forward shift operator in (2.12) gives

$$qx(k) = Ax(k) + Bu(k), \qquad (2.14)$$

$$y(k) = Cx(k) + Du(k) \qquad (2.15)$$

Taking Z-transforms in (2.14), (2.15) for zero initial conditions yields

$$Y(z) = G(z)U(z) \qquad (2.16)$$

where

$$G(z) = C(zI - A)^{-1}B + D \qquad (2.17)$$

The role of discrete transfer functions (2.16) and (2.17) in describing the dynamic behavior of linear systems parallels that of transfer functions for continuous-time systems.

2.3.2   Discrete-time frequency responses

Consider a single-input–single-output discrete-time system described by the transfer function $G(z)$. Assume that the input $u(k)$ of the system is obtained by sampling with period T_s of the continuous-time sinusoidal signal

$$u(t) = \sin(\omega t) \qquad (2.18)$$

Replacing t by kT_s, we see that the input to the discrete-time system is a sinusoidal sequence of the form

$$u(k) = \sin(\Omega k) \qquad (2.19)$$

where $\Omega = \omega T_s$. Then, it is possible to show that the forced system output $y(k)$ is another sinusoidal sequence with the same frequency of the form

$$y(k) = M(\Omega)\sin(\Omega k + \phi) \qquad (2.20)$$

where

$$G(e^{j\Omega}) = M(\Omega)e^{j\phi(\Omega)} \qquad (2.21)$$

Hence, the frequency response plots of a discrete-time system are plots of the magnitude $M(\Omega)$ and the phase $\phi(\Omega)$ of the discrete-time transfer function evaluated at $z = e^{j\Omega}$. These plots are periodic, because $e^{j\Omega}$ is periodic in Ω with period 2π.

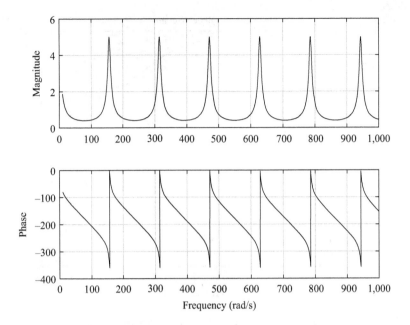

Figure 2.5   Periodicity of the frequency response of a discrete-time system

Since the frequency responses are symmetric about $\Omega = \pi$, the frequency range of Ω from 0 to π is sufficient to completely specify the frequency response of a discrete-time system. Therefore, the frequency response is completely specified if $\omega = \Omega/T_s$ varies in the range from 0 to $\pi/T_s = \omega_N$ where $\omega_N = \omega_s/2$ is the Nyquist frequency.

Example 2.5.   Periodic behavior of discrete-time frequency responses
The periodic behavior of the discrete-time frequency responses is illustrated in Figure 2.5 which shows the magnitude and phase responses of a first-order lag

$$G(s) = \frac{0.5}{0.25s + 1}$$

whose input is sampled with period $T_s = 0.04$ s, the output being reconstructed by using zero-order hold. The corresponding discrete-time transfer function is (to six digits)

$$G(z) = \frac{0.739281}{z - 0.852144}$$

The Bode diagram of the discrete-time system obtained for frequencies in the range $[0.1, \omega_N = 78.5398]$ is shown in Figure 2.6.   ❑

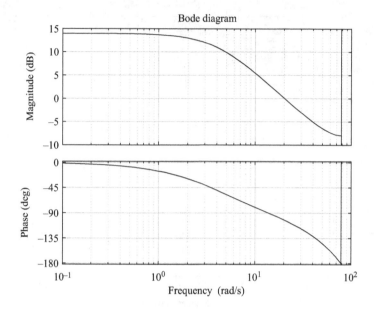

Figure 2.6   Bode diagram a discrete-time system

2.3.3   Discretization of continuous-time models

After finding a linearized description of a nonlinear plant, the next step is to determine a discrete-time linearized model for a given sampling period T_s. This model is used in the design of a discrete-time controller which is implemented in the embedded control system or in the hardware-in-the-loop simulation of the closed-loop system.

The discrete-time equivalent of the continuous-time model (2.5) has the form

$$
\begin{aligned}
x(k+1) &= A_d x(k) + B_d u(k), \\
y(k) &= C_d x(k) + D_d u(k)
\end{aligned}
\tag{2.22}
$$

where $x(k)$, $u(k)$, $y(k)$ are the state vector, control vector, and output vector, respectively, at the sampling instant kT_s and A_d, B_d, C_d, D_d are matrices with appropriate dimensions. These matrices are determined from the corresponding matrices of the continuous-time model (2.5) and depend on the sampling period T_s and the implemented discretization method. Systematic presentation of discretization methods for continuous-time models may be found in the references given in the end of this chapter. In this book, we shall make use of the *zero-order hold discretization method* (Figure 2.7) which produces a discrete-time model with matrices

$$
A_d = e^{AT_s}, \quad B_d = \int_0^{T_s} e^{A\sigma} d\sigma B, \quad C_d = C, \quad D_d = D.
$$

This method is characterized by the fact that the pole p of the continuous-time model is mapped into the pole e^{pT_s} of the discretized model.

Figure 2.7   Discretization of a continuous-time plant

It is possible to show that the matrices A_d and B_d of the discrete-time model may be found by computing a matrix exponential. Let

$$Z = \begin{bmatrix} A & B \\ 0_{m \times n} & 0_{m \times m} \end{bmatrix} T_s$$

is a block $(n + m) \times (n + m)$ matrix. Then

$$\begin{bmatrix} A_d & B_d \\ 0 & I_m \end{bmatrix} = e^Z.$$

The derivation of discrete-time model of a nonlinear plant may be done in MATLAB in two ways. The first opportunity is to discretize the continuous-time linear model obtained by some of the linearization methods, considered in the previous section. This can be done by using the function c2d. The second opportunity is to use the function dlinmod which derives directly the linearized discrete-time model from the continuous-time nonlinear plant model. Both opportunities will be considered briefly below.

The function c2d has an option to implement five discretization methods, namely,

Method	Description
'zoh'	Zero-order hold on the inputs (default method)
'foh'	First-order hold (linear interpolation of inputs)
'impulse'	Impulse-invariant discretization
'tustin'	Bilinear (Tustin) approximation
'matched'	Matched pole-zero method (for SISO systems only)

Example 2.6. Discretization of cart–pendulum system

Given the nominal continuous-time linearized model of the cart–pendulum system, derived in Example 2.2, the discrete-time model for sampling interval $T_s = 0.01$ s is obtained as

```
[Ad,Bd] = c2d(A,B,Ts)
```

The discrete-time model has a state vector

$$x(k) = [p(k) \; \theta(k) \; \dot{p}(k) \; \dot{\theta}(k)]^T$$

and matrices

$$A_d = \begin{bmatrix} 1 & 3.2815 \times 10^{-4} & 9.9695 \times 10^{-3} & 9.7135 \times 10^{-8} \\ 0 & 1.0016 \times 10^{0} & -9.2424 \times 10^{-5} & 1.0005 \times 10^{-2} \\ 0 & 6.5578 \times 10^{-3} & 9.9390 \times 10^{-1} & 3.0358 \times 10^{-5} \\ 0 & 3.1691 \times 10^{-1} & -1.8470 \times 10^{-2} & 1.0015 \times 10^{0} \end{bmatrix},$$

$$B_d = \begin{bmatrix} 7.8521 \times 10^{-4} \\ 2.3771 \times 10^{-3} \\ 1.5689 \times 10^{-1} \\ 4.7506 \times 10^{-1} \end{bmatrix}, \quad C_d = \begin{bmatrix} 1 & 0 & 0 & 0 \\ 0 & 1 & 0 & 0 \end{bmatrix}, \quad D_d = \begin{bmatrix} 0 \\ 0 \end{bmatrix}.$$

The discrete-time transfer functions in respect to p and θ are determined from the discrete-time state-space model by using the function tf. As a result, one obtains

$$H_{pu}(z) = \frac{0.0007852z^3 - 0.0007891z^2 - 0.0007843z + 0.0007835}{z^4 - 3.997z^3 + 5.988z^2 - 3.984z + 0.9938},$$

$$H_{\theta u}(z) = \frac{0.002377z^2 - 4.938e - 06z - 0.002372}{z^3 - 2.997z^2 + 2.991z - 0.9938},$$

respectively.

The zeros of H_{pu} are

$$z_1 = -0.99796, \quad z_2 = 1.05594, \quad z_3 = 0.94691$$

and the zeros of $H_{\theta u}$ are

$$z_1 = 1.0, \quad z_2 = -0.99792.$$

Note that these transfer functions have $n - 1$ zeros, where n is the order of the corresponding transfer function. This is a common property of the discretized models with zero order hold method [43, Section 13.3]. ❑

In what follows we compare the properties of the discretization methods implementing zero-order hold, first-order hold, and Tustin approximation, respectively. Using the continuous-time model of the cart–pendulum system in respect to the output p, the function c2d produces three different discrete-time models using the methods "zoh", "foh", and "tustin".

The magnitude and phase plots of the discretized models are compared in Figures 2.8 and 2.9, respectively. As it is seen from the figures, the Tustin and zero-order hold method tend to produce large magnitude errors near to the Nyquist frequency while the zero-order hold method introduces large phase lag in the same frequency region. These approximation errors may be reduced to some extent by increasing the sampling frequency. Independently on the relatively worse properties of the zero-order hold method it is widely used in practice due to the simplicity of sampling and hold technical implementation. Note that the method "tustin" may be used in conjunction with *frequency prewarping* which allows to remove the frequency distortion at a specific frequency, see for details [3, Section 6].

As noted previously, a discrete-time plant model may be derived directly from the continuous-time nonlinear model by using the function dlinmod which is a

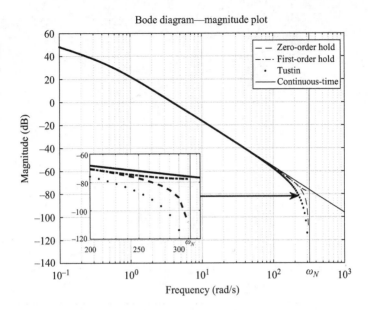

Figure 2.8 Magnitude plots of discretized models

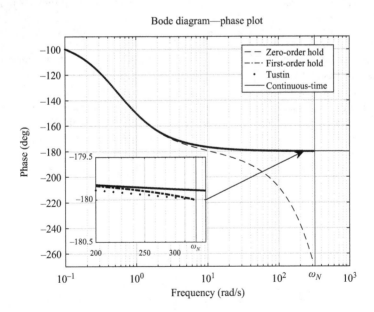

Figure 2.9 Phase plots of discretized models

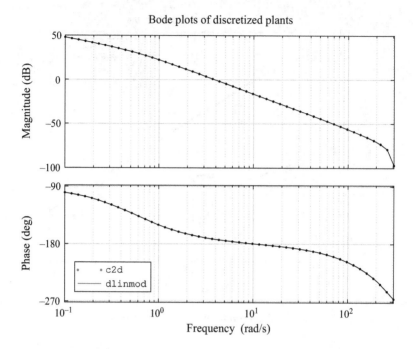

Figure 2.10   Comparison of Bode plots of discrete-time models obtained by c2d *and* dlinmod

discrete-time counterpart of the linearization function linmod. It is appropriate to use the function dlinmod if a Simulink model of the continuous-time or mixed continuous-time and discrete-time plant is available.

In Figure 2.10, we show the Bode plots of the discrete-time models of cart–pendulum system obtained by the functions c2d (using the default option "zoh") and dlinmod. As it is seen, the plots of both models coincide which shows that these functions are interchangeable.

In some cases it might be necessary for a given discrete-time model to determine an equivalent continuous-time model or to resample a discrete-time model to a new sample frequency. This can be done by using the functions d2c and d2d, respectively.

2.3.4   Discretization of time delay systems

Consider the continuous-time system with time-delay

$$\frac{d}{dt}x(t) = Ax(t) + Bu(t - \tau). \tag{2.23}$$

If the time-delay τ is larger than the sampling period T_s, then τ can be represented as

$$\tau = (d - 1)T_s + \zeta, \qquad 0 < \zeta \le T_s$$

where d is a positive integer. In such a case, the system (2.23) is discretized as [21]

$$
\begin{bmatrix}
x(kT_s + T_s) \\
u(kT_s - (d-1)T_s) \\
\vdots \\
u(kT_s - T_s) \\
u(kT_s)
\end{bmatrix}
=
\begin{bmatrix}
A_d & B_{d1} & B_{d0} & \cdots & 0 \\
0 & 0 & I_m & \cdots & 0 \\
\vdots & \vdots & \vdots & \ddots & \vdots \\
0 & 0 & 0 & \cdots & I_m \\
0 & 0 & 0 & \cdots & 0
\end{bmatrix}
\begin{bmatrix}
x(kT_s) \\
u(kT_s - dT_s) \\
\vdots \\
u(kT_s - 2T_s) \\
u(kT_s - T_s)
\end{bmatrix}
$$

$$
+
\begin{bmatrix}
0 \\
0 \\
\vdots \\
0 \\
I_m
\end{bmatrix}
u(kT_s)
\tag{2.24}
$$

where

$$A_d = e^{AT_s},$$

$$B_{d0} = \int_0^{T_s - \tau} e^{A\sigma} d\sigma B,$$

$$B_{d1} = e^{A(T_s - \tau)} \int_0^{\tau} e^{A\sigma} d\sigma B.$$

The discrete-time state-space model (2.24) has $d \times m$ extra states representing the past control inputs

$$u(kT_s - dT_s), \ldots, u(kT_s - 2T_s), u(kT_s - T_s)$$

which are used to describe the delay. That is why to specify the system state it is necessary to store the past d control vectors over a time interval equal to the time delay. For large τ and small T_s, the order of the model (2.24) may increase very much. Nevertheless, this model is very convenient, since it incorporates the time delay into the standard discrete-time state-space representation.

If the time delay τ is less than the sampling interval T_s than the model (2.24) reduces to

$$
\begin{bmatrix}
x(kT_s + T_s) \\
u(kT_s)
\end{bmatrix}
=
\begin{bmatrix}
A_d & B_{d1} \\
0 & 0
\end{bmatrix}
\begin{bmatrix}
x(kT_s) \\
u(kT_s - T_s)
\end{bmatrix}
+
\begin{bmatrix}
B_{d0} \\
I_m
\end{bmatrix}
u(kT_s)
\tag{2.25}
$$

2.3.5 Choice of the sampling period

The sampling period T_s of a discrete-time system depends on several factors, including closed-loop performance and technology capabilities considerations. The decreasing of the sampling period makes the behavior of the discretized model closer to this of the continuous-time model but increases the requirements to the speed of the digital devices (A/D and D/A converters, controller, etc.). On the other hand, the increasing of the sampling period creates problems related to the behavior of the system between the sampling instants.

An approximate lower bound on the sampling frequency in the case of zero-order hold can be determined as follows. The continuous-time transfer function of the zero-order hold is [21]

$$G_{zoh} = \frac{1}{s}(1 - e^{-sT_s}).$$

For small sampling periods, this transfer function is approximated as

$$\frac{1 - e^{-sT_s}}{s} \approx \frac{1 - 1 + sT_s - (sT_s)^2/2 + (sT_s)^3/6 - \cdots}{sT_s} = 1 - \frac{sT_s}{2} + \frac{s^2 T_s^2}{6} - \cdots$$

The first two terms correspond to the Taylor series expansion of $\exp^{-sT_s/2}$. In this way, for small T_s, the zero-order hold can be approximated by a time delay of half a sampling interval. Hence, the zero-order hold decreases the phase margin by

$$\phi = \frac{\omega_c T_s}{2}$$

where ω_c is the crossover frequency (in rad/s) of the continuous-time system.

Assuming that the phase margin can be decreased by $\Delta\phi$, it follows that the sampling period must satisfy

$$T_s < \frac{\Delta\phi}{\omega_c}.$$

This inequality gives a Nyquist frequency

$$F_N > \frac{\omega_c}{(2\Delta\pi)}. \tag{2.26}$$

If, for instance, $\Delta\phi = 5° = 0.087$ rad, then according to (2.26) it follows that the Nyquist frequency must be 36 times higher than the crossover frequency. Note that this value increases in the presence of antialiasing or notch filters, see [3, Section 6] for details. When considering the closed-loop performance, instead of crossover frequency one should take the closed-loop bandwidth (defined in Section 3.2).

As shown by an example in [21, Chapter 3], the reachability and observability of a discretized linear systems may be lost for some low sampling frequencies, although the corresponding continuous-time system is both controllable and observable. This may happen when the discrete-time transfer function has common poles and zeros. Since poles and zeros are functions of the sampling period, a small change of T_s can make the system reachable and/or observable again. It should be noted that the contemporary processors make possible implementation of very high sampling rates which avoids the danger of reachability and/or observability lost.

2.3.6   Discretization of nonlinear models

Sometimes, it is necessary to have a discrete-time equivalent of the nonlinear plant model (2.3) for some sampling interval T_s. For example, when simulating a closed-loop discrete-time system it might be much faster to obtain the system transient response in discrete time provided the nonlinear continuous-time plant (2.3) is replaced by its discrete-time equivalent

$$x(k+1) = \phi(x(k), u(k)),$$
$$y(k) = \eta(x(k), u(k))$$

(2.27)

for a fixed sampling interval T_s. The discrete-time equivalent

$$x(k+1) = \phi(x(k), u(k))$$

(2.28)

of the continuous-time state equation

$$\frac{d}{dt}x(t) = f(t, x, u)$$

can be determined by using some scheme for numerical integration of differential equations which ensures sufficient accuracy with small number of computational operations. For this purpose, it is appropriate to implement efficiently the Bogacki–Shampine method [28, Chapter 7] which is described as follows.

Assuming that the discrete-time input $u(k)$ is obtained by using a zero-order hold, the approximate solution $x(k+1)$ at the instant $t(k+1) = t(k) + h$, where $h = T_s$ is computed by using the equations

$$s_1 = f(t(k), x(k), u(k)),$$
$$s_2 = f(t(k) + \tfrac{h}{2}, x(k) + \tfrac{h}{2}s_1, u(k)),$$
$$s_3 = f(t(k) + \tfrac{3}{4}h, x(k) + \tfrac{3}{4}hs_2, u(k)),$$
$$t(k+1) = t(k) + h,$$
$$x(k+1) = x(k) + \tfrac{h}{9}(2s_1 + 3s_2 + 4s_3).$$

(2.29)

The block diagram of the nonlinear discrete-time model (2.29) is shown in Figure 2.11. (The discretization of the algebraic equation for the output $y(k)$ is straightforward.) The model is easily implemented in Simulink. Its usage in simulating a nonlinear helicopter control system is demonstrated in Chapter 6.

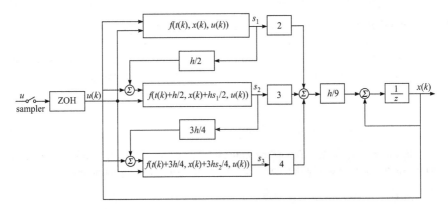

Figure 2.11   Block diagram of the nonlinear discrete-time model

2.4 Stochastic modeling

MATLAB functions used in this section

Function	Description
kalman	Design of Kalman filters for continuous-time and discrete-time systems
lyap	Solution of Lyapunov equation

2.4.1 Stochastic linear systems

A time-invariant continuous-time linear system with stochastic inputs can be represented as

$$\dot{x}(t) = Ax(t) + Gv(t),$$
$$y(t) = Cx(t) + w(t), \tag{2.30}$$

where $v(t)$ and $w(t)$ are random processes. The stochastic process $v(t)$ is called the *process noise* and the stochastic process $w(t)$ is called the *measurement noise*. At each time, the state $x(t)$ and the output $y(t)$ are also stochastic processes. If $v(t)$ and $w(t)$ are Gaussian random processes, then $x(t)$ and $y(t)$ also are Gaussian processes.

Assuming that the model is constructed so that $v(t)$ and $w(t)$ are stationary white noises, the mean and the covariance of the random variables $v(t)$ and $w(t)$ are denoted as

$$m_v = E\{v(t)\} = 0, \tag{2.31}$$

$$E\{v(t+\tau)v(t)^T\} = Q\delta(\tau), \tag{2.32}$$

$$m_w = E\{w(t)\} = 0, \tag{2.33}$$

$$E\{w(t+\tau)w(t)^T\} = R\delta(\tau). \tag{2.34}$$

It is frequently assumed that the process and measurement noise are independent on the current and previous state and independent on each other.

Consider the propagation of state and output means and variances of the continuous-time system (2.30). The initial conditions of the system are

$$E\{x(t_0)\} = m_0, \tag{2.35}$$

$$E\{x(t_0)x(t_0)^T\} = P(t_0). \tag{2.36}$$

The mean $m_x(t) = E\{x(t)\}$ of the state and the mean $m_y(t) = E\{y(t)\}$ of the output propagate according to the deterministic dynamics of the system and are given by

$$\dot{m}_x(t) = Ax(t), \quad m_x(t_0) = m_0, \tag{2.37}$$

$$m_y(t) = Cm_x(t). \tag{2.38}$$

It is possible to prove that the state covariance matrix

$$P(t) = C_x(t,t) = E\left\{[x(t) - m_x(t)][x(t) - m_x]^T\right\}$$

is the solution of matrix differential equation

$$\dot{P}(t) = AP(t) + P(t)A^T + GQG^T \tag{2.39}$$

with initial condition $P(t_0)$.

Equation (2.39) is particularly useful to find the steady-state covariance of a stable time-invariant system. In such case, it reduces to the dual matrix algebraic Lyapunov equation

$$AP + PA^T + GQG^T = 0. \tag{2.40}$$

If the pair (A, G) is stabilizable, then the solution of the Lyapunov equation is positive semidefinite (see Appendix B, Section B.4).

The variance of the output of (2.30) is given by

$$C_y(t,t) = E\left\{[y(t) - m_y(t)][y(t) - m_y(t)]^T\right\} = CP(t)C^T + R. \tag{2.41}$$

A time-invariant discrete-time linear system with stochastic inputs is described by

$$\begin{aligned} x(k+1) &= A_d x(k) + G_d v(k), \\ y(k) &= C_d x(k) + w(k), \end{aligned} \tag{2.42}$$

The mean and the covariance of the random variables $v(k)$ and $w(k)$ are denoted as

$$m_{v(k)} = E\{v(k)\} = 0, \tag{2.43}$$

$$E\left\{v(k)v_j^T\right\} = Q_d \delta(k-j), \tag{2.44}$$

$$m_{w(k)} = E\{w(k)\} = 0, \tag{2.45}$$

$$E\left\{w(k)w_j^T\right\} = R_d \delta(k-j) \tag{2.46}$$

where $\delta(k)$ is the Kronecker delta function.

The mean $m_{x(k)} = E\{x(k)\}$ of the state is a solution of the linear difference equation

$$m_{x(k+1)} = A_d m_{x(k)}, \quad m_{x_0} = E\{x_0\}. \tag{2.47}$$

The discrete-time state covariance matrix is defined as

$$P(k) = E\left\{(x(k) - m_{x(k)})(x(k) - m_{x(k)})^T\right\}. \tag{2.48}$$

It is possible to prove that the state covariance at time $k+1$ is the solution of matrix difference equation

$$P(k+1) = A_d P(k)A_d^T + G_d Q_d G_d^T. \tag{2.49}$$

The steady-state covariance matrix of a stable discrete-time system is obtained as the solution of the discrete-time Lyapunov equation (or *Stein equation*)

$$P = A_d P A_d^T + G_d Q_d G_d^T. \tag{2.50}$$

If the pair (A_d, G_d) is stabilizable, then P is a positive semidefinite matrix (see Appendix B, Section B.4).

2.4.2 Discretization of stochastic models

In most cases instead of with a deterministic model, the control plant is described by the stochastic time-invariant model

$$\frac{d}{dt}x = Ax + Bu + Gv,$$
$$y = Cx + w \tag{2.51}$$

where x is the state vector, y is the measurement vector and v, w are the process noise and measurement noise, respectively; A, B, C are constant matrices with appropriate dimensions. It is assumed that v and w are stationary independent zero-mean Gaussian white noises with covariances

$$E\left\{v(t)v(\tau)^T\right\} = V\delta(t - \tau), \quad E\left\{w(t)w(\tau)^T\right\} = W\delta(t - \tau)$$

where V, W are known covariance matrices.

For a given sampling interval T_s, it is necessary to find a discrete-time equivalent of the model (2.51) in the form

$$x(k + 1) = A_d x(k) + B_d u(k) + v(k),$$
$$y(k) = C_d x(k) + w(k) \tag{2.52}$$

where A_d, B_d, C_d are the matrices of the discretized model and $v(k), w(k)$ are discrete-time stochastic processes. The model matrices can be found in the same way as in the discretization of the deterministic model (2.5). For example, in the case of zero-order hold these matrices are found from their continuous-time counterparts by the expressions

$$A_d = e^{AT_s}, \quad B_d = \int_0^{T_s} e^{A\sigma} d\sigma B, \quad C_d = C.$$

It is possible to show [8, Section 9.4.4], [44, Section 2.4.1], [45, Section 6.4.6] that in the case of zero-order hold the processes $v(k), w(k)$ represent discrete white noise sequences with covariances

$$E\left\{v(k)v_j^T\right\} = V_d\delta(j - k), \quad E\left\{w(k)w_j^T\right\} = W_d\delta(j - k) \tag{2.53}$$

where

$$V_d = \int_0^{T_s} e^{A\tau} GVG^T e^{A^T\tau} d\tau, \quad W_d = \frac{W}{T_s}. \tag{2.54}$$

The covariance matrix V_d can be evaluated using the Taylor series approximation for e^{At},

$$e^{At} = I + At + \frac{1}{2!}(At)^2 + \frac{1}{3!}(At)^3 + \cdots$$

which gives (up to the second order in A and third order in T_s)

$$V_d \approx T_s GVG^T + \frac{T_s^2}{2}(AGVG^T + GVG^T A^T)$$
$$+ \frac{T_s^3}{6}(A^2 GVG^T + 2AGVG^T A^T + GVG^T(A^T)^2). \tag{2.55}$$

If T_s is very short compared to the system time constants, then one may use only the first term in (2.55),

$$V_d \approx T_s GVG^T. \tag{2.56}$$

An accurate result for V_d can be found by using the Van Loan algorithm [46]. Let

$$Z = \begin{bmatrix} -A & GVG^T \\ 0 & A^T \end{bmatrix} T_s$$

is a 2-by-2 block triangular matrix, then

$$e^Z = \begin{bmatrix} \Phi_{11} & \Phi_{12} \\ 0 & \Phi_{22} \end{bmatrix}$$

and

$$V_d = \Phi_{22}^T \Phi_{12} \tag{2.57}$$

This method is implemented by the M-file disrw.m which can be found on the MathWorks® site https://www.mathworks.com.

Example 2.7. Discretization of a stochastic system

Let us compute the matrices A_d and V_d for the double integrator

$$A = \begin{bmatrix} 0 & 1 \\ 0 & 0 \end{bmatrix},$$

with

$$G = \begin{bmatrix} 0 \\ 1 \end{bmatrix}, \quad V = 0.1$$

for $T_s = 1$.

Using the MATLAB function expm, one obtains that

$$A_d = \begin{bmatrix} 1 & 1 \\ 0 & 1 \end{bmatrix}$$

The approximation (2.56) produces

$$V_d \approx \begin{bmatrix} 0 & 0 \\ 0 & 1 \end{bmatrix} \times 10^{-1}. \tag{2.58}$$

It is easy to obtain a solution by (2.55) since $A^k = 0$ for $k \geq 2$. As a result, one finds

$$A_d = I_2 + AT_s = \begin{bmatrix} 1 & 1 \\ 0 & 1 \end{bmatrix}$$

and

$$V_d = GVG^T T_s + (AGVG^T + GVG^T A^T) T_s^2/2 + 2AGVG^T A^T T_s^3/6$$

$$= \begin{bmatrix} \frac{1}{3} & \frac{1}{2} \\ \frac{1}{2} & 1 \end{bmatrix} \times 10^{-1}.$$

This result coincides with the exact solution (2.57) and is significantly different from the approximate solution (2.58) obtained from (2.56). □

2.4.3 Optimal estimation

Consider the time-invariant continuous-time stochastic system described by the equations

$$\dot{x}(t) = Ax(t) + Bu(t) + Gv(t), \quad x(0) = x_0 \tag{2.59}$$

$$y(t) = Cx(t) + w(t), \tag{2.60}$$

where state $x(t) \in \mathcal{R}^n$; control input $u(t) \in \mathcal{R}^m$; output $y(t) \in \mathcal{R}^r$; process noise $v(t) \in \mathcal{R}^q$; measurement noise $w(t) \in \mathcal{R}^r$; and A, B, C, G are known constant matrices with appropriate dimensions. The control input $u(t)$ is a deterministic variable. It is assumed that $v(t)$ and $w(t)$ are zero mean Gaussian white noises with covariances

$$E\{v(t+\tau)v(t)^T\} = V\delta(\tau), \tag{2.61}$$

$$E\{w(t+\tau)w(t)^T\} = W\delta(\tau), \tag{2.62}$$

$$E\{v(t+\tau)w(t)^T\} = 0 \tag{2.63}$$

where V, W are known constant positive definite matrices. Equation (2.63) implies that $v(t)$ and $w(t)$ are uncorrelated. We further assume that the pair (C, A) is detectable (observable) and the pair (A, G) is stabilizable (controllable).

The *estimation problem* for the system (2.59), (2.60) is to determine an "optimal" approximation $\hat{x}(t)$ of the state $x(t)$ using the noisy measurement vector $y(t)$ and the deterministic input $u(t)$. An optimal in stochastic sense estimate can be obtained by using an auxiliary linear system called *optimal filter*.

Define the *estimation error* as

$$e(t) := x(t) - \hat{x}(t). \tag{2.64}$$

The stochastic system

$$\dot{\hat{x}}(t) = A\hat{x}(t) + Bu(t) + K[y(t) - C\hat{x}(t)],$$
$$\hat{x}(0) = \hat{x}_0; \hat{x}(t) \in \mathcal{R}^n, K \in \mathcal{R}^{n \times r}, \tag{2.65}$$

which minimizes the quadratic cost

$$J = \lim_{t \to \infty} E\{e^T(t)e(t)\} = \lim_{t \to \infty} \text{Tr}[E\{e(t)e(t)^T\}], \tag{2.66}$$

is called *continuous-time optimal filter* (or Kalman–Bucy filter) for the system (2.59), (2.60). The vector $\hat{x}(t)$ is called *optimum estimate* of the state $x(t)$ of (2.59), the vector

and K is the optimum filter gain matrix. The gain matrix multiplies the *residual* $z(t) = y(t) - C\hat{x}(t)$.

Equation (2.65) may be rewritten in the form

$$\dot{\hat{x}}(t) = (A - KC)\hat{x}(t) + Bu(t) + Ky(t), \tag{2.67}$$

which shows that the stability of the filter is determined by the properties of the matrix $A - KC$. Since the pair (C, A) is detectable, there exists a matrix K such that the filter is stable, i.e., $A - KC$ is a stable.

Subtracting (2.65) from (2.59) and taking into account (2.60), one obtains that the estimation error is a solution of the differential equation

$$\dot{e}(t) = (A - KC)e(t) + Gv(t) - Kw(t), \quad e(0) = x_0 - \hat{x}_0. \tag{2.68}$$

Hence, the filter stability means stability of the estimation error.

According to (2.40), the covariance matrix P of the estimation error $e(t)$ for $t \to \infty$ is the solution of Lyapunov equation

$$(A - KC)P + P(A - KC)^T + GVG^T + KWK^T = 0 \tag{2.69}$$

and the quadratic cost has the value

$$J = \text{Tr}(P). \tag{2.70}$$

It is possible to show that the minimum of J is obtained if the gain matrix K and the initial condition $\hat{x}_0$ are determined from [47]

$$K = PC^T W^{-1} \tag{2.71}$$

and

$$\hat{x}_0 = m_{x_0} = E\{x_0\}, \tag{2.72}$$

respectively, where the matrix $P \in \mathscr{R}^{n \times n}$ is the unique and positive semidefinite solution of the *matrix algebraic Riccati equation*

$$AP + PA^T + GVG^T - PC^T W^{-1} CP = 0. \tag{2.73}$$

Furthermore, if the pair (A, G) is controllable, then the matrix P is positive definite.

Taking into account (2.71), (2.73) reduces to the Lyapunov equation (2.69) which shows that the covariance of the steady state estimation error $e(t)$ is given by

$$\lim_{t \to \infty} E\{e(t)e^T(t)\} = P. \tag{2.74}$$

The diagonal elements of the matrix P can be used to assess the covariances of the steady-state components of state estimate $\hat{x}(t)$.

Since the matrix P is chosen as the positive semidefinite solution of (2.73) and since the pair (A, G) is stabilizable, it follows by (2.69) that the Kalman–Bucy filter is stable (the eigenvalues of $A - K^*C$ have negative real parts).

Having in mind that $v(t)$ and $w(t)$ are zero mean, one obtains from (2.68) that

$$\lim_{t \to \infty} m_{e(t)} = 0. \tag{2.75}$$

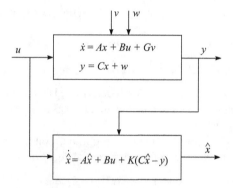

Figure 2.12   Continuous-time system with Kalman–Bucy filter

This shows that the estimate $\hat{x}(t)$ is unbiased, i.e.,

$$\lim_{t\to\infty} E\{\hat{x}(t)\} = \lim_{t\to\infty} E\{x(t)\}.$$

The residual $z(t)$ satisfies the relationship

$$z(t) = Ce(t) + w(t) \tag{2.76}$$

and hence is a zero mean white noise whose covariance tends to

$$P_{z(t)} = CPC^T + W \tag{2.77}$$

as $t \to \infty$.

The Kalman–Bucy filter is the best filter that uses a linear combination of the measurements. It is possible to prove that even if the noise is not Gaussian, the Kalman filter is still the optimal linear filter.

A block diagram of continuous-time system with Kalman–Bucy filter is shown in Figure 2.12.

The Kalman–Bucy filter (2.65) is designed in MATLAB using the function `kalman`. The function `kalman` designs a Kalman filter of a system with the general description

$$\dot{x}(t) = Ax(t) + Bu(t) + Gv(t),$$

$$y(t) = Cx(t) + Hv(t) + w(t)$$

where the process noise $v(t)$ and the measurement noise $w(t)$ can be correlated.

Assume now that a time-invariant stochastic system is described by the difference state and output equations

$$x(k+1) = A_d x(k) + B_d u(k) + G_d v(k),$$
$$y(k) = C_d x(k) + w(k) \tag{2.78}$$

where $x(k) \in \mathscr{R}^n$ is the state vector, $y(k) \in \mathscr{R}^r$ is the measurement vector, $v(k) \in \mathscr{R}^q$ is the process noise, $w(k) \in \mathscr{R}^r$ is the measurement noise and A_d, B_d, C_d, G_d are constant matrices with appropriate dimensions. It is assumed that $v(k)$ and $w(k)$ are stationary independent zero-mean Gaussian white noises with covariances

$$E\left\{v(k)v(j)^T\right\} = V\delta(j-k), \quad E\left\{w(k)w(j)^T\right\} = W\delta(j-k)$$

where V, W are known covariance positive definite matrices.

The discrete-time optimal state estimator (*Kalman filter*) of the stochastic system (2.78) is described by the difference equation

$$\hat{x}(k+1) = A_d\hat{x}(k) + B_d u(k) + K(y(k+1) - C_d A_d\hat{x}(k) - C_d B_d u(k)). \quad (2.79)$$

Equation (2.79) may be rewritten in the form

$$\hat{x}(k+1) = (I - KC_d)A_d\hat{x}(k) + (I - KC_d)B_d u(k) + Ky(k+1) \quad (2.80)$$

which shows that the stability of the filter is determined by the matrix $(I - KC_d)A_d$.

Provided that the matrix pair (A_d, G_d) is stabilizable and the matrix pair (C_d, A_d) is detectable, the optimal gain matrix K of the Kalman filter (2.79) is determined as

$$K = PC_d^T(C_d PC_d^T + W)^{-1} \quad (2.81)$$

where P is the positive semidefinite solution of the discrete matrix algebraic Riccati equation

$$A_d PA_d^T - P + G_d VG_d^T - A_d PC_d^T(C_d PC_d^T + W)^{-1}C_d PA_d^T = 0. \quad (2.82)$$

In such case, the Kalman filter is stable (the eigenvalues of $(I - KC_d)A_d$ are inside the unit circle), i.e., $E\left\{\hat{x}(k)\right\} \to E\left\{x(k)\right\}$ as far as $k \to \infty$ and the mean squared estimation error is given by

$$E\left\{(x(k) - \hat{x}(k))(x(k) - \hat{x}(k))^T\right\} = P. \quad (2.83)$$

The elements of the matrix P can be used to assess the covariances of the state estimate $\hat{x}(k)$, similarly to the continuous-time case.

Examples of Kalman filter application are given in Section 2.7 and in Chapter 4.

2.5   Plant identification

MATLAB files used in this section

File	Description
`ident_black_box_model`	Black box model identification of cart–pendulum system
`ident_gray_box_model`	Gray-box model identification of of cart–pendulum system
`clp_PID_controller.slx`	Simulink model of the cart–pendulum system with manually tuned PID controller
`clp_PID_validation.slx`	Simulink model of the cart–pendulum system used for plant model validation

In case when a plant description is not available, one may use experimentally obtained input and output data to determine an appropriate plant model. This can be done by some identification method such as the methods described in Appendix D. In this section, we shall illustrate the usage of black box and gray box identification methods in case of unstable system.

2.5.1   Identification of black box model

Example 2.8.   Black box identification of cart–pendulum system
Consider the cart–pendulum system presented in Example 2.1. Assume that we do not know the model structure and parameters. The goal is to obtain a linear black box model which sufficiently well describes plant dynamics in upper pendulum position. The identified plant is inheritably unstable which requires to design a controller which stabilizes the system and then performing closed-loop identification experiment. There are many specific methods which deal with closed-loop identification, but some of methods described in Appendix D are also appropriate, provided that input signal guarantees persistent excitation of sufficient order and that the exact structure of the model is in the model set [48]. Such methods are prediction error method and subspace method that uses an ARX-estimation-based algorithm to compute the weighting matrices [49]. They are implemented in various functions of System Identification Toolbox™.

First, we design the closed-loop identification experiment. Input–output data are obtained by the Simulink model `clp_PID_controller.slx`. The block scheme of identification experiment is presented in Figure 2.13. The plant dynamics is simulated by the nonlinear model (2.1), which is implemented in Simulink subsystem block "nonlinear pendulum–cart model." Furthermore, to make identification more realistic the 12-bit encoders are modeled by the quantizer blocks, which corrupts the

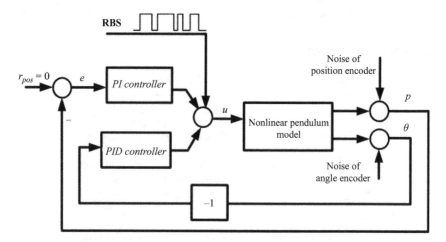

Figure 2.13   Block scheme of closed-loop identification experiment

outputs with noise. Cart–pendulum system is stabilized by two outputs, one input discrete-time PID controller

$$
\begin{aligned}
u(k) = K_{p_{pos}} &\left(e(k) + \frac{T_{d_{pos}} N_{pos}(q-1)}{(1+N_{pos}T_s)q-1} e(k) \right) \\
&- K_{p_\theta} \left(\theta(k) + K_{i\theta} \frac{T_s}{(q-1)} \theta(k) + \frac{T_{d\theta} N_\theta(q-1)}{(1+N_\theta T_s)q-1} \theta(k) \right),
\end{aligned}
\tag{2.84}
$$

where $e(k) = r_{pos}(k) - p(k)$ is the cart position error, $r_{pos}(k) = 0$ is the cart position reference, $K_{p_{pos}} = -0.02$, $T_{d_{pos}} = 0.15$, $N_{pos} = 2$ are the parameters of PD controller for cart position control, $K_{p_\theta} = 1.5$, $K_{i_\theta} = 0.4$, $T_{d_\theta} = 0.0066$, $N_\theta = 3$ are the parameters of PID controller for pendulum angle stabilization. The controller parameters are tuned by trial and error method. The random binary signal (RBS) is added to the controller output which provides persistent excitation of identification input signal. RBS is obtained from filtered through relay white Gaussian noise. The amplitude of RBS is chosen to be ± 10 percent, so the control signal stays in linear region and the pendulum tilts enough (4–5 degrees) without falling. The input–output data sampled by $T_s = 0.01$ s is obtained by the command lines

```
Ts = 0.01;
u_max = 0.5;
unc_lin_model
%
%Set uncertain parameters to their nominal values
val_all = [];
%
%Simulation
sim('clp_PID_controller')
datae = iddata([position.signals.values(10001:20000), ...
                theta.signals.values(10001:20000)], ...
                control.signals.values(10001:20000),0.01);
datae.OutputName = {'Cart position','Pendulum angle'}
datae.InputName = {'Control'}
datav = iddata([position.signals.values(20001:21000), ...
                theta.signals.values(20001:21000)], ...
                control.signals.values(20001:21000),0.01)
datav.OutputName = {'Cart position','Pendulum angle'}
datav.InputName = {'Control'}
```

Data is separated into two data sets—datae and datav. The first one consists of 10,000 samples and is used for model parameter estimation and the second consists of 1,000 samples and is used for model validation. Next steps are to check persistence excitation level of input signal and to plot data sets (Figures 2.14 and 2.15). These actions are done by the command lines

```
pexcit(datae,10000)
figure(1)
idplot(datae),grid
title('Estimation data')
figure(2)
idplot(datav),grid
title('Validation data')
```

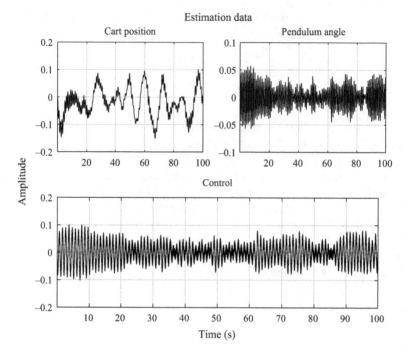

Figure 2.14   Estimation data set

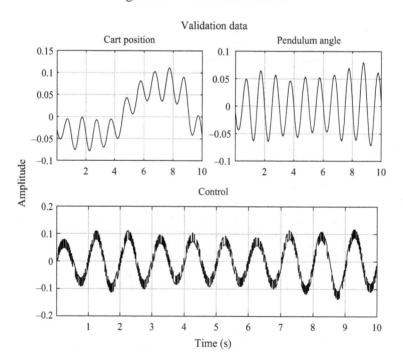

Figure 2.15   Validation data set

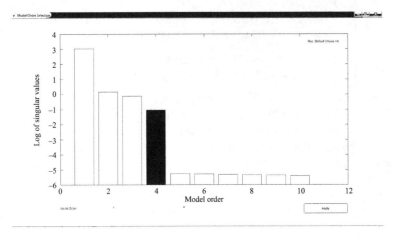

Figure 2.16   Hankel singular values of estimated models

The excitation level of input signal is 2,500, which means that we can estimate up to 2,500 parameters from estimation data set.

As mentioned in Appendix D, if we have not a priori information about model type and structure, the estimation procedure should be started with sufficiently simple black box model with few design parameters. Such model is state-space model (D.65) with free parametrization. It is especially useful in case of single-input–multiple-output plant and has only one design parameter—the model order. Assuming that the possible model order is between 1 and 10, we form the model set of ten state-space models. The order of the best model from this set is determined by MATLAB function n4sid with the command line

```
modeln4sid = n4sid(datae,1:10)
```

The resulting plot of Hankel singular values is depicted in Figure 2.16. As it is seen, the best model order is equal to 4. Then, the fourth order state-space model is estimated by

```
modeln4sid = n4sid(datae,4,'N4weight','SSarx')
```

The obtained model is in innovation form and has matrices

$$
A = \begin{bmatrix}
1 & 0.01495 & 6.34 \times 10^{-5} & -1.94 \times 10^{-5} \\
3.84 \times 10^{-3} & 0.9939 & -1.91 \times 10^{-3} & -0.02123 \\
-0.0039 & -0.05686 & 1.008 & -0.02123 \\
0.04218 & -0.0151 & -0.1407 & 0.9942
\end{bmatrix},
$$

$$
B = \begin{bmatrix}
-0.0682 \\
-1.911 \\
0.3206 \\
-5.202
\end{bmatrix}
$$

$$C = \begin{bmatrix} -0.05427 & 1.536 \times 10^{-3} & 1.38 \times 10^{-5} & -2.63 \times 10^{-6} \\ 6.19 \times 10^{-3} & 7.54 \times 10^{-5} & 0.0204 & 8.03 \times 10^{-4} \end{bmatrix},$$

$$D = \begin{bmatrix} 0 \\ 0 \end{bmatrix}, K = \begin{bmatrix} -5.731 & -29.69 & 2.268 & -151.3 \\ -3.22 \times 10^{-3} & -0.1949 & 15.8 & -76.61 \end{bmatrix}$$

The values of metrics (D.67) and (D.70), defined in Appendix D, are $MSE = 2.406 \times 10^{-7}$, $FPE = 7.833 \times 10^{-17}$. Note that using a SSARX weighting scheme in function n4sid allows obtaining unbiased estimates in case of closed-loop identification. The next step is to perform validation tests of estimated state-space model. The test of residuals is done by the command lines

```
figure(3)
resid(modeln4sid,datav),grid
title('Residuals correlation of n4sid model')
figure(4)
resid(datav,modeln4sid,'fr'),grid
title('Residuals frequency response of n4sid model')
e=resid(datav,modeln4sid);
figure(5)
plot(e),grid
title('Residuals of n4sid model')
```

The results from whitening test and independence test are shown in Figure 2.17. The frequency response of estimated high-order finite impulse response (FIR) model

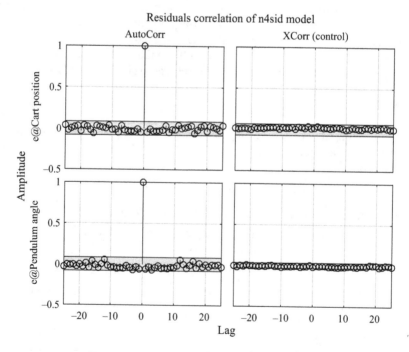

Figure 2.17   *Residual test of* modeln4sid

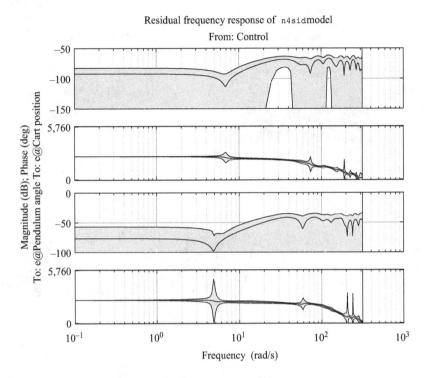

Figure 2.18   Residual to input signal frequency response

between control signal and residuals along with 99 percent confidence region is depicted in Figure 2.18. The residuals plot is shown in Figure 2.19. As can be seen from Figure 2.17, the obtained model passes both tests, which means that the model captures sufficiently well plant dynamics and the noise model is "good." Figure 2.18 shows that there is not significant dynamics between the input signal and residuals in the whole interested frequency range.

It is seen from Figure 2.19 that the residual from cart position is 10,000 times smaller than the actual cart position and the residual from pendulum angle is 50 times smaller than the actual pendulum angle.

The simulation of modeln4sid and true nonlinear pendulum–cart model in closed-loop operation is done by the Simulink model clp_PID_validation .slx. The comparison between identified model outputs and true nonlinear model outputs is shown in Figures 2.20 and 2.21. The values of metric (D.71) for cart position and pendulum angle are $FIT_p = 70.6077$ percent and $FIT_\theta = 94.0585$ percent, which is sufficiently good result. The estimated by subspace algorithm state-space model describes plant dynamics very well and passes all validation tests. Then, the question is, "Can we describe plant dynamics sufficiently well with simpler model?". Now we can try to estimate model with simpler noise description, such as ARX model (D.47).

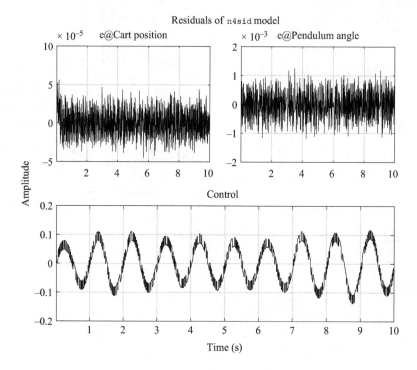

Figure 2.19   Residuals of `model n4sid`

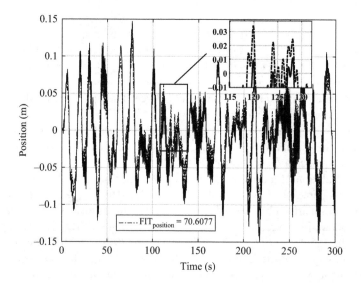

Figure 2.20   Simulated cart position of `model n4sid` *and true nonlinear model*

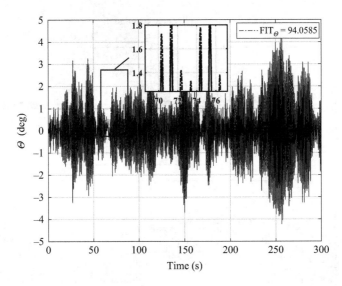

Figure 2.21 *Simulated pendulum angle of* `modeln4sid` *and true nonlinear*
model

The orders of all polynomials are chosen to be 2, so that the obtained model will be
of fourth order. The estimation and validation is done by the command lines

```
modelarx = arx(datae,'na',[2 2;2 2],'nb',[2;2],'nk',[1;1])
figure(6)
resid(modelarx,datav),grid
e = resid(datav,modelarx);
figure(7)
plot(e), grid
title('Residuals of ARX model')
modelv = ss(modelarx)
sim('clp_PID_pendulum_validation')
figure(8), plot(position_validate(:,1),position_validate(:,2),'r', ...
        position_validate(:,1), position_validate(:,3),'b'), grid
FIT_position = 100*(1 - norm(position_validate(:,3) - ...
            position_validate(:,2))/norm(position_validate(:,3) ...
            - mean(position_validate(:,3))))
legend([char('FIT_{position} = '), num2str(FIT_position)])
xlabel('Time [s]'), ylabel('Position [m]')
figure(9), plot(theta_validate(:,1), theta_validate(:,2), 'r', ...
            theta_validate(:,1), theta_validate(:,3), 'b'), grid
FIT_theta = 100 * (1 - norm(theta_validate(:,3) - theta_validate(:,2)) ...
        /norm(theta_validate(:,3) - mean(theta_validate(:,3))))
legend([char('FIT_{\Theta} = '), num2str(FIT_theta)])
xlabel('Time [s]'), ylabel('\Theta [deg]')
```

The results from various tests are shown in Figures 2.22–2.25. As can be seen
from Figure 2.22, the obtained model does not pass whitening and independence tests,

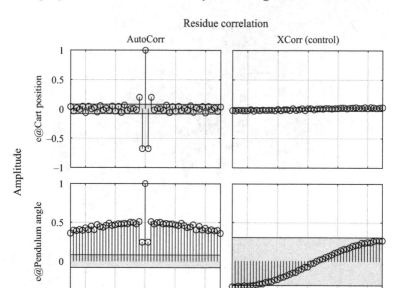

Figure 2.22   Residual test of `modelarx`

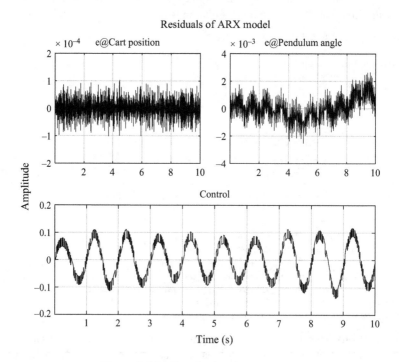

Figure 2.23   Residuals of `modelarx`

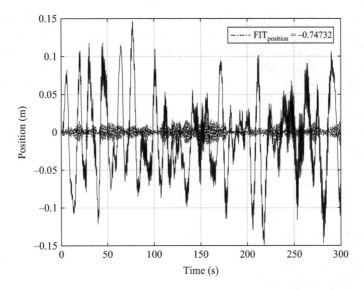

Figure 2.24 Simulated cart position of `modelarx` *and true nonlinear model*

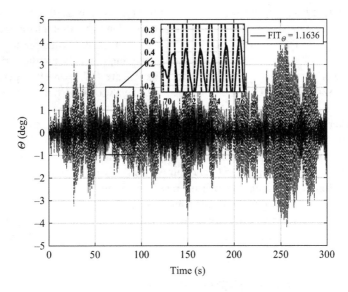

Figure 2.25 Simulated pendulum angle of `modelarx` *and true nonlinear model*

which means that the model does not describe well plant dynamics and the noise model is not adequate. The maximum values of cart position residual and pendulum angle residual are two times greater than the residuals obtained for `modeln4sid` (see Figure 2.23). The fits (D.71) from Appendix D between nonlinear model outputs and `modelarx` outputs are $FIT_p = -0.74732$ percent and $FIT_\theta = 1.1636$ percent. The negative value of cart position fit means that the obtained ARX model fits data no better than a straight line with slope equal to the mean of the data. The fit of pendulum angle is very "poor." In general, the performance of ARX model is "bad."

Further improvement of identification results can be obtained by estimation of state-space model with MATLAB function `ssest`, which implements prediction error method. The `modeln4sid` is used as initial estimate of dynamics in function `ssest`. The estimated model is validated by the same tests as these applied for `modeln4sid`. The estimation and validation is done by the command lines

```
init_sys = modeln4sid
opt = ssestOptions('SearchMethod','lm');
modelssest = ssest(datae,init_sys,'Ts',0.01,opt)
figure(10)
resid(modelssest,datav), grid, ...
    title('Residuals correlation of ssest model')
figure(11)
resid(datav,modelssest,'fr'), grid, ...
    title('Residuals frequency response of ssest model')
e = resid(datav,modelssest);
figure(12)
plot(e),grid,title('Residuals of ssest model')
```

The results from various tests are shown in Figures 2.26–2.30. Obtained model passes the whitening and independence tests. The residuals are similar to ones obtained for `model4sid`. The values of mean square error and Akaike's final prediction error are $MSE = 2.251 \times 10^{-7}$, $FPE = 7.208 \times 10^{-17}$, which are closer to the corresponding quantities obtained for `modeln4sid`. The fits are $FIT_p = 81.028$ percent and $FIT_\theta = 92.5249$ percent. Obtained fit of pendulum position is 11 percent greater than one of `modeln4sid`.

Another useful validation test is to compare the frequency response of estimated models with the response of the discrete model from Example 2.6. The results are shown in Figure 2.31. It is seen that the magnitudes of estimated state-space model are close to these of the linearized pendulum model, whereas the magnitude of ARX model is different, which again confirms results from other tests.

As a result we obtain two adequate state-space models. They pass all validation tests and describe very well pendulum dynamics. The model obtained by prediction error method gives better fit of the cart position. Both models can be used for controller design.   ❑

2.5.2 Identification of gray-box model

The gray box identification approach is illustrated by the following example.

Example 2.9. Gray box identification of cart–pendulum system
Consider the linearized continuous-time model of cart–pendulum system presented in Example 2.2. Assume that we do not know the values of cart mass M kg, pendulum moment of inertia I kg m^2, dynamic cart friction coefficient f_c N s/m and rotational friction coefficient f_p N ms/rad. The goal is to estimate the unknown parameters and

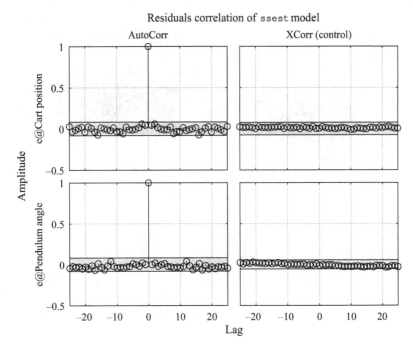

Figure 2.26   Residual test of `model` `ssest`

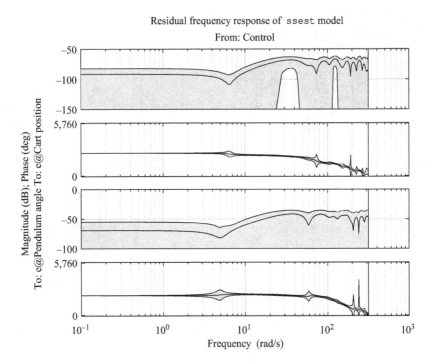

Figure 2.27   Residual to input signal frequency response

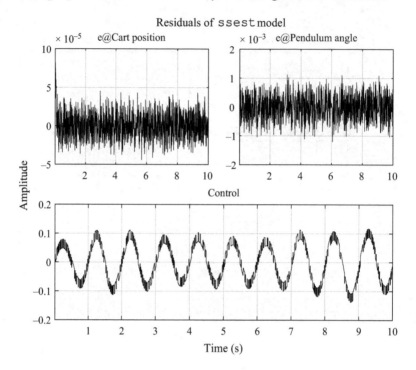

Figure 2.28   *Residuals of* model ssest

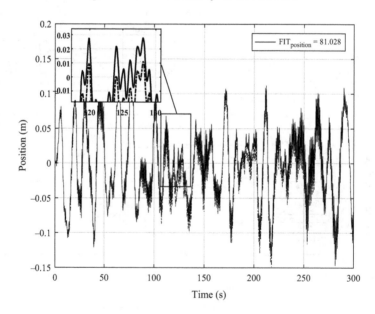

Figure 2.29   *Simulated cart position of* model ssest *and true nonlinear model*

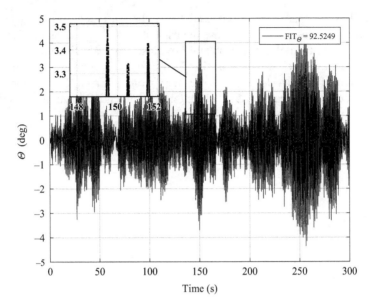

Figure 2.30   Simulated pendulum angle of modelssest *and true nonlinear model*

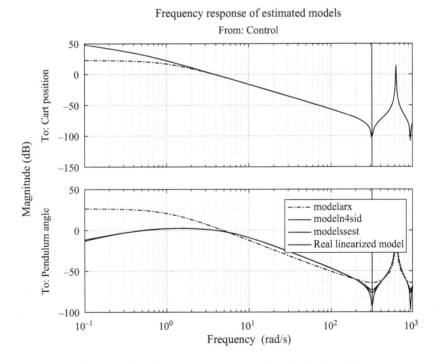

Figure 2.31   Frequency responses of estimated models

noise model such that the gray-box model describes sufficiently well dynamics in upper pendulum position. We will use the estimation and validation data sets from Example 2.8. Note that this data is obtained by closed-loop identification experiment and the excitation level of input signal is 2,500. To estimate continuous time linear gray-box model by MATLAB, first, we must write a MATLAB function, which describes the model dynamics as set of first-order deferential equations. This is done by the command lines

```
function [A,B,C,D] = pendulum_lin_model1(par,Ts)
%
%parameters to be identified
M = par(1); %kg
f_c = par(2); %N s/m
f_p = par(3); % N m s/rad
I = par(4); % kg m^2
%
% parameters that will not be estimated
theta_trim = 0.0;
m = 0.104; % kg
g = 9.81;   % m/s^2
l = 0.174;  % m
I_t = I + m*l^2;   % kg m^2
M_t = M + m;       % kg
den = M_t*I_t - m^2*l^2*cos(theta_trim);
k_F = 12.86;       % N
%%
u_max = 0.5;
 A = [0 0 1 0;
      0 0 0 1;
      0 m^2*l^2*g/den -f_c*I_t/den  -f_p*l*m/den;
      0 M_t*m*g*l/den -f_c*m*l/den  -f_p*M_t/den];
%
B = [0;
     0;
     I_t/den;
     l*m/den]*k_F;
%
C = [1 0 0 0;
     0 1 0 0];
%
D = [0;
     0];
```

The next step is to write a main function, which creates gray-box model and estimates its parameters. The estimation is done by MATLAB function greyest which uses the prediction error method for estimation of state-space model in innovation form

$$\dot{x}(t) = A(\theta)x(t) + B(\theta)u(t) + Ke(t)$$
$$y(t) = C(\theta)x(t) + e(t)$$

(2.85)

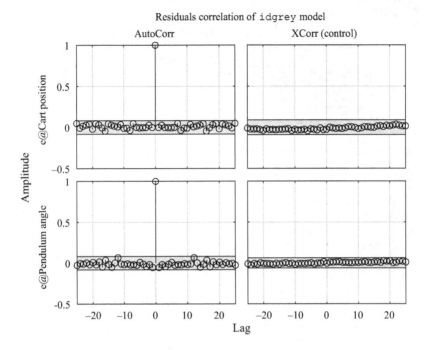

Figure 2.32   Residual test of gray-box model

where matrices *A*, *B*, and *C* are parameterized in the MATLAB function `pendulum_lin_model` and disturbance model matrix is free parameterized. Estimation is done be the command lines

```
par = [0.4; 0.8; 0.000001; 0.000001]
pendulum = idgrey('pendulum_lin_model1', par, 'c');
opt = greyestOptions;
opt.DisturbanceModel = 'estimate'
opt.Searchmethod = 'lm'
m_idgrey = greyest(datae,pendulum,opt)
```

The continuous time gray-box model is created by the MATLAB function `idgrey`. The initial values of unknown parameters are stored in the vector `par`. Here we set search algorithm to be the Levenberg–Marquardt method. After estimation, the validation tests should be performed. First, the comparison between estimated parameters and real ones are performed $\hat{I} = 0.02837$, $I = 0.00283$, $\hat{M} = 0.7679$, $M = 0.768$, $\hat{f}_c = 0.50004$, $f_c = 0.5$, $\hat{f}_p = 5.34 \times 10^{-5}$, $f_p = 6.65 \times 10^{-5}$. It is seen that the values of estimated parameters are very close to their physical values.

The results from whitening and independence tests are shown in Figure 2.32. The frequency response of estimated high-order FIR model between control signal and residuals along with 99 percent confidence region is depicted in Figure 2.33. As can be seen, the obtained model passes both tests, which means that the model capture sufficiently well the plant dynamics and the disturbance model is "good." The cart

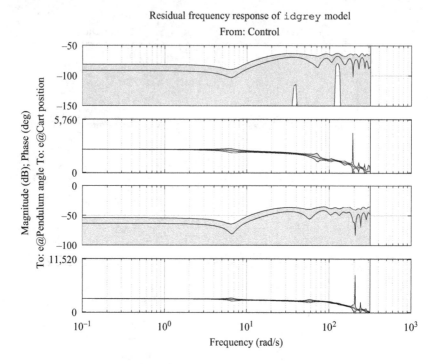

Figure 2.33   Residual to input signal frequency response

position residual showed in Figure 2.34 is approximately 10,000 times smaller than the actual cart position and the pendulum angle residual is 50 times smaller than the actual pendulum angle, which again indicates that the model captures very well plant dynamics.

The comparison between gray-box model outputs and real nonlinear model outputs is shown in Figures 2.35 and 2.36. The values for cart position fit and pendulum angle fit are $FIT_p = 79.55$ percent and $FIT_\theta = 92.27$ percent, which, as expected, are close to the fits obtained for black box state-space model estimated by prediction error method.

The comparison between frequency response of gray-box model and this of the linearized pendulum model is shown in Figure 2.37. It is seen that the magnitudes of estimated model are very close to these of linearized pendulum model in the whole frequency range.   ❑

As a result of gray box identification, we obtain physical parameterized state-space model whose parameters are very close to their real values. This model passes all validation tests and describes very well pendulum dynamics. The disturbance model is very good too which gives opportunity to design both a controller and optimal stochastic estimators such as Kalman filter or H_∞ filter.

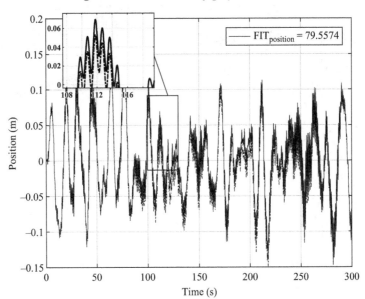

Figure 2.34   Residuals of gray-box model

Figure 2.35   Simulated cart position of gray box and real nonlinear model

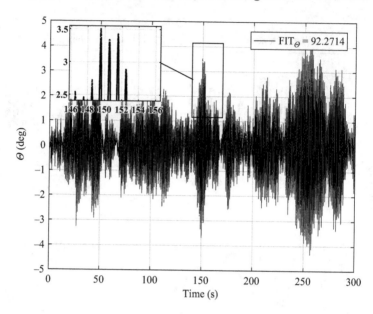

Figure 2.36   Simulated cart position of gray box and real nonlinear model

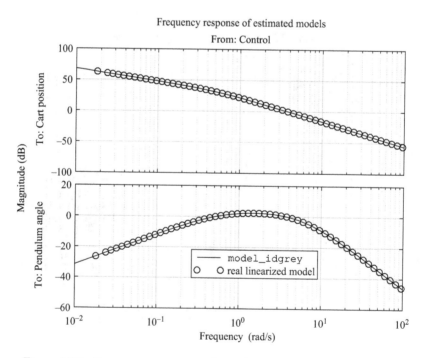

Figure 2.37   Frequency responses of gray box and real linearized models

2.6 Uncertainty modeling

MATLAB files used in this section

File	Description
`par_pendulum_cart`	Uncertain parameters of cart–pendulum system
`uncertain_pendulum_cart`	Uncertain model of cart–pendulum system
`uncertainty_analysis`	Analysis of the effect of uncertain parameters
`unc_model_simulink`	Deriving uncertain model from Simulink model
`unc_dscr_pendulum_cart`	Uncertain discrete-time model of cart–pendulum system
`unc_pendulum_cart.slx`	Simulink nonlinear model of uncertain cart–pendulum system
`uncertain_model_ident`	Identification of uncertain model

The mathematical models of control systems are rarely known exactly. In the general case, there exists some *uncertainty* in system model due to the lack of sufficient knowledge about its operation. The uncertainty in control plants can be divided into two categories: uncertain disturbance signals and dynamic perturbations. The former includes input and output disturbances, sensor noises, actuator noises, and so on. The latter represents the discrepancy between the mathematical model and the actual system dynamics. Typical sources of the discrepancy include unmodeled (usually high-frequency) dynamics, neglected nonlinearities, and system-parameter variations due to environmental changes and elements ageing. These modeling errors may adversely affect the stability and performance of the closed-loop control system which makes necessary to take them into account when building the plant model.

The modeling error is referred to as *structured* or *parametric* uncertainty if it is associated with system parameters which vary in given ranges due to production tolerances or changes of operational conditions. The second type of modeling error is known under the name of *unstructured uncertainty* and corresponds to the case of high-frequency dynamics which is usually neglected during the model development, for instance time delays.

In this section, we present several techniques to describe structured and unstructured uncertainty in MATLAB and Simulink and show how to build uncertain state-space models of continuous-time and discrete-time plants. The corresponding models are used in the next chapters to analyze the robustness of uncertain closed-loop systems and to design controllers ensuring robust stability and robust performance of uncertain plant systems.

2.6.1 Structured uncertainty models

The structured uncertainty models are characterized by the presence of *uncertain parameters* which can be expressed in the form

$$k = \bar{k}(1 + p_k \delta_k) \tag{2.86}$$

Table 2.2   Cart–pendulum uncertain parameters

Parameter	Description	Nominal value	Units	Tolerance (%)
M	Equivalent cart mass	0.768	kg	±10.0
I	Pendulum moment of inertia	2.83×10^{-3}	kg m^2	±20.0
f_c	Dynamic cart friction coefficient	0.5	N s/m	±20.0
f_p	Rotational friction coefficient	6.65×10^{-5}	N ms/rad	±20.0

where $\bar{k}$ is the *nominal value* of the parameter k, p_k is the *relative uncertainty* of k and δ_k is a real or complex uncertainty which is scaled so that $|\delta_k| \leq 1$. The uncertain real parameters are represented in Robust Control Toolbox™ using the function `ureal` and the uncertain complex parameters by `ucomplex` [50, Chapter 1]. These parameters may be used in arbitrary regular expressions involving the elementary arithmetic operations and may participate in matrix elements like the usual MATLAB variables. They are implemented in both continuous-time and discrete-time models. Random samples of these parameters are obtained by the function `usample` and uniformly gridded samples are generated by the function `gridureal`. Such samples are used in the Monte-Carlo simulation of uncertain systems.

Example 2.10. Uncertain model of cart–pendulum system

Further on, we shall make use of an uncertain cart–pendulum system model, which is obtained under the assumption that the system parameters M, I, f_c, and f_p vary in some intervals. These variations are due to changes of the operational conditions or may be a result of lack of data or lack of sufficiently accurate models of some phenomena. Specifically, we assume that the change of the equivalent cart mass M is ±10 percent, the uncertainty in the value of pendulum moment of inertia is ±20 percent and the errors in determining the coefficients f_c and f_p are also ±20 percent.

In Table 2.2, we give the values determining the uncertain cart–pendulum parameters M, I, f_c, and f_p.

The known cart–pendulum system parameters and the trim conditions are set in MATLAB in the usual way. The uncertain parameters are entered by the lines

```
M = ureal('M',0.768,'Percentage',10);        % kg
I = ureal('I',2.83*10^(-3),'Percentage',20);  % kg m^2
f_c = ureal('f_c',0.5,'Percentage',20);       % N s/m
f_p = ureal('f_p',6.65*10^(-5),'Percentage',20); % N m s/rad
```

and the variables depending on the uncertain parameters are computed as

```
I_t = I + m*l^2;                              % kg m^2
M_t = M + m;                                  % kg
den = M_t*I_t - m^2*l^2*cos(theta_trim);
```

The uncertain nonlinear model (2.2) of the cart–pendulum system is implemented by the Simulink file `unc_pendulum_cart.slx`. The uncertain parameters are modeled by Uncertain State Space blocks. Using the Uncertainty

`Value` field of such block, one may specify to use the nominal value, random value or to sample the uncertain variable in the prescribed range. This file can be used to simulate the model for various values in the uncertainty range (Monte-Carlo simulation) or to derive linearized uncertain plant model.

Consider now how to build uncertain linear model of the cart–pendulum system. The matrices of the uncertain linear state-space model are set in MATLAB exactly in the same way as in the case of fixed parameters (see Section 2.2) entering the lines

```
A = [0      0                 1           0
     0      0                 0           1
     0 m^2*l^2*g/den -f_c*I_t/den -f_p*l*m/den
     0 M_t*m*g*l/den -f_c*m*l/den -f_p*M_t/den];
%
B = [   0
        0
     I_t/den
     l*m/den]*k_F;
%
C = [1  0  0  0
     0  1  0  0];
%
D = [0
     0];
```

The uncertain state-space model of the cart–pendulum system is obtained using the function `ss` by the line

```
G_ulin = ss(A,B,C,D)
```

As a result, one obtains the uncertain state-space (`uss`) object

```
G_ulin =

  Uncertain continuous-time state-space model with 2 outputs,
                                        1 inputs, 4 states.
  The model uncertainty consists of the following blocks:
    I: Uncertain real, nominal = 0.00283, variability = [-20,20]%,
                                            10 occurrences
    M: Uncertain real, nominal = 0.768, variability = [-10,10]%,
                                            10 occurrences
    f_c: Uncertain real, nominal = 0.5, variability = [-20,20]%,
                                            2 occurrences
    f_p: Uncertain real, nominal = 6.65e-05, variability = [-20,20]%,
                                            2 occurrences
```

It is seen from the report that the parameters M and I occur ten times in the model while f_c and f_p occur only two times. Note that the large number of uncertain parameter occurrences makes difficult the analysis and especially the design of uncertain control systems.

The uncertain model G_{ulin} can be used to determine the time-domain and frequency domain plant characteristics using the corresponding MATLAB functions implemented in the case of fixed parameter models.

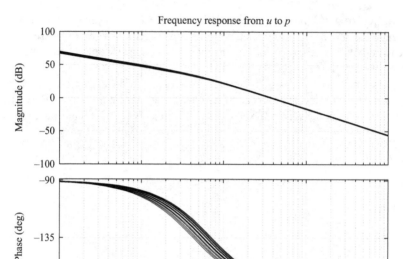

Figure 2.38   Bode plot of the uncertain model with output p

In Figures 2.38 and 2.39, we display the Bode plots of the uncertain model in respect to the outputs p and θ, which are obtained by the lines

```
omega = logspace(-2,2,50);
figure(1)
bode(G_ulin(1,1),omega)
figure(2)
bode(G_ulin(2,1),omega)
```

The usual single curves of these plots are replaced by families of curves obtained for 20 random values of the uncertain parameters set in the prescribed ranges.

It is instructive to consider the variations of plant "gain" as a result of the changes of uncertain parameters. As a measure of the relative change in the gain, it is appropriate to use the quantity

$$\frac{\bar{\sigma}(G_{ulin}(j\omega) - \bar{G}_{ulin}(j\omega))}{\bar{\sigma}(\bar{G}_{ulin}(j\omega))}$$

where $G_{ulin}(s)$ is the uncertain plant transfer function, $\bar{G}_{ulin}(s)$ is the nominal plant transfer function, and $\bar{\sigma}(G_{ulin}(j\omega))$ is the maximum singular value of the frequency response $G(j\omega)$ for a given frequency ω.

The relative changes of the plant gain to the individual variations in M, I, f_c, f_p are represented in Figures 2.40–2.43, respectively. It is seen from the figures that the relative changes in plant gain due to the uncertainties in M and I are comparable

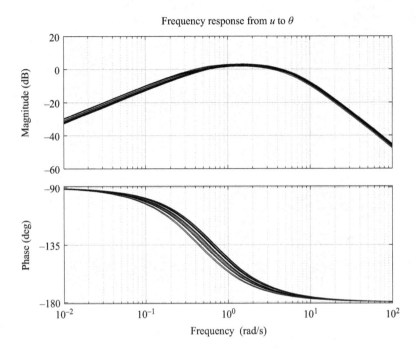

Figure 2.39   Bode plot of the uncertain model with output θ

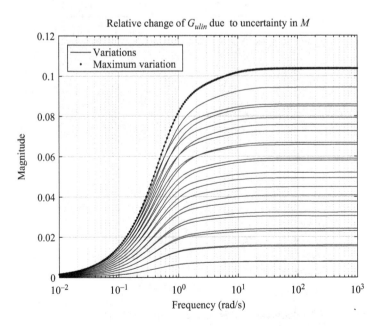

Figure 2.40   Relative change in plant gain due to the uncertainty of M

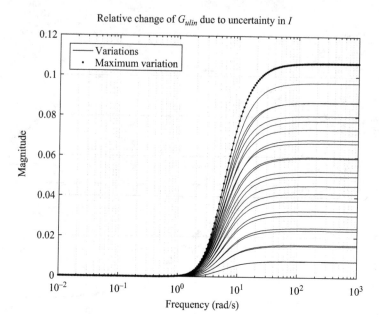

Figure 2.41   Relative change in plant gain due to the uncertainty of I

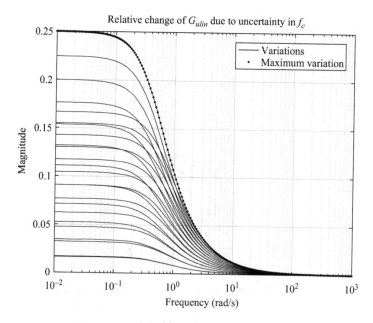

Figure 2.42   Relative change in plant gain due to the uncertainty of f_c

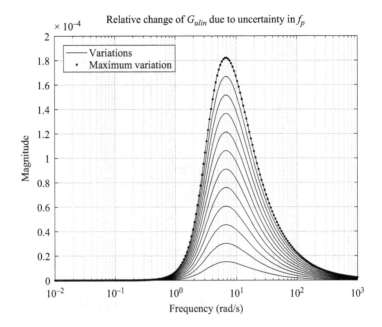

Figure 2.43 Relative change in plant gain due to the uncertainty of f_p

in size but the changes due to the variations of M appear in wider high-frequency range. The changes due to the uncertainties in f_c and f_p have different effects. The changes due to f_c are in the low frequency range, while the changes due to variations in f_p have a maximum around 6 rad/s. The effect of the friction coefficient f_p is 10^4 times smaller in comparison to the effect of other uncertain parameters and may be neglected to simplify the plant uncertainty model.

In Figure 2.44, we show the changes of the plant gain due to the changes in all four uncertain parameters. It is seen that the joint effect of the four uncertainties do not exceed 25 percent and is stronger in the low-frequency range. This is typical for the case of parametric uncertainties. ❐

2.6.2 Representing uncertain models by LFT

Consider the connection represented in Figure 2.45. It is described by the linear system of equations

$$\begin{bmatrix} z \\ y \end{bmatrix} = \begin{bmatrix} 0 & \bar{k} \\ p_k & \bar{k} \end{bmatrix} \begin{bmatrix} w \\ u \end{bmatrix}$$
$$w = \delta_k z \tag{2.87}$$

Performing the matrix-vector multiplication in the first equation of the system (2.87) and eliminating w, z, one obtains

$$y = \bar{k}(1 + p_k \delta_k)u$$

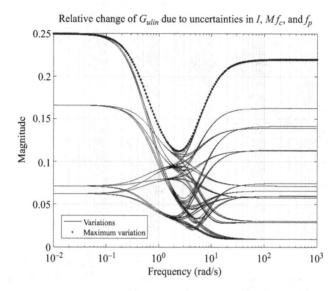

Figure 2.44   *Relative change in plant gain due to the uncertainties of four parameters*

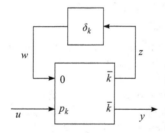

Figure 2.45   *Representation of uncertain parameter by an upper LFT*

which corresponds to the uncertain parameter representation (2.86). In this way, an uncertain real parameter can be represented by the connection shown in Figure 2.45 which is a particular case of the *upper linear fractional transformation (LFT)* shown in Figure 2.46. The connection, represented in this figure, consists of a fixed part described by the known transfer function matrix M and uncertain matrix Δ which may have a special structure.

Let the interconnection transfer function matrix M be partitioned as

$$M(s) = \begin{bmatrix} M_{11} & M_{12} \\ M_{21} & M_{22} \end{bmatrix}$$

according to the dimensions of the vectors u, y, w, z. Then, the LFT shown in Figure 2.46 is described by

$$y = F_U(M, \Delta)u \qquad\qquad (2.88)$$

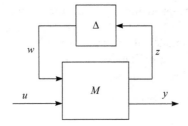

Figure 2.46   Upper linear fractional transformation

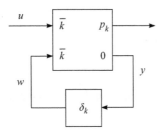

Figure 2.47   Representation of uncertain parameter by a lower LFT

where

$$F_U(M, \Delta) =: [M_{22} + M_{21}\Delta(I - M_{11}\Delta)^{-1}M_{12}]$$

provided that $I - M_{11}\Delta$ is invertible. The notation F_U prompts that the *upper* loop of M is closed by Δ.

The representation of the uncertain parameter

$$k = \bar{k}(1 + p_k\delta_k)$$

by the interconnection shown in Figure 2.45 is not unique. This parameter can also be represented by the interconnection shown in Figure 2.47 which is a particular case of the *lower LFT*.

The LFTs are defined in MATLAB by the function lft, and the matrices M and Δ may be retrieved from a given LFT by the function lftdata.

The LFT is universal tool for representing uncertain systems. For example, plants with several real uncertain parameters can be represented by the LFT, shown in Figure 2.46, where the matrix Δ is a diagonal matrix whose nontrivial elements are the uncertainties of the respective elements. The LFT may be also used to represent other type of uncertain models.

2.6.3   *Deriving uncertain state-space models from Simulink® models*

An uncertain state-space model can be derived from the nonlinear Simulink model of the uncertain plant using the function ulinearize. For this purpose, it is necessary to specify the linearization input and output points of the Simulink model using getlinio command.

In the case of cart–pendulum system, the derivation of uncertain state-space model is done by the M-file `unc_model_simulink` making use of the nonlinear uncertain model of the system implemented by the Simulink file `unc_pendulum_cart.slx`. The derivation of uncertain model is done by the lines

```
% Create operation point specifications for the model
val_all = [];
op = operspec('unc_pendulum_cart')
%
% Get linearization I/O settings for the model
open_system('unc_pendulum_cart')
io = getlinio('unc_pendulum_cart');
%
%Linearization of the model
G_unum = ulinearize('unc_pendulum_cart',op,io)
```

The obtained state-space model G_{unum} is the same as the model G_{ulin} obtained from the linear state-space model, up to the ordering of state vector components.

2.6.4   Unstructured uncertainty models

In the case of unstructured uncertainty, the magnitude of perturbations is limited, but the uncertainty is not associated with specific plant parameters. The unstructured uncertainty can be used to represent the effect of several perturbations of different kind which are lumped into a single perturbation block Δ. The unstructured uncertainty can be described in different ways, depending on the specific application. The most frequently used descriptions are given below.

    1. **Additive uncertainty** (Figure 2.48).
    The uncertain plant is described as

$$G(s) = \bar{G}(s) + W_a(s)\Delta_a(s) \tag{2.89}$$

where $\bar{G}(s)$ is the nominal plant model and the weighting transfer function matrix W_a is chosen so that $\bar{\sigma}(\Delta_a(j\omega)) \leq 1$ for each ω in the frequency range of interest. The scaling of Δ_a is convenient in the robust stability analysis and design of the respective closed-loop system.

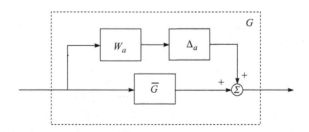

Figure 2.48   Plant with additive uncertainty

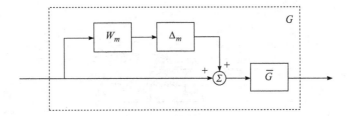

Figure 2.49   Plant with input multiplicative uncertainty

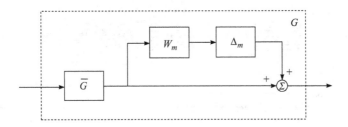

Figure 2.50   Plant with output multiplicative uncertainty

This uncertainty represents the absolute error between the actual dynamics and the nominal model.

2. **Input multiplicative uncertainty** (Figure 2.49).
The perturbed plant is described by

$$G(s) = \bar{G}(s)(I + W_m(s)\Delta_m(s)) \tag{2.90}$$

where $W_m(s)$ is a weighting transfer function matrix chosen so that $\bar{\sigma}(\Delta(j\omega)) \leq 1$.

3. **Output multiplicative uncertainty** (Figure 2.50).
In this case, the plant is described by

$$G(s) = (I + W_m(s)\Delta_m(s))\bar{G}(s) \tag{2.91}$$

Note that for single-input–single-output plants, models (2.90) and (2.91) coincide.

The input and output multiplicative uncertainties are convenient for using since they are related to the relative rather than to absolute modeling error.

The unstructured (complex) uncertainty models are built using the function `ultidyn`. The uncertain linear, time-invariant dynamics object `ultidyn` represents an unknown stable linear system whose only known attribute is a magnitude bound on its frequency response. The nominal value of this uncertain object always has a zero transfer function matrix. Every `ultidyn` element is interpreted as a continuous-time system. However, when such element is an uncertain element of an uncertain state-space model (`uss`), then the time-domain characteristic of the element

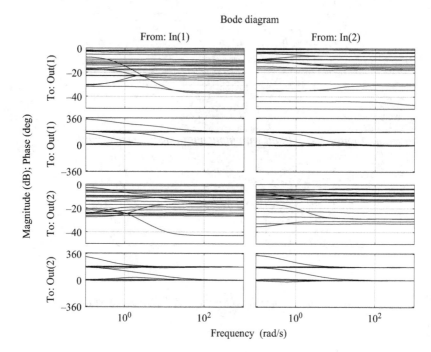

Figure 2.51   Bode plot of an uncertain `ultidyn` *element*

is determined from the time-domain characteristic of the system. Random samples of `ultidyn` element are obtained by the function `usample`.

In Figure 2.51, we show the Bode plots corresponding to the four transfer functions of a two-input–two-output uncertain LTI object Δ generated by the line

```
delta = ultidyn('delta',[2 2])
```

The unstructured uncertainty models (2.89)–(2.91) can be implemented for different purposes. Frequently, they are used to approximate the effect of several uncertain parameters in order to simplify the plant model. Example of such approximation is given in Chapter 4, Example 4.6. Other important application of the unstructured uncertainty models is to describe unmodeled dynamics, as described below.

Consider the set of plants

$$G(s) = \bar{G}(s)f(s),$$

where $\bar{G}(s)$ is a fixed (and known) transfer function. We desire to neglect the term $f(s)$ (which may be a fixed transfer function or belongs to an uncertainty set) and to represent $G(s)$ by multiplicative uncertainty with a nominal model $\bar{G}$ in the form

$$G(s) = \bar{G}(s)(1 + W_m(s)\Delta_m(s)),$$

where $\bar{\sigma}(\Delta_m(j\omega)) \leq 1$.

Since

$$\frac{G(s) - \bar{G}(s)}{\bar{G}(s)} = f(s) - 1,$$

then the magnitude frequency response of the relative uncertainty due to the neglect of the dynamics of $f(s)$, is

$$\frac{|G(j\omega) - \bar{G}(j\omega)|}{|\bar{G}(j\omega)|} = |f(j\omega) - 1|.$$

From this expression, one obtains that

$$|W_m(j\omega)| = \max \left| \frac{G(j\omega) - \bar{G}(j\omega)}{\bar{G}(j\omega)} \right| = \max |f(j\omega) - 1|.$$

This procedure is illustrated by the following example, in which neglected time delay is represented by a multiplicative uncertainty.

Example 2.11. Approximation of uncertain time delay by multiplicative uncertainty

Given is the plant $G = \bar{G}(s)e^{-\tau s}$, where $0 \leq \tau \leq 0.1$ and $\bar{G}(s)$ does not depend on τ. Our desire is to represent the plant by a multiplicative uncertainty and nominal model $\bar{G}(s)$. For this aim, first we compute the magnitude response of the relative error

$$|f(j\omega) - 1| = |e^{-j\omega\tau} - 1| = \sqrt{(\cos(\omega\tau) - 1)^2 + \sin(\omega\tau)^2}$$

for values of τ between 0 and 0.1 and determine an upper bound `max_err` on this error. This is done by the commands

```
nfreq = 100;
omega = logspace(-1,3,nfreq);
max_err = zeros(1,nfreq);
for tau = 0:0.005:0.1;
    for i = 1:nfreq
        om = omega(i);
        sys_err(i) = sqrt((cos(om*tau)-1)^2 + sin(om*tau)^2);
    end
    for i = 1:nfreq
        max_err(i) = max(sys_err(i),max_err(i));
    end
end
```

The frequency responses obtained are shown in Figure 2.52. Next, we find a second-order stable and minimum-phase approximation of the multiplicative uncertainty using the commands

```
ord = 2;                      % approximation order
sys = frd(max_err,omega);     % creates frd object
Wm = fitmagfrd(sys,ord);      % fits the frequency response
```

As a result, for the weighting transfer function $W_m(s)$, one obtains

```
1.967 s^2 + 52.9 s + 1.21
-------------------------
  s^2 + 38.42 s + 535.5
```

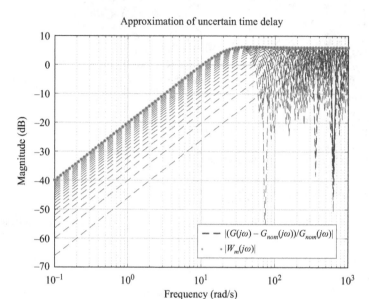

Figure 2.52   Approximation of uncertain time delay

Once the nominal model $\bar{G}$ is set, the uncertain plant model is found by the lines

```
delta = ultidyn('delta',[1 1]);
G_unc = G_nom*(1 + Wm*delta)
```
❒

2.6.5   Mixed uncertainty models

In practice, it is customary to have the presence of both structured and unstructured uncertainty in the plant model. This situation is referred to as *mixed uncertainty*.

The appearance of structured and unstructured uncertainty in the mixed uncertainty model leads to a general model of an uncertain plant. All uncertain parts of this model are combined in a block-diagonal matrix Δ. This matrix has two type of blocks: *repeated scalar* blocks $\delta_1 I_{r_1}, \ldots, \delta_s I_{r_s}$ which corresponds to structured uncertainties and *full* blocks $\Delta_1, \ldots, \Delta_f$ corresponding to unstructured uncertainties. The block

$$
\delta_i I_{r_i} =
\begin{bmatrix}
\delta_i & 0 & \ldots & 0 \\
0 & \delta_i & \ldots & 0 \\
\vdots & \vdots & \ddots & \vdots \\
0 & 0 & \ldots & \delta_i
\end{bmatrix}
$$

contains scalar uncertainty δ_i which is repeated r_i times, $i = 1, \ldots, s$. That is why, the scalar block dimensions are equal to $r_1, \ldots, r_s$. The full blocks $\Delta_1, \ldots, \Delta_f$ have dimensions $m_1, \ldots, m_f$, respectively.

The uncertain block Δ containing all uncertainties is defined as

$$\Delta = \left\{ \begin{bmatrix} \begin{bmatrix} \delta_1 I_{r_1} & & & \\ & \ddots & & \mathbf{0} \\ & & \delta_s I_{r_s} & \\ & & & \Delta_1 \\ \mathbf{0} & & & & \ddots \\ & & & & & \Delta_f \end{bmatrix} \end{bmatrix} : \delta_i \in \mathbb{C}, \ \Delta_j \in \mathbb{C}^{m_j \times m_j} \right\}, \tag{2.92}$$

where $\sum_{i=1}^{s} r_i + \sum_{j=1}^{f} m_j = n$ and n is the dimension of Δ. The set of all matrices Δ is defined as $\Delta \subset \mathbb{C}^{n \times n}$. The parameters δ_i of the repeated scalar blocks can be real numbers only, if the corresponding uncertainties are real. Note that in (2.92), all scalar block appear first which is done to simplify the notation. The corresponding functions in Robust Control Toolbox may work even with nonsquare blocks as well as arbitrary order of blocks.

To build models with mixed sensitivity, it is convenient to use the functions `sysic` or `iconnect`. Examples of using the function `sysic` are given in Chapter 4.

2.6.6 Discretization of uncertain models

Consider the linear continuous-time uncertain plant described by

$$\begin{bmatrix} z(s) \\ y(s) \end{bmatrix} = M \begin{bmatrix} w(s) \\ u(s) \end{bmatrix} \tag{2.93}$$

$$w(s) = \Delta z(s) \tag{2.94}$$

where

$$M = \begin{bmatrix} A & B_1 & B_2 \\ C_1 & D_{11} & D_{12} \\ C_2 & D_{21} & D_{22} \end{bmatrix}$$

is the transfer function matrix of the fixed part of the system and the uncertainty Δ has the form shown in (2.92). According to (2.93) and (2.94), the uncertain continuous-time plant model can be represented as the upper LFT

$$G(s) = F_U(M, \Delta). \tag{2.95}$$

Denoting by $u(k), y(k)$, the sampled signals of u, y with sampling period $T_s > 0$, the discrete-time equivalent of the uncertain plant model (2.95) is defined as

$$G_d(z) = F_U(M_d, \Delta_d). \tag{2.96}$$

A commonly used approach to discretize (2.95), called *Full ZOH approach* [51], is to apply sampling and zero-order holds on all signals of (2.93) as shown in Figure 2.53, assuming that the hold and sampling devices are perfectly synchronized with the sampling period. This setting implies that (2.93) is discretized as stand-alone system disregarding (2.94) which is equivalent to discretize the fixed part transfer function

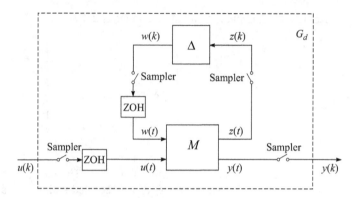

Figure 2.53   Discretization of uncertain plant

matrix $M(s)$ for the specified sampling period T_s. As a result of the discretization, one obtains a discrete-time transfer function $M_d(z)$. The discrete-time uncertain model is then determined as

$$G_d(z) = F_U(M_d, \Delta) \tag{2.97}$$

where Δ is interpreted now as a discrete-time uncertainty. Such interpretation is possible in Robust Control Toolbox since the `ureal`, `ucomplex`, and `ultidyn` objects are considered as continuous-time or discrete-time objects depending on the nature of the uncertain system within which they are uncertain elements.

The Full ZOH approach leads to a simple discretization method for uncertain models which is sufficiently accurate for small values of T_s. More accurate but more complicated methods are presented in [51].

Example 2.12. Discretization of uncertain cart–pendulum system model

The discrete-time model DG_{ulin} of the uncertain cart–pendulum system model G_{ulin}, obtained in Example 2.6, is computed in the case of zero-order hold by the lines

```
Ts = 0.01;
[M,Delta] = lftdata(G_ulin);
Md = c2d(M,Ts);
DG_ulin = lft(Delta,Md)
```

As noted previously, the functions `ureal`, `ucomplex`, and `ultidyn` generate uncertain objects which can be implemented in continuous-time as well as in discrete-time models. This explains the usage of the same variable `Delta` in both continuous-time and discrete-time models.

The Bode plots of continuous-time and discrete-time cart–pendulum system models in respect to the position p and pendulum angle θ, obtained for 30 random samples of the uncertainty, are compared in Figures 2.54 and 2.55, respectively. It is seen from the figures that the Bode plot of the discretized model is close to the Bode plot of the continuous-time one except for frequencies near to the Nyquist frequency. The large errors near to this frequency are due to the properties of the method ("zoh") used for

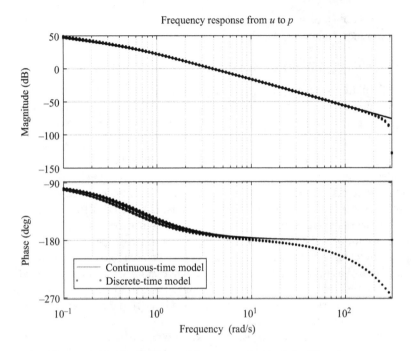

Figure 2.54 Bode plots of the uncertain model with output p

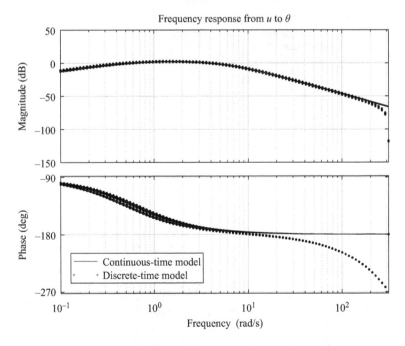

Figure 2.55 Bode plots of the uncertain model with output θ

discretization of fixed model part. If instead of "zoh" method one uses "tustin", the phase errors are reduced significantly. ❏

2.6.7 Deriving uncertainty models by identification

Uncertain plant model may be determined on the basis of the results, obtained by model identification. Example 2.13 illustrates the derivation of uncertainty model of an unstable plant using the functions of System Identification Toolbox.

Example 2.13. Deriving uncertain model by identification
Consider the derivation of uncertainty model of cart–pendulum system, using the results from black box identification of the system, presented in Example 2.8. The condition of unbiased parameter estimates obtained by the identification guarantees that the exact parameter values are contained in the confidence intervals of parameter estimates with probability close to 1. This allows to derive parametric uncertainty model with scalar uncertainties. However, since the number of estimated parameters is usually large, the implementation of structured uncertainty model is not practical. That is why, it is preferable to derive unstructured uncertainty model. Based on parameter 3σ confidence intervals, maximum relative deviations from the nominal model in the frequency domain can be obtained.

Assume that the model `modeln4sid` is obtained as shown in Example 2.8. Then, the maximum deviations of the model frequency responses from u to θ are obtained and plotted by the following command lines:

```
w = logspace(-1,log10(pi/Ts),500);
[MAG,PHASE,W,SDMAG,SDPHASE] = bode(modeln4sid,w);
MAGt2 = []; MAGt2(1:length(w)) = MAG(2,1,:);
SDMAGt21 = []; SDMAGt2(1:length(W)) = SDMAG(2,1,:);
relmag21 = 3*SDMAGt2./MAGt2;
figure(1)
semilogx(w,relmag21,'r'),grid
xlabel('Frequency (rad/s)')
ylabel('Relative uncertainty')
title('Relative magnitude uncertainty')
xlim([0.1 pi/0.01])
%
% Uncertain magnitude and phase plots (3sigma confidence intervals)
mag_nom_theta = reshape(MAG(2,1,:),size(MAG(2,1,:),3),1);
mag_max_theta = reshape(MAG(2,1,:)+3*SDMAG(2,1,:), ...
                                  size(MAG(2,1,:),3),1);
mag_min_theta = reshape(MAG(2,1,:)-3*SDMAG(2,1,:), ...
                                  size(MAG(2,1,:),3),1);
%
phase_nom_theta = reshape(PHASE(2,1,:),size(PHASE(2,1,:),3),1);
phase_max_theta = reshape(PHASE(2,1,:)+3*SDPHASE(2,1,:), ...
                                  size(PHASE(2,1,:),3),1);
phase_min_theta = reshape(PHASE(2,1,:)-3*SDPHASE(2,1,:), ...
                                  size(PHASE(2,1,:),3),1);
%
figure(2)
```

```
semilogx(w,mag_min_theta,'b--',w,mag_nom_theta,'m-', ...
          w,mag_max_theta,'r-.'), grid
xlabel('Frequency (rad/s)')
ylabel('Magnitude')
title('Magnitude response from u to \theta')
legend('Minimum magnitude','Nominal magnitude','Maximum magnitude')
xlim([0.1 pi/0.01])
%
figure(3)
semilogx(w,phase_min_theta,'b--',w,phase_nom_theta,'m-', ...
          w,phase_max_theta,'r-.'), grid
xlabel('Frequency (rad/s)')
ylabel('Phase (deg)')
title('Phase response from u to \theta')
legend('Minimum phase','Nominal phase','Maximum phase')
xlim([0.1 pi/0.01])
```

The frequency plot of the relative magnitude uncertainty of the model is shown in Figure 2.56. Clearly, the model has larger uncertainties in the low-frequency and high-frequency ranges. Note that the bound of relative uncertainty is conservative which may lead to pessimistic results in the analysis and design.

In Figures 2.57 and 2.58, we show the uncertainty bounds of the model frequency responses. These bounds can be used to derive unstructured uncertainty model used for robustness analysis and design. For this aim, the bounds are approximated by optimization procedure with shaping filter which is represented by transfer function of corresponding order. Example of such model is implemented in Chapter 7.　❑

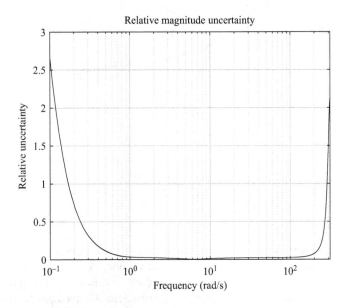

Figure 2.56　Relative magnitude uncertainty

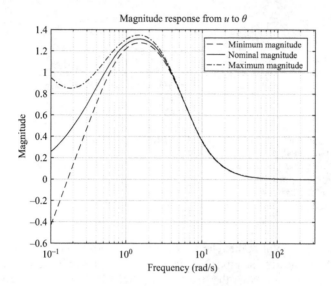

Figure 2.57   Uncertain magnitude response

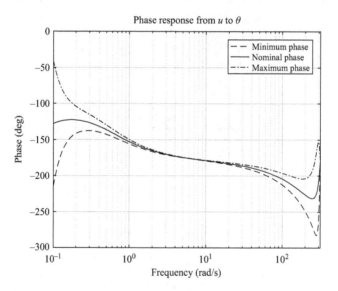

Figure 2.58   Uncertain phase response

It is necessary to point out that using different estimation sets it is possible to obtain uncertainty models with different uncertainty bounds. This indicates that additional optimization may be necessary in order to obtain a model with tight uncertainty bounds. The alternative is to use special identification methods for uncertainty models, see for instance [52–54].

2.7 Sensor modeling

MATLAB files used in this section

File	Description
gyro_model	Model of gyro noise
gyro_fun	Function used in optimization of gyro model
sim_gyro_model.slx	Simulink model of gyro noise
accel_model	Model of accelerometer noise
accel_fun	Function used in optimization of accelerometer model
sim_accel_model.slx	Simulink model of accelerometer noise
filtering_design	Kalman filter design for sensor fusion
filtering_sim	Kalman filtering simulation
sim_fusion.slx	Simulink model of sensor fusion

This section is devoted to the modeling of sensor stochastic errors which affect the performance of embedded closed-loop control systems. Among the large variety of sensors, we choose to present the modeling of microelectromechanical systems (MEMS) sensors intended to measure motion variables like angular rates and accelerations. The reason behind this is that the MEMS inertial sensors (gyroscopes and accelerometers) have lots of applications in embedded low-cost navigation and control systems. A common disadvantage of these sensors are the significant errors which accompany the corresponding measurements. This necessitates the development of adequate error models which may be used to achieve acceptable measurement accuracy with the aid of appropriate filtering. Note that triaxial gyroscopes and accelerometers may be combined in a single device named *Inertial Measurement Unit* (IMU) used in navigation and motion control.

The inertial sensor errors consist of deterministic and stochastic parts. The deterministic part includes bias offset, scale factor error, axis nonorthogonality, axis misalignment and so on. These errors are removed from row measurements by the corresponding calibration techniques, see for instance Section 3.3 [55]. The stochastic part contains random errors (noises) which cannot be removed from the measurements and should be modeled as stochastic processes. The stochastic sensor errors comprises in-run bias variation (bias instability), rate random walk and angular random walk (for gyroscopes), and acceleration random walk and velocity random walk (for accelerometers).

The gyro and accelerometer stochastic models are build using a variety of techniques. The most widely used tools are the power spectral density (PSD) in the frequency-domain and the Allan variance in the time-domain. Note that it is desirable to keep the model order as low as possible since the model is used in the design of optimal estimators based on the sensor measurements.

In what follows we present a simple methodology that allows to develop stochastic discrete-time models of MEMS gyro and accelerometer noises. The methodology is based on the frequency-domain and time-domain characteristics of the sensors noises and is illustrated for the case of Analog Devices Tri-Axis IMU *ADIS16405*. It is shown that the gyroscope and accelerometer noises may be represented by similar second-order models which are appropriate for usage in development of navigation and control systems.

2.7.1   Allan variance

Consider the application of the Allan variance to the time-domain analysis of random processes.

The Allan variance provides a means of identifying various noise terms in the original data set [56,57].

Assume that a quantity $\Theta(t)$ is measured at discrete time moments $t = kT_s, k = 1, 2, \ldots, L$ with a sample time T_s. The average value of Θ between times $t(k)$ and $t(k) + \tau$ where $\tau = mT_s$, is given by

$$\hat{\Omega}(k)(\tau) = \frac{\Theta(k + m) - \Theta(k)}{\tau}$$

where $\Theta(k) = \Theta(kT_s)$.

The Allan variance is defined as

$$\sigma^2(\tau) = \frac{1}{2}\left\langle (\hat{\Omega}(k + m) - \hat{\Omega}(k))^2 \right\rangle$$
$$= \frac{1}{2\tau^2} \left\langle (\Theta(k + 2m) - 2\Theta(k + m) + \Theta(k))^2 \right\rangle \tag{2.98}$$

where $\langle\ \rangle$ is the ensemble average.

The Allan variance is estimated as follows

$$\sigma^2(\tau) = \frac{1}{2\tau^2(L - 2m)} \sum_{k=1}^{L-2m} (\Theta(k + 2m) - 2\Theta(k + m) + \Theta(k))^2. \tag{2.99}$$

The computation of Allan variance of a given random process corresponds to the filtration of this process by using a filter whose bandwidth depends on the value of τ. This makes possible to examine different types of random processes by varying τ. The Allan variance is usually plotted as a function of τ on a log–log plot.

Consider the application of Allan variance to some simple random processes used to describe the stochastic errors in different devices.

The Allan variance of the white noise is given by

$$\sigma^2(\tau) = \frac{\sigma_x^2}{\tau}.$$

The Allan variance of the random walk is given by

$$\sigma^2(\tau) = \frac{\sigma_w^2 \tau}{3}.$$

The Allan variance of the first-order Gauss–Markov process is determined as [57]

$$\sigma^2(\tau) = \frac{\sigma_x^2}{\tau} \left[1 - \frac{T}{2\tau}(3 - 4e^{-\frac{\tau}{T}} + e^{-2\tau/T}) \right] \tag{2.100}$$

This expression leads to the following conclusions. If τ is much longer than the correlation time T, it is found that

$$\sigma^2(\tau) \Rightarrow \frac{\sigma_x^2}{\tau}, \quad \tau \gg T$$

which is the Allan variance for white noise with strength σ_x^2. For τ much smaller than the correlation time, (2.100) reduces to

$$\sigma^2(\tau) \Rightarrow \frac{\sigma_x^2}{T^2}\frac{\tau}{3}, \quad \tau \ll T$$

which is the Allan variance for random walk with driving noise of strength σ_x^2/T^2.

2.7.2 Stochastic gyro model

The MEMS gyroscope noise typically consists of the following terms:

- The *bias* is a time-variant error, which is independent of the angular rate and may be represented as a sum of static and dynamic components. The static component, also known as the fixed bias, turn-on bias or bias repeatability, comprises the run-to-run variation of each gyroscope bias plus the residual fixed bias remaining after sensor calibration. It is constant throughout an IMU operating period, but varies from run to run. The dynamic component, also known as the in-run bias variation or *bias instability*, varies over periods of order a minute and also incorporates the residual temperature-dependent bias remaining after sensor calibration. The bias instability is a low-frequency process whose origin are components susceptible to random flickering. The intensity of bias instability is inversely proportional to the frequency ($1/f$ noise) and is characterized by nonstationary autocorrelation function. The bias instability is typically expressed in deg/h, where 1 deg/h = 4.848137×10^{-6} rad/s.
- *Rate random walk (RRW)*. This is a rate error due to white noise in angular acceleration. The RRW is a nonstationary process whose intensity increases linearly with time. This error is typically expressed in deg/h/$\sqrt{h}$ or in deg/h^2/$\sqrt{Hz}$, where 1 deg/h/$\sqrt{h}$ = 8.080228×10^{-8} rad/s/$\sqrt{s}$ and 1 deg/h^2/$\sqrt{Hz}$ = 1.346705×10^{-9} rad/s/$\sqrt{s}$.
- *Angular random walk (ARW)*. This is an angular error which is due to the white noise in angular rate. The ARW is described by the first-order differential equation

$$\frac{d\theta}{dt} = w(t)$$

where w is white input driving noise. The strength of this noise is typically expressed in deg/$\sqrt{h}$ or in deg/h/$\sqrt{Hz}$ where 1 deg/$\sqrt{h}$ = 2.90888×10^{-4} rad/$\sqrt{s}$ and 1 deg/h/$\sqrt{Hz}$ = 4.848137×10^{-6} rad/$\sqrt{s}$.

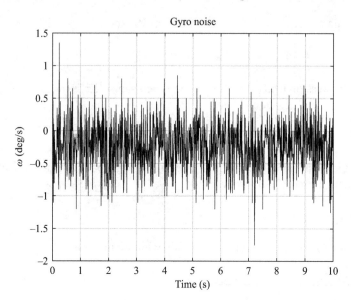

Figure 2.59   Output gyro noise

- *Quantization error.* This is an error representing the rate quantization noise due to the analog-to-digital conversion. This error is expressed in deg/s.

  Other gyro noise terms are described in detail in [57].
  The derivation of stochastic discrete-time model of gyro noise is illustrated by Example 2.14.

Example 2.14. Derivation of stochastic gyro noise model
The model is obtained using the M-file gyro_model. For this aim, we use $L = 10^6$ samples of one of the gyro outputs of *ADIS16405* IMU [58] measured at rest with frequency $f_s = 100$ Hz during the period of 10,000 s. After removing a small constant bias of -0.2071 (deg/s) we found that the noise standard deviation is (to six digits)

$$\sigma_x = 0.382981 \ (\text{deg/s}).$$

The first 1,000 samples of the centered output gyro noise are shown in Figure 2.59.
To determine the individual terms of the gyro noise, we study its frequency domain and time domain characteristics. In the frequency domain, we make use of the one-sided PSD of gyro noise, determined by the function pwelch from MATLAB and shown in Figure 2.60. The PSD is computed for frequencies up to 50 Hz. To find the different error components, the PSD is approximated by straight lines of different slopes. (The coefficient 2 comes from the fact that the one-sided PSD is twice the two-sided PSD used in the approximation.) Based on the PSD approximation, we conclude that there are three noise components, namely bias instability, angular random walk and rate random walk characterized by the parameters B, N

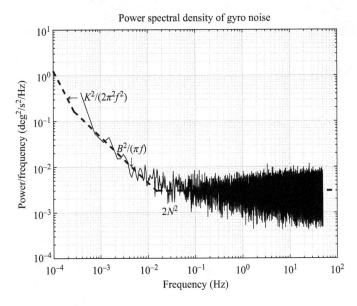

Power spectral density of gyro noise

Figure 2.60   Spectral density of gyro noise

and K, respectively. Since the gyro has 14-bit resolution, its quantization error is negligible.

In the low-frequency range, the PSD is approximated by the term

$$\frac{2K^2}{\omega^2}$$

for some constant coefficient K, which indicates the presence of random walk process. This term can be presented as output of an integrator with white noise driving input of strength K^2.

In the frequency range between 10^{-3} and 10^{-2} Hz, the PSD is approximated by the term

$$\frac{2B^2}{\omega}$$

which cannot be presented as an output of a shaping filter with rational transfer function and white noise driving input. Further on, an appropriate approximation of bias instability is done by first-order Markov process generated as the output of first-order lag with white noise input.

The third section of the PSD is approximated by a horizontal line which shows the presence of a white noise in the measured rate, corresponding to the angle random walk. This noise is of strength N^2.

To determine the parameters B, N, and K precisely, we use the Allan variance of gyro noise, computed with overlapping estimates, and shown in Figure 2.61.

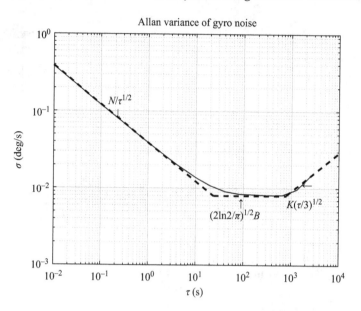

Figure 2.61 Allan variance of gyro noise

The Allan variance of the white noise corresponding to the angular random walk has a slope of $-1/2$ on the log-log plot of $\sigma(\tau)$ versus τ. The numerical value of N can be obtained by reading the slope line at $\tau = 1$.

The Allan variance of the bias instability is represented by a horizontal line drawn at

$$\sigma(\tau) = \sqrt{\frac{2ln2}{\pi}}B$$

on the log–log plot of $\sigma(\tau)$ versus τ [57]. This allows to find simply the value of B.

The Allan variance of random walk is represented by a slope of $+1/2$ on the log–log plot of $\sigma(\tau)$ versus τ. The value of K can be read off the slope line at $\tau = 3$.

As a result, one obtains the values

- $B = 1.20 \times 10^{-2}$ deg/s $= 43.2$ deg/h,
- $K = 5.00 \times 10^{-4}$ deg/s/$\sqrt{s} = 0.2324$ deg/h/$\sqrt{h}$,
- $N = 3.88 \times 10^{-2}$ deg/$\sqrt{s} = 2.328$ deg/$\sqrt{h}$

which are used in the PSD approximation of gyro noise, shown in Figure 2.60.

The stochastic modeling of gyro noise consists in determining appropriate shaping filters with random noise driving inputs whose outputs approximate the respective gyro noise components with sufficient accuracy. For this aim, we shall use a simple optimization algorithm which minimizes the difference between the gyro noise PSD and PSD of the approximate model. In accordance with the analysis presented, the approximate noise model is taken in the form

$$\tilde{x} = x^{bias} + x^{rrw} + x^{arw}, \tag{2.101}$$

where x^{bias}, x^{rrw}, and x^{arw} stand for bias instability, rate random walk, and angular random walk driving noise, respectively.

An approximation of the bias instability is obtained by the output of a discrete-time first-order filter

$$x(k+1)^{bias} = a_d x(k)^{bias} + b_d v(k), \tag{2.102}$$

where a_d, b_d are filter coefficients and v is an input white noise. This filter is obtained by discretization of the first-order Gauss–Markov process

$$T\dot{x}(t) + x(t) = v(t),$$

where T is the correlation time. The correlation time is determined by an optimization procedure along with the input noise strengths. The coefficients a_d, b_d are determined as

$$a_d = e^{-\Delta T/T}, \quad b_d = \int_0^{\Delta T} e^{-\tau/T} d\tau,$$

where $\Delta T = T_s$ is the sampling period.

The random process x^{rrw} is taken as

$$x^{rrw}(z) = DF(z)w(z) \tag{2.103}$$

where

$$DF(z) = \frac{\Delta T}{z-1}$$

is the transfer function of discrete-time integrator and w is an input white noise.

Finally, the driving input of angular random walk is modeled by the discrete-time white noise x^{arw}.

The discrete-time white noises v, w, and x^{arw} are obtained as

$$v = K_{g_1} r_1, \quad w = K_{g_2} r_2, \quad x^{arw} = K_{g_3} r_3$$

where K_{g_1}, K_{g_2}, K_{g_3} are unknown coefficients and r_1, r_2, r_3 are zero mean pseudorandom sequences with standard deviation equal to 1 and length 10^6 generated by the function randn. Note that the parameters K_{g_1}, K_{g_2}, K_{g_3} represent the strengths of the corresponding input driving noises.

In this way, the gyro noise model is characterized by the four unknown parameters T, K_{g_1}, K_{g_2}, and K_{g_3}. They are determined by an iterative procedure, minimizing the difference $\tilde{P}_x - P_x$ where $\tilde{P}_x$ is the PSD of $\tilde{x}$, as defined in (2.101), and P_x is the PSD of the gyro noise x obtained experimentally. The optimal values of these parameters are found by the function lsqnonlin from MATLAB which implements a nonlinear least squares minimization method. The initial values of the unknown parameters are taken as

$$T = 30, \quad K_{g_1} = 1, \quad K_{g_2} = 1, \quad K_{g_3} = \sigma_x.$$

The one-sided PSD of the approximated gyro noise is determined by the function pwelch with overlapping sections of the same length. Best results are obtained when the length of these sections is taken equal to 50,000. As a result, after 110 iterations one obtains the following parameter values (rounded to six significant digits)

$$T = 27.9780 \text{ s},$$

$$K_{g_1} = 0.890636 \text{ (deg/s)}\sqrt{s}, \ K_{g_2} = 0.00687997 \text{ (deg/s)}/\sqrt{s},$$

$$K_{g_3} = 0.376935 \text{ deg}/\sqrt{s}.$$

Since the length of the approximation noise is taken sufficiently large (10^6 samples), the optimal parameters obtained are statistically reliable which is verified by the noise model simulation. Note, however, that the correlation time T may be very sensitive to the changes of initial conditions.

The discrete-time gyro noise model thus obtained may be represented in the form

$$\tilde{x}(k) = x(k)^{bias} + x(k)^{rrw} + x(k)^{arw}, \tag{2.104}$$

where

$$x(k+1)^{bias} = a_d x(k)^{bias} + b_d K_{g_1} \eta(k)^{bias}, \tag{2.105}$$

$$x(k+1)^{rrw} = x(k)^{rrw} + \Delta T K_{g_2} \eta(k)^{rrw}, \tag{2.106}$$

$$x(k)^{arw} = K_{g_3} \eta(k)^{arw} \tag{2.107}$$

where η^{bias}, η^{rrw}, η^{arw} are white noises of unit strength.

Equations (2.104)–(2.107) are represented in the standard form

$$\begin{aligned} x(k+1) &= Ax(k) + Gv(k), \\ y(k) &= Cx(k) + Hv(k) \end{aligned} \tag{2.108}$$

where $x(k) = [x(k)^{bias}, \ x(k)^{rrw}]^T$, $y(k) = \tilde{x}$, $v(k) = [\eta(k)^{bias}, \ \eta(k)^{rrw}, \ \eta(k)^{arw}]^T$, and

$$A = \begin{bmatrix} a_d & 0 \\ 0 & 1 \end{bmatrix}, \quad G = \begin{bmatrix} b_d K_{g_1} & 0 & 0 \\ 0 & \Delta T K_{g_2} & 0 \end{bmatrix},$$

$$H = \begin{bmatrix} 0 & 0 & K_{g_3} \end{bmatrix}, \quad C = \begin{bmatrix} 1 & 1 \end{bmatrix}.$$

The model is of second order and, its output y is expressed in degrees per second. If it is necessary to express the model output in radians per second, then the coefficients K_{g_1}, K_{g_2}, and K_{g_3} should be multiplied by $pi/180$.

As it is seen from Figure 2.62, the PSD of the model noise, generated according to (2.101), fits well the PSD of the measured noise.

The Simulink model of the gyro noise, build upon (2.104)–(2.107), is shown in Figure 2.63.   ❑

2.7.3   Stochastic accelerometer model

Similarly to gyro noise, the MEMS accelerometer noise may be represented as a sum of the following terms.

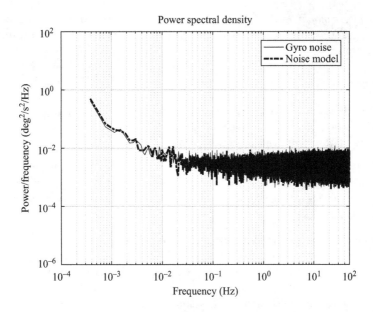

Figure 2.62   Spectral density of gyro noise and model

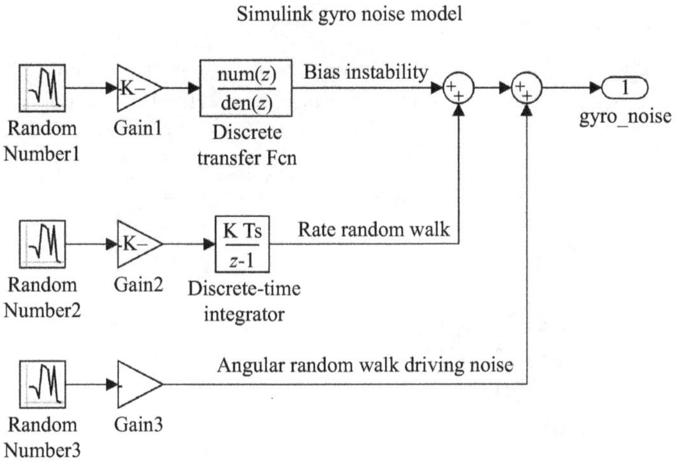

Simulink gyro noise model

Figure 2.63   Simulink® gyro model

- *Bias instability.* The accelerometer bias instability is typically expressed in mg ($10^{-3} \times g$), where $g = 9.80665$ m/s^2 is the earth gravitation unit.
- *Acceleration random walk (ARW).* This is an acceleration error due to white noise in jerk. The ARW is typically expressed in mg/$\sqrt{h}$ where 1 mg/$\sqrt{h} = 1.634442 \times 10^{-4}$ m/s^2/$\sqrt{s}$.

- *Velocity random walk (VRW)*. This is a velocity error due to white noise in acceleration measurements. The VRW is described by the first-order differential equation

$$\frac{dV}{dt} = w(t)$$

where V is the velocity and w is white acceleration noise. The strength of this noise is typically expressed in m/s/$\sqrt{h}$, where 1 m/s/$\sqrt{h}$ = 0.0166667 m/s/$\sqrt{s}$.
- *Quantization error*. This error is expressed in g.

To develop a model of the accelerometer noise one can implement the same methodology as in the case of gyroscopes. This is due to the fact that the PSD and Allan variance of the MEMS accelerometer noise have properties similar to the respective gyro characteristics.

Example 2.15. Accelerometer noise model

The derivation of noise model is done by the M-file `accel_model`.

In the given case we use 10^6 samples of one of the accelerometer outputs of the *ADIS16405* Inertial Measurement Sensor measured at rest with frequency $f_s = 100$ Hz (Figure 2.64). The standard deviation of the zero mean noise is found to be $\sigma_{x_a} = 3.247324 \times 10^{-3}$ g.

The accelerometer noise is modeled as

$$\tilde{x}_a = x_a^{bias} + x_a^{arw} + x_a^{vrw}, \qquad (2.109)$$

where x_a^{bias}, x_a^{arw}, and x_a^{vrw} are the accelerometer bias instability, acceleration random walk, and velocity random walk driving noise, respectively. The unknown model

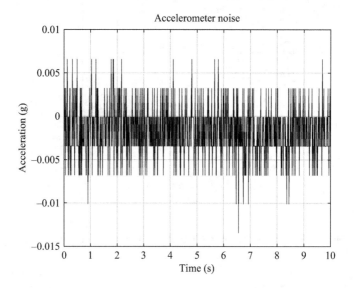

Figure 2.64   Output accelerometer noise

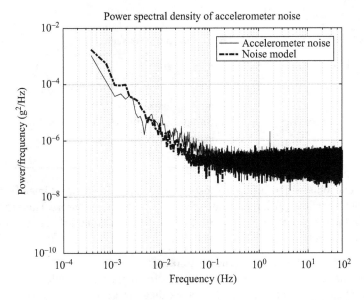

Figure 2.65   Spectral density of accelerometer noise and model

parameters are the time constant T_a of the bias instability, the strength K_{a_1} of bias instability driving noise, the strength K_{a_2} of the acceleration random walk, and the strength K_{a_3} of the velocity random walk. As a result of the optimization procedure, after 120 iterations, one obtains

$$T_a = 4819.36 \text{ s},$$
$$K_{a_1} = 2.69875 \text{ g}\sqrt{s}, \ K_{a_2} = 0.00206606 \text{ g}/\sqrt{s}, \ K_{a_3} = 0.00303117 \text{ gs}/\sqrt{s}.$$

The output of accelerometer model (2.109) is measured in g. To obtain the output in m/s^2 it is necessary to multiply the coefficients K_{a_1}, K_{a_2}, and K_{a_3} by 9.80665.

The accelerometer noise PSD and the model noise PSD are shown in Figure 2.65. Better results can be obtained by using higher order shaping filters but this may complicate the noise model. ❒

We note at the end of this section that in some cases, it might be justified to use simplified models of gyroscope and accelerometer noises in order to reduce the overall system order and thus to reduce the embedded controller order. For instance, in low cost navigation systems, the gyroscope noise is frequently modeled using only the rate random walk and angular random walk terms and the accelerometer noise is modeled taking only the velocity random walk term. Corresponding noise models can be obtained by the procedure described using only the desired terms.

2.7.4 *Sensor data filtering*

MATLAB files used in this section

File	Description
filtering_design	Design of Kalman filter for single-axis attitude estimation
filtering_sim	Simulation of the single-axis attitude estimation system
sim_fusion.slx	Simulink simulation model

The data obtained by different sensors may be fused applying Kalman filtering to obtain better accuracy in comparison to the usage of separate sensors. This is achieved thanks to the properties of the Kalman filter and the fact that different sensors have independent stochastic errors.

Example 2.16. Single-axis attitude estimation
Consider the problem of single-axis attitude estimation using attitude angle measurements and gyro rate information. The block diagram of the sensor fusion model is shown in Figure 2.66. The gyro output

$$\tilde{\omega} = \omega + \Delta\omega$$

consists of the reference signal ω and the additive gyro noise $\Delta\omega$. The attitude measurements are obtained by a sensor whose output

$$\tilde{\theta} = \theta(k) + \Delta\theta$$

is corrupted by the error $\Delta\theta$. The gyro and angle measurements are used as inputs of a Kalman filter which produces the optimal attitude estimate $\hat{\theta}$. The quantity $\Delta\hat{\theta} = \theta - \hat{\theta}$ represents the attitude estimation error.

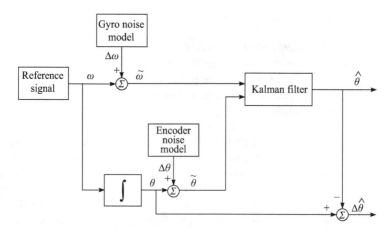

Figure 2.66   Block diagram of the sensor fusion model

Assume that the gyro noise satisfies the model given by (2.108). Hence, the angle rate $\omega = \dot{\theta}$ is related to the gyro output $\tilde{\omega}$ by the equation

$$\dot{\theta} = \tilde{\omega} - x^{bias} - x^{rrw} - x^{arw} \tag{2.110}$$

where x^{bias} is bias instability, x^{rrw} is rate random walk, and x^{arw} is angular random walk driving noise. Note that $x(k)^{bias}$, $x(k)^{rrw}$, and $x(k)^{arw}$ are measured in degrees.

Equation (2.110) is discretized in the form

$$\theta(k+1) = \theta(k) + \Delta T(\tilde{\omega}(k) - x(k)^{bias} - x(k)^{rrw} - x(k)^{arw}) \tag{2.111}$$

where the bias and rate random walk are solutions of the difference equations

$$x(k+1)^{bias} = a_d x(k)^{bias} + b_d K_{g_1} \eta(k)^{bias}, \tag{2.112}$$

$$x(k+1)^{rrw} = x(k)^{rrw} + \Delta T K_{g_2} \eta(k)^{rrw}, \tag{2.113}$$

respectively.

The attitude measurements are written in the form

$$\tilde{\theta} = \theta(k) + \eta^{enc} \tag{2.114}$$

where the encoder error η^{enc} represents a white quantization noise. Assuming that an 8-bit encoder is used, the intensity of the quantization noise is equal to

$$\sigma_{enc}^2 = q^2/12, \quad q = 360/2^8,$$

where the coefficient 360 corresponds to noise measured in degrees.

Equations (2.111)–(2.114) are written in the standard form

$$\begin{aligned} x(k+1) &= Ax(k) + Bu(k) + Gv(k), \\ y(k) &= Cx(k) + w(k) \end{aligned} \tag{2.115}$$

where

$$\begin{aligned} x(k) &= [\theta(k), \ x(k)^{bias}, \ x(k)^{rrw}]^T, \\ u(k) &= \tilde{\omega}(k), \\ y(k) &= \tilde{\theta}(k), \\ v(k) &= [\eta(k)^{bias}, \ \eta(k)^{rrw}, \ \eta(k)^{arw}]^T, w(k) = \eta(k)^{enc} \end{aligned}$$

and

$$A = \begin{bmatrix} 1 & -\Delta T & -\Delta T \\ 0 & a_d & 0 \\ 0 & 0 & 1 \end{bmatrix}, \quad B = \begin{bmatrix} \Delta T \\ 0 \\ 0 \end{bmatrix},$$

$$G = \begin{bmatrix} 0 & 0 & -\Delta T K_{g_3} \\ b_d K_{g_1} & 0 & 0 \\ 0 & \Delta T K_{g_2} & 1 \end{bmatrix}, \quad C = \begin{bmatrix} 1 & 0 & 0 \end{bmatrix}.$$

The noises v and w have covariances given by

$$E\{v(k)v_j^T\} = V\delta(j-k), \quad E\{w(k)w_j^T\} = W\delta(j-k),$$

where $V = I_3$ and $W = \sigma_{enc}^2$.

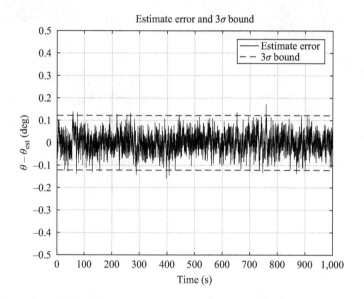

Figure 2.67   Estimate error and bounds

Equation (2.115) is used to design a third-order Kalman filter

$$\hat{x}(k+1) = A\hat{x}(k) + Bu(k) + K(y(k+1) - CA\hat{x}(k) - CBu(k)) \qquad (2.116)$$

which determines the optimal estimate $\hat{x}(k)$ of $x(k)$ based on the noisy measurements $u(k) = \tilde{\omega}$ and $y(k) = \tilde{\theta}$. The steady-state value

$$P = E\left\{(x(k) - \hat{x}(k))(x(k) - \hat{x}(k))^T\right\}.$$

of the variance of $x(k) - \hat{x}(k)$ is given by the positive definite solution P of the associated Riccati equation

$$APA^T - P + V - APC^T(CPC^T + W)^{-1}CPA^T = 0.$$

Therefore, the predicted standard deviation σ_{err} of the estimate error $\Delta\hat{\theta} = \theta(k) - \hat{\theta}(k)$ can be determined as the square root of the element $P(1, 1)$. This gives $\sigma_{err} = 0.0404785$. Note that the encoder error is characterized by standard deviation $\sigma_{enc} = 0.405949$ which is more than ten times larger than σ_{err}.

The system (2.115) along with the Kalman filter (2.116) are modeled in Simulink using the file `sim_fusion.slx`. Synthetic measurements are created using a true angular rate $\dot{\theta} = 0.015$ deg/s. The standard deviation of the actual estimation error is found as $\sigma_{err} = 0.0402275$ which is slightly smaller than the predicted value.

A plot of the attitude-angle error $\theta - \hat{\theta}$ and the 3σ bound equal to ± 0.121435 are shown in Figure 2.67. It is seen that the Kalman filter provides filtered estimates whose errors are indeed bounded by the theoretical 3σ bounds.   ❏

2.8 Notes and references

The process of mathematical modeling is essential in science and engineering. The development of control system models is a large area of scientific research and there exists a vast literature on this subject, see for instance Bishop [2]; de Silva [59]; Egeland and Gravdahl [60]; Fishwick [61]; Golnaraghi and Kuo [62]; Isermann [20]; Karnopp, Margolis, and Rosenberg [63]; Kulakowski, Gardner, and Shearer [64]; Ljung and Glad [65]; Spong, Hutchinson, and Vidyasagar [66]. The plant models should be developed so that to take into account the structure and parameter uncertainties keeping at the same time model order low to simplify the controller design.

The detailed derivation of equations describing the motion of cart–pendulum system can be found in many sources, see for instance Boubaker and Iriarte [67] and Ogata [68, Chapter 3].

The linearization of nonlinear models is considered in almost every textbook on control theory, see Goodwin, Graebe, and Salgado [43, Chapter 3]; Ogata [68, Chapter 3]; Franklin, Powell, and Emami-Naeini [69, Chapter 2]; Golnaraghi and Kuo [62, Chapter 4] where this technique is described in details.

The discretization of continuous-time systems and properties of various discretization methods are considered in depth in the corresponding chapters of Franklin, Powell, and Workman [8]; Goodwin, Graebe, and Salgado [43]; Hendricks, Jannerup, and Sørensen [45]. Various considerations about the choice of sampling frequency may be found in [21, Chapter 8], [8, Chapter 11], [43, Chapter 12], [45, Appendix D], [4, Chapter 2], [26, Chapter 15].

The theory of stochastic control is presented in several books, among them Åström [70]; Bryson and Ho [71]; Maybeck [72]; Hendricks, Jannerup, and Sørensen [45, Chapter 6]; Speyer and Chung [73]. The discrete-time Kalman filter is proposed by R.E. Kalman [74] and is one of the most versatile tools of modern control engineering. The theory of Kalman filter is presented in lots of books, among them Anderson and Moore [75]; Åström [70]; Bryson and Ho [71]; Crassidis and Junkins [76]; Gibbs [77]; Lewis, Xie, and Popa [44]; Simon [78] and Speyer and Chung [73]. The book of Grewal and Andrews [79] deserves special attention since it considers in detail the application of Kalman filtering using MATLAB. A very readable introduction to the theory of Kalman filter is given by Athans [80, Chapter 13].

As noted in Bittanti and Garatti [81], control science is basically a model based discipline and the performance of control is determined by the accuracy of the model representing data. Building uncertain system models is an important step in the design of robust control systems. Unfortunately, the derivation of uncertain plant models may be much more difficult in comparison with modeling of plants with fixed structure and parameters.

A detailed discussion of the properties of LFT and their using to model uncertain systems can be found in [82, Chapter 10].

Different components of the gyro and accelerometer systematic and random errors are described in detail in Aggarwal et al. [55], Groves [83], Titterton and

Weston [84], IEEE Standard 952-1997 [57]. Techniques for building models of MEMS sensor noises are presented in Maybeck [72], Farrell [85], Siouris [86]. Instead of using the PSD, sometimes it is preferred to exploit the autocorrelation function of the noise in order to obtain first-order Gauss–Markov or higher order auto-regressive models. However, as noted in [57], the correlation methods are very model sensitive and not well suited for higher order processes.

Chapter 3

Performance requirements and design limitations

The controller design aims to provide a control signal (control action) which will force the plant to behave in a desired fashion, i.e., to change the output according to a set of performance requirements. In classical control, the performance specifications are given in terms of desired time-domain and frequency-domain measures, such as step response specifications (overshoot, rise time, settling time), frequency response specifications (bandwidth, crossover frequency, resonance frequency, resonance damping) and relative stability in terms of gain margin (GM) and phase margin (PM). Further on, significant specifications are given in terms of disturbance attenuation and noise suppression. A very important requirement is to achieve the necessary performance in the presence of different plant uncertainties, i.e., to ensure *robustness* of the closed-loop system.

In this chapter, we consider briefly some important issues concerning the performance requirements to closed-loop linear systems and the fundamental design limitations in achieving the control aims. The performance specifications are formulated in continuous-time due to the clear physical interpretation in this case. First, we present the relatively simple case of single-input–single-output (SISO) systems which are well studied in the classical control theory. The trade-offs in the design of such systems are shown in some details. Then, we discuss the more complicated case of multiple-input–multiple-output (MIMO) systems whose performance is investigated by using the singular value plots of certain closed-loop transfer function matrices. An important issue in controller design is the closed-loop system performance in presence of different uncertainties. At the end of the chapter, we present some elements of the contemporary approach to the robustness analysis of uncertain linear systems based on the small gain theorem and structured singular value (SSV).

3.1 SISO closed-loop systems

MATLAB® function used in this section

Function	Description
loopsens	Sensitivity functions of closed-loop system

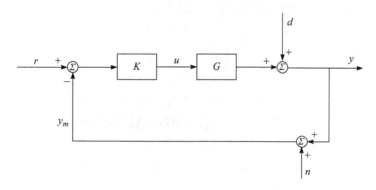

Figure 3.1   Block diagram of a SISO system

The typical block diagram of a negative feedback SISO system is shown in Figure 3.1. It consists of a control plant with transfer function G and one degree-of-freedom controller with transfer function K. The controller has only one input which determines its name. The closed-loop system is a subject to reference r, disturbance d, and measurement noise n. The input to the controller $K(s)$ is $r - y_m$, where $y_m = y + n$ is the measured output.

In the more general case, the disturbance is applied to a separate plant input which reflects its action at some internal point of the plant. Since the plant is linear, its output may be represented in the form

$$y = G_u(s)u + G_d(s)d,$$

where G_u and G_d are the corresponding transfer functions in respect to control and disturbance.

In this way, the system is reduced to the form, shown in Figure 3.2, which means that this system can be represented again by block diagram in Figure 3.1, modifying the disturbance in appropriate way.

The control action (plant input) is obtained by the following equation:

$$u = K(s)(r - y - n). \tag{3.1}$$

The control aim is to obtain a signal u (i.e., to design a controller K), such that the system error

$$e := y - r$$

remains small in the presence of disturbance d and noise n.

The exogenous actions to the system have different character. The reference r and the disturbance d have usually deterministic character and their spectrum is in the low-frequency range. The measurement noise n is a random process and its spectrum as a rule is in the high-frequency range. The determination of these actions is done on the base of plant functioning analysis and may require some laboratory experiments.

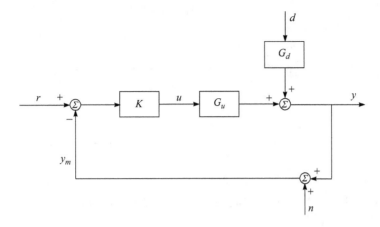

Figure 3.2 Disturbance conversion to plant output

The reasons for using feedback control are

1. Uncertainty in signals—uncertain disturbance and inaccurate statistical characteristics of noise.
2. Uncertainty in the plant model.
3. Plant instability.

The third reason follows from the fact that an unstable plant can be stabilized only by feedback.

In the general case, the usage of feedback reduces the model uncertainty that involves signal uncertainty and plant model uncertainty.

For the system shown in Figure 3.1, the plant model is described by the following equation:

$$y = Gu + d. \tag{3.2}$$

The substitution of (3.1) in (3.2) yields

$$y = GK(r - y - n) + d$$

or

$$(I + GK)y = GKr + d - GKn.$$

Hence, the closed-loop response is

$$y = \underbrace{\frac{GK}{I + GK}}_{T} r + \underbrace{\frac{1}{I + GK}}_{S} d - \underbrace{\frac{GK}{I + GK}}_{T} n. \tag{3.3}$$

The system error is given by

$$e = y - r = -Sr + Sd - Tn, \tag{3.4}$$

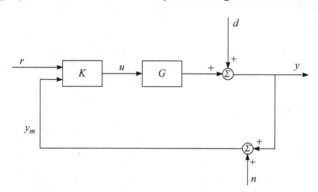

Figure 3.3   Two degree-of-freedom controller

where one utilizes the fact that $T - 1 = -S$. The corresponding plant input is

$$u = KS(r - d - n). \tag{3.5}$$

Further on, we use the following notation:

$L = GK$ — transfer function of the open-loop system,

$S = \dfrac{1}{I + GK} = \dfrac{1}{I + L}$ — sensitivity function,

$T = \dfrac{GK}{I + GK} = \dfrac{L}{I + L}$ — complementary sensitivity function.

Equation (3.3) may be rewritten as

$$y = Tr + Sd - Tn. \tag{3.6}$$

Clearly, S is the closed-loop transfer function from the disturbance to the output, while T is the closed-loop transfer function from reference to the output. The transfer function—T may be considered as the noise transfer function. The term *complementary sensitivity* for T comes from the identity

$$S + T = 1. \tag{3.7}$$

The open-loop transfer function L and the sensitivity functions S and T are determined from transfer functions G and K in MATLAB by using the function `loopsens`.

Better closed-loop performance may be achieved by using the so-called two degree-of-freedom controllers. Typical two degree-of-freedom controller is shown in Figure 3.3. It has separate inputs for the reference r and measured output y_m and one output u. The linear two degree-of-freedom controller may be separated in two blocks,

$$K = \begin{bmatrix} K_r & K_y \end{bmatrix},$$

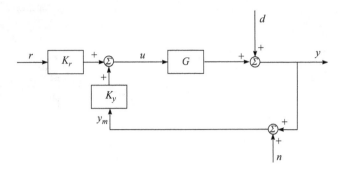

Figure 3.4 Two degree-of-freedom controller with prefilter and feedback

where K_y is the feedback part and K_r is a reference prefilter. As a result, one has that

$$u = K_r r + K_y y_m. \tag{3.8}$$

The closed-loop system structure in this case is shown in Figure 3.4. The feedback is used to reduce the uncertainty effect (disturbances and model errors), while the prefilter ensures the necessary tracking accuracy.

The closed-loop transfer functions can be modified easily in the case of two degree-of-freedom controller taking into account the relationship (3.8).

3.2 Performance specifications of SISO systems

MATLAB functions used in this section

Function	Description
loopmargin	Stability margins of closed-loop system
bode	Bode frequency response
nyquist	Nyquist frequency response

3.2.1 Time-domain specifications

The standard specifications of time-domain performance of SISO system in case of step reference are illustrated in Figure 3.5. The most important time-domain characteristics are as follows:

- *Overshoot:* The first peak value of step response, divided by the steady-state value y_{ss}. In case of unit step response, the overshoot is given by the value of M and is expressed in percent. Typically, it shouldn't exceed 30 percent.
- *Rise time (t_r):* The time required for the output signal to reach 90 percent of its steady-state value. Should be as small as possible.

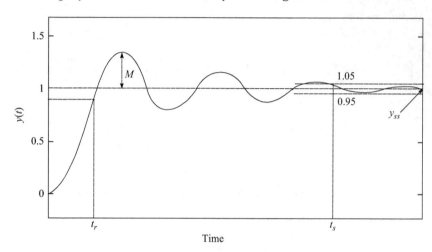

Figure 3.5   Specification of time-domain performance

- *Settling time (t_s):* The minimum period of time after which the unit step response remains within 5 percent of its steady-state value. Should be as small as possible.
- *Steady-state error (offset):* The difference between the steady-state value of output response and its reference value. Should be as small as possible.

3.2.2   Frequency-domain specifications

Important performance specifications in the frequency domain are the GM and PM, maximum peaks of sensitivity functions, crossover, and bandwidth frequencies.

Let $L(s)$ be the transfer function of the open-loop system with negative feedback. The typical Bode plot of the open-loop system, illustrating the notion of GM and PM, is shown in Figure 3.6.

The *GM* is defined as

$$\text{GM} = \frac{1}{|L(j\omega_{180})|}, \tag{3.9}$$

where *phase crossover frequency* ω_{180} is the frequency for which the phase frequency response of $L(j\omega)$ crosses the axis $-180°$, i.e.,

$$\angle L(j\omega_{180}) = -180°. \tag{3.10}$$

The GM is the factor, by which the open-loop gain $|L(j\omega)|$ may be increased before the closed-loop system becomes unstable. Usually, it is required that GM $> 2 = 6$ dB.

The *PM* is defined as

$$\text{PM} = \angle L(j\omega_c) + 180°, \tag{3.11}$$

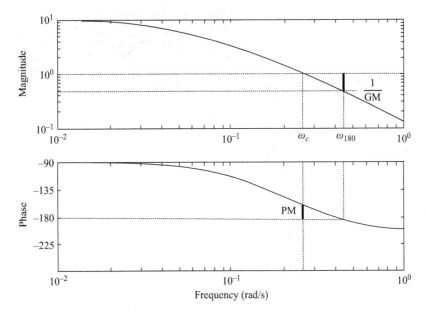

Figure 3.6 Bode plot of open-loop system with PM and GM indicated

where *the gain crossover frequency* ω_c is the frequency, for which $|L(j\omega)|$ crosses for the first time 1 from above, i.e.,

$$|L(j\omega_c)| = 1.$$

The PM shows what amount of negative phase (phase lag) can be added to $L(s)$ at frequency ω_c before the phase at this frequency to become $-180°$, which corresponds to closed-loop instability. Usually, it is required that PM be greater than $30°$.

An alternative to classical GM and PM is the *disk margin*. The disk margin is the largest region of the complex plane such that for all gain and phase variations inside the region the nominal closed-loop system is stable. The definition of the disk margin is illustrated in Figure 3.7 which shows the Nyquist plot of a third-order open-loop system with transfer function

$$L(s) = \frac{0.22s + 11}{0.28s^3 + 1.437s^2 + 6.6s + 18}.$$

The cross-points denoted by GM and PM indicate the classical GM and PM for the system, while the points denoted by disk gain margin (DGM) and disk phase margin (DPM) indicate the disk gain and disk PMs, respectively. The disk margins provide a lower bound on classical GM and PM [50]. Both classical and disk stability margins may be determined using the Robust Control Toolbox™ function `loopmargin`.

In case of uncertain system model, it is more appropriate to use the worst case GM and PM (see Section 3.8).

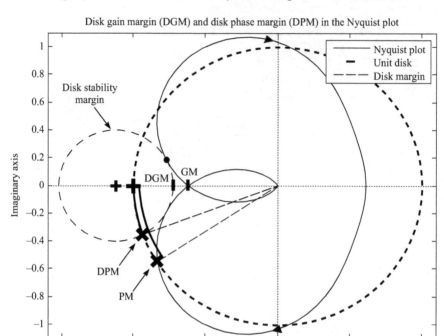

Figure 3.7   Disk gain and phase margins

The GM and PM are related to the maximum peaks of the sensitivity and complementary sensitivity functions defined as

$$M_S = \max_{\omega} |S(j\omega)|, \quad M_T = \max_{\omega} |T(j\omega)|.$$

It is possible to show [87] that for $M_T = 2$ (6 dB), it is guaranteed that GM ≥ 1.5 and PM $\geq 29°$ and similar relationships exist involving M_S. Larger values of M_S and M_T indicate poor performance as well as poor robustness.

The closed-loop bandwidth characterizes the transient response speed. Considering the closed-loop system as a low-pass filter, larger bandwidth means faster response since high-frequency signals pass more quickly on the system output. However, as it is shown in Section 3.3, the increasing of the bandwidth also increases noises effect and parameter variations.

The closed-loop bandwidth ω_B is the highest frequency at which $|T(j\omega)|$ crosses $1/\sqrt{2} = 0.707$ *(≈ -3 dB) from above.*

Alternative definition of the bandwidth may be given by using the sensitivity function S instead of T.

3.3   Trade-offs in the design of SISO systems

MATLAB functions used in this section

Function	Description
loopmargin	Stability margins of closed-loop system
pole	Transfer function poles
zero	Transfer function zeros and gain
bodemag	Bode magnitude plot

In this section, we discuss in brief the limitations on the performance of SISO systems which impose trade-offs in controller design.

3.3.1   Limitations on S and T

As shown by (3.7), the sensitivity S and the complementary sensitivity T of a SISO system satisfy

$$S + T = 1.$$

In the ideal case, S should be small for small tracking error and disturbance effect, and T should be small to reduce the measurement noise effect. Unfortunately, S and T cannot be made small simultaneously. For each frequency ω either $|S(j\omega)|$ or $|T(j\omega)|$ should be greater than or equal to 0.5. This means that the shaping of the frequency responses of S and T should be done so as to achieve a compromise between the disturbance attenuation and noise suppression.

The limitations on the sensitivity functions are clarified by the following result:

(Bode's integral formula). *Suppose that the open-loop transfer function $L(s)$ has at least two more poles than zeros and let $S(s)$ is the sensitivity function. If the open-loop transfer function has poles $\{p_i : i = 1, \ldots, N_p\}$ in the right half-plane, then for closed-loop stability the sensitivity function must satisfy*

$$\int_0^\infty \ln |S(j\omega)| d\omega = \int_0^\infty \ln \frac{1}{|1 + L(j\omega)|} d\omega = \pi \sum_i^{N_p} \mathrm{Re}(p_i), \qquad (3.12)$$

where $\mathrm{Re}(p_i)$ denotes the real part of the pole p_i.

Equation (3.12) shows that if the sensitivity function is made small at some frequencies it should increase at other frequencies so that the integral of $\ln |S(j\omega)|$ to remain constant. This implies that the controller design may be considered as a redistribution of the disturbance attenuation over different frequencies. If the disturbance attenuation is improved in some frequency range, it will worsen in other, a property which is frequently called *waterbed effect*. If $N_p > 0$, then the sensitivity reduction area ($\ln |S(j\omega)| < 0$) is less than the sensitivity increase area ($\ln |S(j\omega)| > 0$) by an amount proportional to the sum of the distances from the right half-plane poles to the imaginary axis. This indicates that a portion of the loop gain which could otherwise contribute to sensitivity reduction must instead be used to pull the unstable poles into the left half plane [88].

Similarly to (3.12), it is possible to prove the following result in respect to the complementary sensitivity function T:

$$\int_0^\infty \frac{\ln|T(j\omega)|}{\omega^2} d\omega = \pi \sum_j^{N_z} \frac{1}{\text{Re}(z_j)}, \qquad (3.13)$$

where N_z is the number of right half-plane zeros. On the basis of (3.12) and (3.13), one may conclude that fast right half-plane poles (larger $\text{Re}(p_i)$) are worse than slow ones and slow right half-plane zeros (smaller $\text{Re}(z_j)$) are worse than fast ones.

3.3.2   Right half-plane poles and zeros

The plant poles and zeros, especially those in the right half-plane (unstable poles and nonminimum phase zeros), may have a strong impact on the closed-loop system dynamics. As it is well known from the classical root locus analysis, when the open-loop gain tends to infinity, then the closed-loop poles tend to open-loop zeros. That is why the presence of nonminimum phase zeros causes instability for large gains. Very dangerous are the small in module nonminimum phase zeros which limits strongly the closed-loop bandwidth and hence the speed of closed-loop response.

The analysis of the closed-loop dynamics leads to the following conclusions [43].

- If the absolute value of the real part of the dominant closed-loop poles is greater than the smallest nonminimum phase open-loop zero, then the time response is characterized by large undershoot. *That is why the closed-loop bandwidth should be set less than the smallest nonminimum phase zero.*
- If the absolute value of the real part of the dominant closed-loop poles is greater than the absolute value of the smallest stable open-loop zero, then significant overshoot will occur.
- If the absolute value of the real part of the dominant closed-loop poles is less than the absolute value of the largest unstable open-loop pole then a large overshoot will occur or the system error will change its sign quickly. *That is why the closed-loop bandwidth should be set greater than the real part of each unstable pole.*
- If the absolute value of the real part of the dominant closed-loop poles is greater than the absolute value of the smallest stable open-loop pole, then an overshoot will again occur.

The above considerations imply that the closed-loop bandwidth should lie between the larger unstable open-loop pole and the smallest nonminimum phase open-loop zero, i.e.,

$$\max_{\text{Re}(p_k)>0} |p_k| < \omega_B < \min_{\text{Re}(z_j)>0} |z_j|.$$

If this relationship is not satisfied, then the closed-loop system may have bad response.

Example 3.1. Influence of plant poles and zeros

Consider a plant model given by

$$G(s) = \frac{s - z}{s(s - p)}.$$

The closed-loop poles are assigned to $\{-1, -1, -1\}$. Then, the corresponding controller is

$$K(s) = K_c \frac{s - z_c}{s - p_c}.$$

The controller parameters are given in Table 3.1 for four cases of plant pole-zero configurations. The unit step responses for these cases are shown in Figure 3.8.

From the responses obtained, it is possible to derive the following conclusions.

Case 1 (Very small stable zero) It is seen that the transient response has a large overshoot.

Case 2 (Nonminimum phase zero, stable pole) There is a large undershoot, due to the nonminimum phase zero. There is also a small overshoot due to the stable pole at -0.5.

Case 3 (Nonminimum phase zero, small unstable pole) First, there is large undershoot due to the nonminimum phase zero, then one can see significant overshoot due to the unstable pole.

Case 4 (Small nonminimum phase zero, large unstable pole) In this case, the undershoot is due to the nonminimum phase zero and the overshoot is due to the unstable pole. The overshoot is larger in comparison with Case 3, which is due to the fact that the unstable pole is on the right of the nonminimum phase zero. ☐

It is necessary to note that open-loop poles depend on the intrinsic dynamics of the system and can be affected by changing the plant construction. Unlike to this, the open-loop zeros depend on how the sensors and actuators are coupled to the states. The zeros can thus be affected by moving the sensors and actuators or by adding sensors and actuators [89].

Table 3.1   Controller parameters for different pole-zero configurations

	Case 1	Case 2	Case 3	Case 4
	$p = -0.5$	$p = -0.5$	$p = 0.2$	$p = 0.5$
	$z = -0.1$	$z = 0.5$	$z = 0.5$	$z = 0.2$
K_c	20.6250	-3.7500	-18.8000	32.5000
p_c	18.1250	-6.2500	-22.0000	29.0000
z_c	-0.4848	-0.5333	-0.1064	0.1538

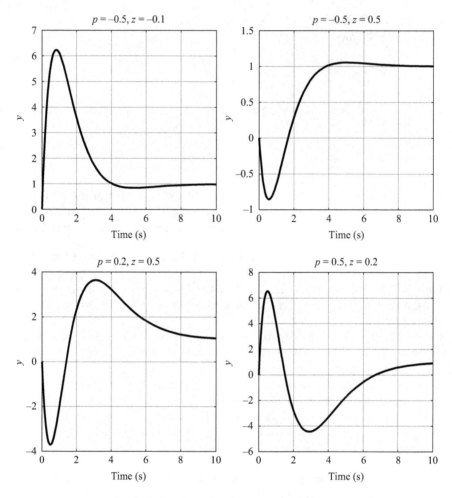

Figure 3.8   Plant output for four pole-zero configurations

3.3.3   Limitations imposed by time delays

The effect of a plant time delay on the closed-loop performance is similar to the effect of a nonminimum phase zero.

Consider a plant $G(s)$ which involves a time-delay $e^{-\tau s}$ and has not nonminimum phase zeros. From the Padé approximation of the time delay,

$$e^{-\tau s} \approx \frac{1 - \tau/2s}{1 + \tau/2s},$$

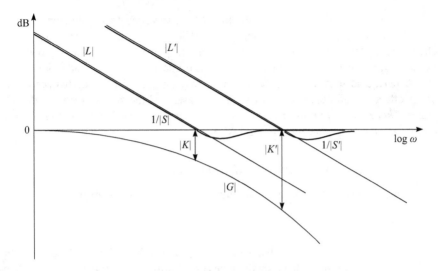

Figure 3.9   Noise amplification due to bandwidth increasing

it follows that the approximated plant has a nonminimum phase zero at $2/\tau$. According to the recommendations given above, the closed-loop bandwidth should satisfy

$$\omega_c < \frac{1}{\tau}. \tag{3.14}$$

3.3.4   Limitations imposed by measurement noise

As it was previously shown, the effect of measurement noise on the system output and control action is given by

$$y = -Tn, \tag{3.15}$$

$$u = -KSn \tag{3.16}$$

It is seen from (3.15) that the deleterious effect of the noise on the system output may be attenuated if $|T(j\omega)|$ is small in the frequency domain where $|n(j\omega)|$ is significant. *Since the noise is typically significant in the high-frequency range, it imposes an upper limit on the closed-loop bandwidth.*

It should be pointed out that the effect of the measurement noise on the plant input may be much stronger than its effect on the system output. This is illustrated in Figure 3.9, which displays typical characteristics of G and L.

It follows from (3.16) that in the low-frequency range where $|L| = |GK|$ is much greater than 1, the noise $n_u \approx -(1/G)n$ in the control action does not depend on the controller K. However, in the high-frequency range where $|L|$ is much less than 1, $n_u \approx -Kn$ and if the controller gain is large this may lead to large amplification of the noise in control action.

It is seen from Figure 3.9 that the increase of the open-loop gain from L to L' is achieved at the price of increasing $|K|$ to $|K'|$, i.e., at the price of larger noise effect on the plant input. The noise intensity and its power increase not only because $|K| > |K'|$, but also because the noise power is proportional to the bandwidth.

When the noise is very large the actuators may saturate which leads to decreasing of the effective gain, the signal distortion increases and the system accuracy deteriorates. *That is why the noise effect on the actuator inputs should be restricted. This limits the open-loop gain by limitation of $|K|$.*

3.3.5   Limitations, imposed by disturbances

According to (3.6), the effect of disturbance on system output is given by

$$y = Sd.$$

Assume that the disturbance d has significant power only in the bandwidth B_d. In such case, it is desirable to have small values of $|S(j\omega)|$ in the range B_d. This implies $S(j\omega) \approx 0$ and $T(j\omega) \approx 1$ in this frequency range.

Therefore, to achieve acceptable closed-loop performance in the presence of disturbances, it is necessary to set a lower limit on the closed-loop bandwidth.

3.3.6   Limitations on control action

In real systems, all actuators have constraints on the maximum output in the form of amplitude or rate saturation. Peaks of the control action occur as a result of fast changes of reference, disturbance or noise. As shown by (3.5), for one degree-of-freedom loops the controller output is given by

$$u = KS(r - d - n).$$

Since $KS = T/G$, we have that

$$u = \frac{T}{G}(r - d - n). \tag{3.17}$$

It follows from (3.17) that if the closed-loop bandwidth is much larger than this of the plant $G(s)$, then the transfer function $K(s)S(s)$ will amplify significantly the high-frequency components in $r(j\omega)$, $d(j\omega)$, and $n(j\omega)$.

Example 3.2. Large magnitude of control action
Consider a plant and controller given by

$$G(s) = \frac{1}{(2s + 1)(0.5s + 1)},$$

$$K(s) = \frac{9(0.4s + 1)}{0.3s + 1}.$$

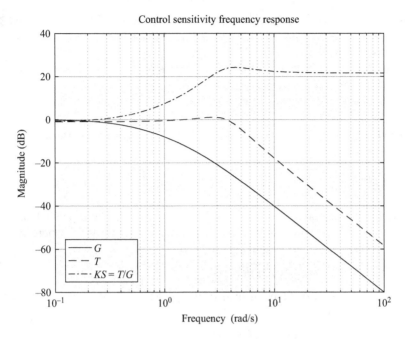

Figure 3.10 Magnitude responses of plant, closed-loop system, and control sensitivity

The sensitivity of control action to reference, disturbance, and noise is determined by the transfer function

$$K(s)S(s) = \frac{T(s)}{G(s)} = \frac{3.6s^3 + 18s^2 + 26.1s + 9}{0.3s^3 + 1.75s^2 + 6.4s + 10}.$$

The magnitude responses of plant, closed-loop system, and control action sensitivity (determined by $|K(j\omega)S(j\omega)| = |T(j\omega)/G(j\omega)|$) are shown in Figure 3.10. It is seen that the closed-loop bandwidth is about ten times larger than the plant bandwidth. This leads to high sensitivity of control in the high-frequency range that results in large picks of the control action in presence of high-frequency components in the reference, disturbance, or noise.   ❏

In this way, we may conclude that to avoid actuator saturation, it is necessary to put an upper limit on the closed-loop bandwidth.

3.3.7 Limitations due to model errors

Usually, the controller design is based on the nominal plant model after that one adds the requirement the resulting closed-loop performance to be insensitive to the difference between the actual and nominal model. This property of the closed-loop system is referred to as *robustness*. The closed-loop robustness in the presence of different kind of uncertainties is investigated in more details in Sections 3.7 and 3.8. Here, we consider in simplified form the effect of unstructured perturbations on the closed-loop dynamics.

As shown in Chapter 2, for a SISO system, the modeling errors due to unstructured uncertainty may be expressed as

$$y(s) = G(s)u(s) = G_{nom}(s)(1 + G_\Delta(s)),$$

where $G_{nom}(s)$ is the *nominal plant model* and $G_{\Delta(s)}$ is the *model multiplicative error* given by

$$G_\Delta(s) = \frac{G_\delta(s)}{G_{nom}(s)} = \frac{G(s) - G_{nom}(s)}{G_{nom}(s)}.$$

The difference between nominal and actual model may be expressed also as a difference between nominal and actual closed-loop sensitivity. If we denote by

$$S_{nom}(s) = \frac{1}{1 + G_{nom}(s)K(s)}$$

the nominal sensitivity function then in the presence of model errors the actual sensitivity one obtains

$$S(s) = S_{nom}(s)S_\Delta(s),$$

where

$$S_\Delta(s) = \frac{1}{1 + T_{nom}(s)G_\Delta(s)}.$$

Usually, the nominal model represents the plant behavior with sufficient accuracy in the low-frequency range, i.e., in the case when the plant input is constant or slowly varying. The modeling accuracy deteriorates as frequency increases since the dynamics features which are neglected in the nominal model, become considerable. This implies that $|G_\Delta(j\omega)|$ will become more substantial with frequency increasing. *Therefore, to achieve acceptable performance in the presence of modeling errors, it is necessary to put upper limit on the closed-loop bandwidth.*

We may summarize that in the controller design of SISO systems, it is necessary to make trade-offs which are determined by the following fundamental limitations.

$S_{nom}(s) = 1 - T_{nom}(s)$
  i.e., the disturbance is rejected only at frequencies for which $|T(j\omega)| \approx 1$.
$y(s) = -T_{nom}(s)n(s)$
  i.e., the measurement noise $n(t)$ is suppressed only at frequencies for which $|T(j\omega)| \approx 0$.
$u(s) = \frac{T_{nom}(s)}{G(s)}(r(s) - d(s) - n(s))$
  i.e., large control actions occur at frequencies where $|T(j\omega)| \approx 1$ but $|G(j\omega)| \ll 1$, which takes place when the closed-loop system is much faster than the plant.

$$S(s) = S_{nom}(s)S_\Delta(s) \text{ where } S_\Delta(s) = \frac{1}{1 + T_{nom}(s)G_\Delta(s)}$$

i.e., fast response to reference and disturbances at frequencies where the model is inaccurate jeopardize the closed-loop stability; note that the relative modeling error $G_\Delta(s)$ usually has magnitude and phase which increase in the high-frequency range.

Forcing the closed-loop to become faster than the nonminimum zeros leads to larger undershoots in the closed-loop response.

It is necessary to stress that these design trade-offs are independent on the methods, used in controller design.

Example 3.3. Tradeoffs in controller design
The following example illustrates some of the tradeoffs in controller design.

Consider the system shown in Figure 3.1 with a second-order plant described by the transfer function

$$G = \frac{K_o}{T_o^2 s^2 + 2\xi T_o s + 1}$$

where $K_o = 10$, $T_o = 1$, and $\xi = 0.5$.

As controllers, we shall make use of three lead compensators with different gains:

$$K_1 = 2 \times \frac{\tau_1 s + 1}{\tau_2 s + 1}, \quad K_2 = 5 \times \frac{\tau_1 s + 1}{\tau_2 s + 1}, \quad K_3 = 10 \times \frac{\tau_1 s + 1}{\tau_2 s + 1}$$

where $\tau_1 = 1$, $\tau_2 = 0.02$.

The Bode plots of the closed-loop systems with the three controllers are shown in Figure 3.11. It is seen that the largest closed-loop bandwidth corresponds to the highest gain controller (the third controller).

The output closed-loop responses corresponding to the controllers K_1, K_2, and K_3 are displayed in Figure 3.12. With the increase of controller gain, the rise time decreases but the overshoot increases as well. Note that further increase of the controller gain may lead to closed-loop instability.

The control actions produced by the three controllers are presented in Figure 3.13. The fastest response for the third controller is obtained at the price of the largest control impulse at the initial moment. Such impulses may lead to saturation of the actuators and bad performance of the closed-loop system.

The magnitude responses corresponding to the output sensitivity functions are shown in Figure 3.14. For the highest gain (the third controller), the corresponding sensitivity function has the smallest value (-40 dB) in the low-frequency range up to 1 rad/s, which means that the disturbances with spectrum in this range will be suppressed 100 times when this controller is used.

The transient responses to sinusoidal disturbance with unit magnitude and frequency 0.2 rad/s are shown in Figure 3.15. The response has a magnitude equal to 0.01 in the case of the third controller, as predicted from the frequency analysis of output sensitivity function.

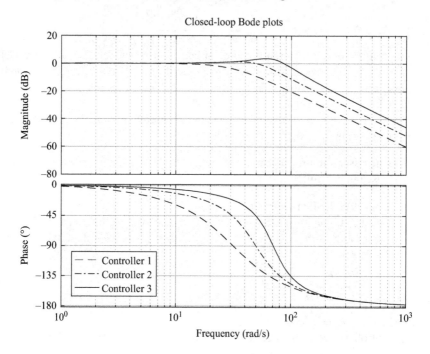

Figure 3.11   Bode plots of the closed-loop systems

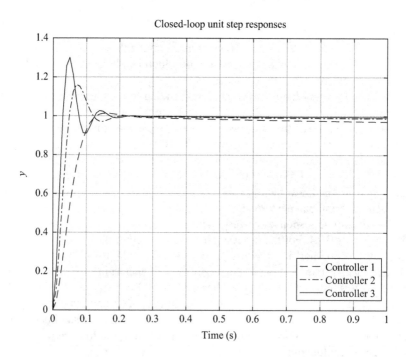

Figure 3.12   Closed-loop transient responses

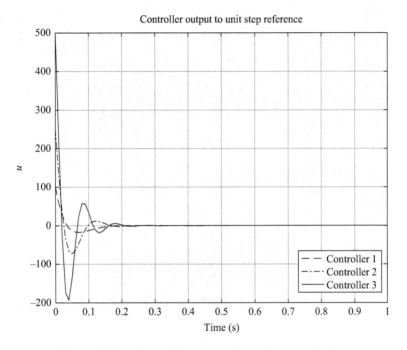

Figure 3.13　Controller outputs for unit step reference

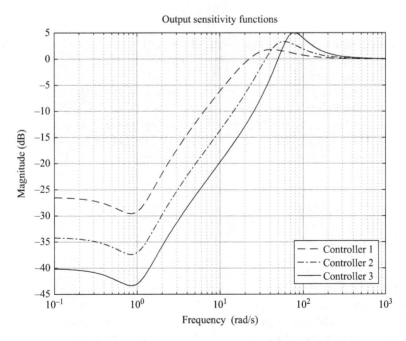

Figure 3.14　Output sensitivity functions

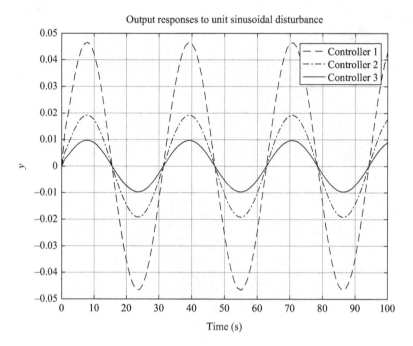

Figure 3.15   Output response to sinusoidal disturbance

Consider finally the effect of measurement noise on the system behavior. The measurement noise is generated with the random number generator rand, has a zero mean value and standard deviation equal to 5.8×10^{-3}.

It is seen from the transient responses, presented in Figure 3.16, that the measurement noise has the largest effect in the case of third controller (highest gain). This follows also from the magnitude responses of the complementary sensitivity function (Figure 3.11), which is equal up to the sign to the noise transfer function.

The measurement noise may lead to large noise at the plant input, so as to cause actuator saturation. For the example under consideration, the transfer functions of the closed-loop "measurement noise—plant input" are given by

$$W_1 = -K_1 S_1, \quad W_2 = -K_2 S_2, \quad W_3 = -K_3 S_3$$

The magnitude responses of this loop for the three controllers are given in Figure 3.17. Clearly, the amplitude of the high-frequency noise at the plant input will increase with increasing of the controller gain.

The closed-loop transient responses at the plant input, due to the measurement noise, are shown in Figure 3.18. For the third controller, the noise intensity at the plant input is several times higher than the intensity of the measurement noise and is significantly higher than the intensity of the noise at system output.   ❐

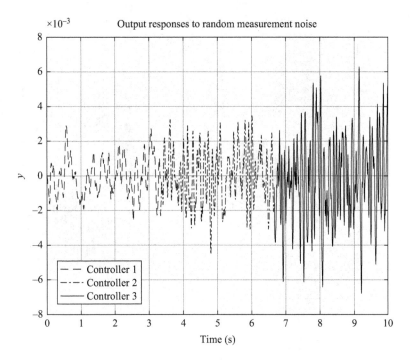

Figure 3.16   Output transient response due to the measurement noise

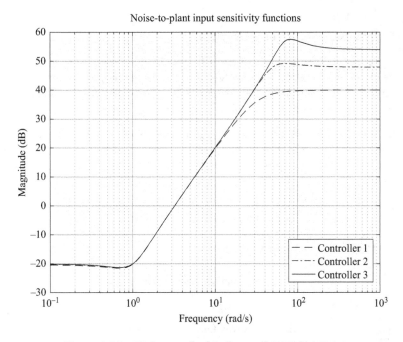

Figure 3.17   Noise-to-plant input sensitivity functions

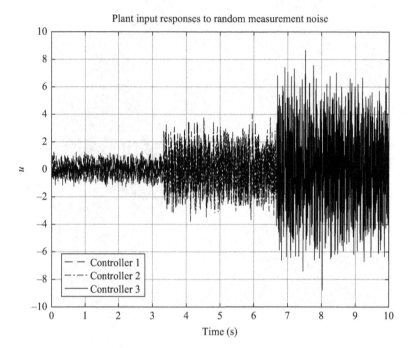

Figure 3.18   Transient responses at the plant input, due to the measurement noise

The results obtained in this example lead to the following conclusions:

- Higher controller gain leads to larger bandwidth and faster transient response of the closed-loop system. The steady-state error for higher gain is smaller.
- Larger bandwidth leads to better suppression of the low-frequency disturbances.
- Larger bandwidth leads to amplification of the high-frequency noises.

It should be stressed that these conclusions hold also for high-order plants.

Unfortunately, higher gain and larger bandwidth may lead to high sensitivity to parameter variations, i.e., to pure robustness. That is why the controller design is a tradeoff between achievement of good performance (good suppression of disturbances and noises) and good robustness (low sensitivity to plant model uncertainty).

3.4   MIMO closed-loop systems

MATLAB function used in this section

Function	Description
loopsens	Sensitivity functions of closed-loop system

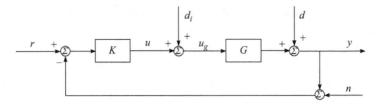

Figure 3.19 Block diagram of a MIMO system

The block diagram of a MIMO negative feedback system is shown in Figure 3.19. The system consists of a MIMO plant G and a controller K upon which act the reference r, the sensor noise n, input plant disturbance d_i, and output disturbance d. In the general case, all signals are represented by vectors and the transfer function matrices G and K have appropriate dimensions.

With an eye on the further considerations, it is convenient to define *the input loop transfer function matrix L_i* and *output loop transfer function matrix L_o*, as

$$L_i = KG, \quad L_o = GK,$$

respectively, where L_i is obtained when breaking the loop at the input of the plant and L_o is obtained when breaking the loop at the output of the plant. *The input sensitivity function* is defined as the transfer function matrix from d_i to u_g,

$$S_i = (I + L_i)^{-1}, \quad u_g = S_i d_i.$$

The *output sensitivity function* is defined as the transfer function matrix from d to y,

$$S_o = (I + L_o)^{-1}, \quad y = S_o d.$$

The matrices of the *input complementary sensitivity* and *output complementary sensitivity* are defined respectively as

$$T_i = I - S_i = L_i(I + L_i)^{-1},$$

$$T_o = I - S_o = L_o(I + L_o)^{-1}.$$

(The term *complementary* is used to emphasize that T is complementary to S, $T = I - S$.) The matrix $I + L_i$ is called *input return difference*, and $I + L_o$ is called *output return difference*.

It is necessary to point out that in the general case, the sensitivity functions and the loop transfer functions in respect to the input and output are different ($S_o \neq S_i$, $T_o \neq T_i$, $L_o \neq L_i$).

As in the SISO case, the determination of the loop transfer functions and closed-loop sensitivity functions in MATLAB may be done by the function `loopsens`.

It is possible to show that the closed-loop system satisfies the following relationships

$$y = T_o(r - n) + S_o G d_i + S_o d, \tag{3.18}$$

$$r - y = S_o(r - d) + T_o n - S_o G d_i, \tag{3.19}$$

$$u = S_i K(r - n) - S_i K d - T_i d_i, \tag{3.20}$$

$$u_g = S_i K(r - n) - S_i K d + S_i d_i. \tag{3.21}$$

Since $S_i K = K S_o$, (3.20) and (3.21) may also be written as

$$u = K S_o(r - n) - K S_o d - T_i d_i,$$

$$u_g = K S_o(r - n) - K S_o d + S_i d_i.$$

Equations (3.18) through (3.21) play fundamental role in analysis and design of MIMO control systems. For instance, (3.18) reveals that the effect of disturbance d on the system output (y) can be made "small" making the output sensitivity function S_o "small." In a similar way, (3.21) shows that the effect of the disturbance d_i on the plant input (u_g) can be made small, making the input sensitivity S_i small.

For MIMO systems, one uses more frequently the matrices L_o, S_o, and T_o, which for brevity sometimes are denoted as L, S, and T. For SISO systems, one has that $L_i = L_o = L$, $S_i = S_o = S$, $T_i = T_o = T$.

For a better performance, a MIMO system may incorporate two degree-of-freedom controller similar to the SISO case.

Consider again the MIMO system shown in Figure 3.19. Let us regroup the external input signals into the feedback loop as w_1 and w_2 and regroup the input signals and controller as e_1 and e_2. Then, the feedback loop with the plant and the controller can be represented as in Figure 3.20.

Assume that the state-space realizations for G and K are stabilizable and detectable. Then the negative feedback system in Figure 3.20 is said to be *internally stable* if the transfer function matrix

$$\begin{bmatrix} I & K \\ -G & I \end{bmatrix}^{-1} = \begin{bmatrix} (I + KG)^{-1} & -K(I + GK)^{-1} \\ G(I + KG)^{-1} & (I + GK)^{-1} \end{bmatrix} \tag{3.22}$$

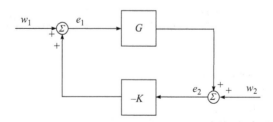

Figure 3.20   Internal stability analysis block diagram

from the inputs (w_1, w_2) to the outputs (e_1, e_2) is stable. The internal stability guarantees that for all bounded inputs (w_1, w_2) the outputs (e_1, e_2) are also bounded.

It should be noted, that for certain controllers, the 2×2 block matrix in (3.22) may not be proper, even though G and K are proper transfer function matrices. This means that for some inputs (w_1, w_2), the algebraic loop cannot be solved in respect to (e_1, e_2) and the feedback loop is said to be *ill-posed*. Controllers with such property are excluded from further consideration. If either G or K is strictly proper, then well-posedness is assured.

To check the internal stability it is necessary and sufficient to test whether each of the four transfer function matrices in (3.22) is stable. In the special case when G and K are stable, the internal stability is checked easily.

Let G and K are stable. Then, the system shown in Figure 3.20 is internally stable if and only if $(I + GK)^{-1}$ is stable, or equivalently

$$\det[I + G(j\omega)K(j\omega)] \neq 0 \text{ for each } \omega. \tag{3.23}$$

Equation (3.23) represents a generalization of the Nyquist criterion to the stability of MIMO systems.

3.5 Performance specifications of MIMO systems

MATLAB functions used in this section

Function	Description
norm(T,inf)	$\mathcal{H}_\infty$-norm of transfer function T

Performance analysis in the MIMO case is more complicated than in the SISO case and makes use of the singular values of matrix frequency responses and the $\mathcal{H}_\infty$ norm of system transfer function.

3.5.1 *Using singular values for performance analysis*

Consider the application of singular values to the performance analysis of MIMO system in Figure 3.19. It follows from (3.18) through (3.21) that the effect of disturbances and noises on the closed-loop dynamics depends on the magnitudes of the transfer function matrices G, K, S_i, S_o, T_i, and T_o. The magnitude of a transfer function matrix G in a given frequency range can be characterized by using the singular values of the corresponding frequency response matrix $G(j\omega)$. (Further on, the argument $j\omega$ is skipped for brevity.) Let the singular value decomposition (SVD) of $n \times m$ matrix M is (see Appendix A)

$$G = U\Sigma V^H \tag{3.24}$$

The column vectors of U, denoted by u_j, are called *output directions* of the plant. They are orthonormal, i.e., they are orthogonal and have unit length,

$$\|u_j\|_2 = \sqrt{|u_{j1}|^2 + |u_{j2}|^2 + \cdots + |u_{jn}|^2} = 1,$$

$$u_j^H u_j = 1, \quad u_i^H u_j = 0, \quad i \neq j.$$

In a similar way, the column vectors of V, denoted by v_j, are orthonormal and are called *input directions*. The input and output directions are related through the singular values. To see this, (3.24) is written as $GV = U\Sigma$, which for the jth column yields

$$Gv_j = \sigma_j u_j, \tag{3.25}$$

where σ_j is the jth singular value of G. In this way, if we consider an *input* in direction v_j, then the *output* is in direction u_j. Since $\|v_j\|_2 = 1$ and $\|u_j\|_2 = 1$, we see that the jth singular value σ_j characterizes the gain of the matrix G in this direction.

Equation (3.25) may be written as

$$\sigma_j(G) = \|Gv_j\|_2 = \frac{\|Gv_j\|_2}{\|v_j\|_2}.$$

The largest gain for any input direction is equal to the maximum singular value

$$\bar{\sigma}(G) = \sigma_1(G) = \max_{d \neq 0} \frac{\|Gd\|_2}{\|d\|_2} = \frac{\|Gv_1\|_2}{\|v_1\|_2}$$

and the smallest gain for any input direction is equal to the minimum singular value

$$\underline{\sigma}(G) = \sigma_k(G) = \min_{d \neq 0} \frac{\|Gd\|_2}{\|d\|_2} = \frac{\|Gv_k\|_2}{\|v_k\|_2}$$

where $k = \min\{n, m\}$. This means that for any input vector d it is fulfilled that

$$\underline{\sigma}(G) \leq \frac{\|Gd\|_2}{\|d\|_2} \leq \bar{\sigma}(G) \tag{3.26}$$

The frequency response plots of $\bar{\sigma}(G)$ and $\underline{\sigma}(G)$ may be considered as a generalization of the magnitude plot of scalar transfer function to the case of transfer function matrix. Indeed, if G is a scalar transfer function, the frequency response of its single singular value coincides with the magnitude response $|G|$. For a given frequency, the maximum and minimum singular values may be considered as the maximum and minimum gains of the corresponding transfer function matrix. These makes them convenient in the performance analysis of MIMO systems. Note that the singular values cannot be used in the stability analysis since they provide information only for the gain but not for the phase.

In MATLAB, the singular value plot of a transfer function matrix is obtained by the function `sigma`.

The singular value plots of the output sensitivity and complementary sensitivity transfer function matrices of a sixth-order two-input–two-output system are shown in Figure 3.21. In the given case, the matrices have two singular values which are the maximum and minimum one.

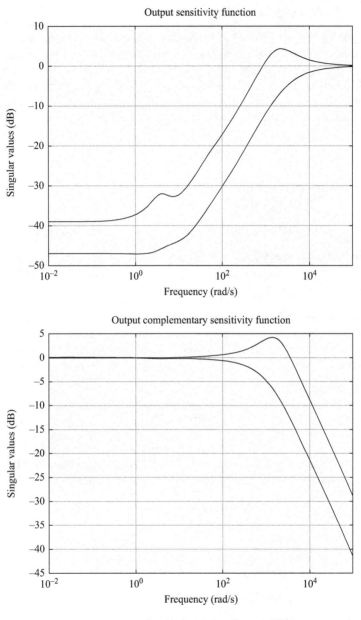

Figure 3.21   Singular value plot of S_o and T_o

The SVD is useful in the analysis and design of systems with nonsquare transfer function matrices (systems with different number of inputs and outputs). If, for instance, we have a plant with more outputs than inputs, then the additional output singular vectors show in which directions the plant cannot be controlled.

3.5.2 $\mathscr{H}_\infty$ *Norm of a system*

Consider the system in Figure 3.22 with stable transfer function matrix G. To assess the system performance, it is useful to have a single number which characterizes the "gain" of G, i.e., the size of the output signal $z(t)$ for an input signal $w(t)$ with unit size. This gain is determined by using an appropriate system norm. For the aim of robustness analysis and design the relevant norm is the so-called $\mathscr{H}_\infty$ norm.

Further on the size of $z(t)$ is evaluated by using the signal 2-norm

$$\|z(t)\|_2 = \sqrt{\sum_i \int_{-\infty}^{\infty} |z_i(\tau)|^2 d\tau}. \tag{3.27}$$

We assume that $w(t)$ is an arbitrary signal which satisfies $\|w(t)\|_2 = 1$.

The $\mathscr{H}_\infty$ norm is defined as

$$\|G(s)\|_\infty := \max_\omega \bar{\sigma}(G(j\omega)) \tag{3.28}$$

where for a fixed frequency ω,

$$\bar{\sigma}(G(j\omega)) = \max_{w\neq 0} \frac{\|z(j\omega)\|_2}{\|w(j\omega)\|_2}$$

is the induced matrix 2-norm of $G(j\omega)$. From system point of view, it follows from (3.28) that the $\mathscr{H}_\infty$ norm is the "peak" of the frequency response characteristics (the peak of the maximum singular value) over the frequency. Note that for a SISO system the maximum peaks of the sensitivity and complementary sensitivity functions,

$$M_S = \max_\omega |S(j\omega)|, \quad M_T = \max_\omega |T(j\omega)|,$$

are related to the $\mathscr{H}_\infty$ norm by the relationships

$$M_S = \|S\|_\infty, \quad M_T = \|T\|_\infty.$$

The $\mathscr{H}_\infty$ norm also has several interpretations in the time domain. This norm is equal to the induced (worst case) 2-norm in the time domain:

$$\|G(s)\|_\infty = \max_{w(t)\neq 0} \frac{\|z(t)\|_2}{\|w(t)\|_2} = \max_{\|w(t)\|_2=1} \|z(t)\|_2. \tag{3.29}$$

In this equation, the worst case input signal $w(t)$ is a sinusoid with frequency ω^* and a direction which gives $\bar{\sigma}(G(j\omega^*))$ as a maximum gain.

The $\mathscr{H}_\infty$ norm satisfies the multiplicative property

$$\|A(s)B(s)\|_\infty \leq \|A(s)\|_\infty \|B(s)\|_\infty \tag{3.30}$$

which has important implications in robust control theory.

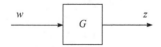

Figure 3.22   System G

The $\mathcal{H}_\infty$ norm also has interpretation in terms of the mathematical expectation of stochastic signals [82, Section 4.5].

The computation of $\mathcal{H}_\infty$ norm of a system $G(s)$ can be done graphically using the singular value plot of $G(j\omega)$ and the definition (3.28). This, however, may lead to errors in case of lightly damped system or when the frequency grid is not sufficiently dense. That is why to compute $\|G(s)\|_\infty$ it is better to use the MATLAB function `norm(G,'inf')`, which implements an iterative state-space procedure.

Example 3.4. $\mathcal{H}_\infty$ norm of a system
Consider a 2 × 2 transfer function matrix

$$G(s) = \begin{bmatrix} \dfrac{10(s+1)}{s^2 + 0.2s + 100} & \dfrac{1}{s+1} \\[2ex] \dfrac{s+2}{s^2 + 0.1s + 10} & \dfrac{5(s+1)}{(s+2)(s+3)} \end{bmatrix}.$$

The singular values of $G(j\omega)$ are shown in Figure 3.23. From the maximum singular value, we obtain that its peak value is about 34 dB, which corresponds to a value of the $\mathcal{H}_\infty$ norm, equal to 50.12. A more accurate computation of the norm is done by the command line

```
norm(G,'inf')
```

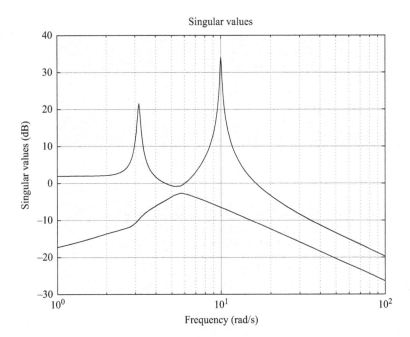

Figure 3.23   Maximum and minimum singular value of $G(j\omega)$

which produces

```
ans =

    50.2496
```

3.5.3 Hankel norm

The Hankel norm of a stable system $G(s)$ is obtained when one applies an input $w(t)$ up to the moment $t = 0$ and measures the output $z(t)$ for $t > 0$, $w(t)$ being chosen so that to maximize the ratio of the 2-norms of these signals:

$$\|G(s)\|_H := \max_{w(t)} \frac{\sqrt{\int_0^\infty \|z(\tau)\|_2^2 d\tau}}{\sqrt{\int_{-\infty}^0 \|w(\tau)\|_2^2 d\tau}}. \tag{3.31}$$

Hankel norm is a type of induced norm from the past inputs to the future outputs.

It is possible to show that the Hankel norm is equal to

$$\|G(s)\|_H = \sqrt{\rho(PQ)}, \tag{3.32}$$

where ρ is the spectral radius (the maximum eigenvalue), P is the controllability Grammian and Q is the observability Grammian. (See Appendix B for the definitions of Grammians.) The corresponding *Hankel singular values* are the positive square roots of the eigenvalues of product PQ,

$$\sigma_i = \sqrt{\lambda_i(PQ)}. \tag{3.33}$$

The Hankel norm and the $\mathcal{H}_\infty$ norm are closely related. It may be shown that

$$\|G(s)\|_H \equiv \sigma_1 \le \|G(s)\|_\infty \le 2 \sum_{i=1}^n \sigma_i. \tag{3.34}$$

In this way, the Hankel norm is always less than (or equal to) $\mathcal{H}_\infty$ norm, which is seen by comparing the definitions (3.29) and (3.31).

3.6 Trade-offs in the design of MIMO systems

MATLAB function used in this section

Function	Description
sigma	Singular value plot of MIMO system

In this section, we analyze the main limitations and trade-offs in the design of MIMO systems in terms of singular values of certain transfer function matrices.

3.6.1   Disturbance rejection

Efficient disturbance rejection at the plant output (y) in the low-frequency range, where d and d_i are significant, requires

$$\bar{\sigma}(S_o) \ll 1 \quad \text{(for disturbance at plant output, } d\text{)}, \tag{3.35}$$

$$\bar{\sigma}(S_o G) \ll 1 \quad \text{(for disturbance at plant input, } d_i\text{)}. \tag{3.36}$$

In a similar way, good disturbance rejection at the plant input (u_g) requires

$$\bar{\sigma}(S_i) \ll 1 \quad \text{(for disturbance at plant input, } d_i\text{)}, \tag{3.37}$$

$$\bar{\sigma}(S_i K) \ll 1 \quad \text{(for disturbance at plant output, } d\text{)}. \tag{3.38}$$

Equations (3.35) and (3.36) imply that

$$\underline{\sigma}(L_o) \gg 1, \quad \underline{\sigma}(GK) \gg 1, \tag{3.39}$$

$$\bar{\sigma}(K^{-1}) \ll 1, \quad \frac{1}{\underline{\sigma}(K)} \ll 1, \quad \underline{\sigma}(K) \gg 1 \tag{3.40}$$

and (3.37) and (3.38) imply that

$$\underline{\sigma}(L_i) \gg 1, \quad \underline{\sigma}(KG) \gg 1, \tag{3.41}$$

$$\bar{\sigma}(G^{-1}) \ll 1, \quad \frac{1}{\underline{\sigma}(G)} \ll 1, \quad \underline{\sigma}(G) \gg 1. \tag{3.42}$$

Conditions (3.39) and (3.41) may be considered as a generalization of the requirement for high loop gain of SISO systems. They are obtained from (3.35) and (3.37) as follows. Since $S_o = (I + GK)^{-1}$, $S_i = (I + KG)^{-1}$, according to (A.20) we have that

$$\frac{1}{\underline{\sigma}(GK) + 1} \leq \bar{\sigma}(S_o) \leq \frac{1}{\underline{\sigma}(GK) - 1}, \quad \text{if } \underline{\sigma}(GK) > 1,$$

$$\frac{1}{\underline{\sigma}(KG) + 1} \leq \bar{\sigma}(S_i) \leq \frac{1}{\underline{\sigma}(KG) - 1}, \quad \text{if } \underline{\sigma}(KG) > 1.$$

For simplicity, conditions (3.40) and (3.42) are obtained under the assumption that the transfer function matrices G and K are invertible. (The same conditions may be obtained in case of rectangular transfer matrices using the pseudoinverse.) This gives

$$\bar{\sigma}(S_o G) = \bar{\sigma}((I + GK)^{-1}G) \approx \bar{\sigma}(K^{-1}) = \frac{1}{\underline{\sigma}(K)} \quad \text{if } \underline{\sigma}(GK) \gg 1,$$

$$\bar{\sigma}(S_i K) = \bar{\sigma}((I + KG)^{-1}K) \approx \bar{\sigma}(G^{-1}) = \frac{1}{\underline{\sigma}(G)} \quad \text{if } \underline{\sigma}(KG) \gg 1.$$

In this way, efficient disturbance attenuation at the plant output (y) requires large output loop gain $\underline{\sigma}(L_o) = \underline{\sigma}(GK) \gg 1$ in the frequency range where d is significant in order to suppress d and large controller gain $\underline{\sigma}(K) \gg 1$ in the frequency range where

d_i is significant to suppress d_i. In a similar way, efficient disturbance rejection at the plant input (u_g) requires large input loop gain $\underline{\sigma}(L_i) = \underline{\sigma}(KG) \gg 1$ in the frequency range where d_i is significant for attenuation of d_i and large plant gain $\underline{\sigma}(G) \gg 1$ in the frequency range where d is significant for attenuation of d. Note that the later condition cannot be changed by the controller design.

To summarize, disturbance rejection imposes a lower bound on the loop gains in the low frequency range. It should be stressed, however, that the loop gains cannot be made arbitrarily high in large frequency range. This makes necessary a trade-off between closed-loop performance and design limitations.

3.6.2   Noise suppression

Another basic requirement to the closed-loop MIMO system performance is the suppression of measurement noises. As mentioned earlier, large values of $\underline{\sigma}(L_o(j\omega))$ over a large frequency range attenuate the effect of the disturbance d but in the same time they increase the effect of n, since $T_o \approx 1$ and the noise gets through over the same frequency range, i.e.,

$$y = T_o(r - n) + S_o G d_i + S_o d \approx (r - n).$$

Usually, the noise n is intensive in the high-frequency range. As in the SISO case, the large loop gain outside of the bandwidth of G, i.e., $\underline{\sigma}(L_o(j\omega)) \gg 1$ or $\underline{\sigma}(L_i(j\omega)) \gg 1$, while $\bar{\sigma}(G(j\omega)) \ll 1$ may amplify greatly the noise in control action (u). In fact, in this case, we have that $T_i \approx I$ and assuming that G is invertible it follows that

$$u = S_i K(r - n - d) - T_i d_i = T_i G^{-1}(r - n - d) - T_i d_i \approx G^{-1}(r - n - d) - d_i.$$

Since for ω such that $\bar{\sigma}(G(j\omega)) \ll 1$ one has

$$\underline{\sigma}(G^{-1}) = \frac{1}{\bar{\sigma}(G)} \gg 1$$

it follows that the disturbance and errors are amplified in u when the frequency range exceeds the bandwidth of G. In a similar way, the controller gain $\sigma(K)$ should be small in the frequency range where the loop gain is small to avoid large control actions and actuator saturation. In such case, $\bar{\sigma}(L_i(j\omega)) \ll 1, S_i \approx I, T_i \approx 0$ and

$$u = S_i K(r - n - d) - T_i d_i \approx K(r - n - d).$$

3.6.3   Model errors

Suppose that the plant model G is perturbed to $(I + \Delta)G$ with Δ stable and assume that the nominal system is stable, i.e., the closed-loop system with $\Delta = 0$ is stable. Then, the perturbed closed-loop system is stable if the polynomial

$$\det(I + (I + \Delta)GK) = \det(I + GK)\det(I + \Delta T_o)$$

has no zeros in the right half of the complex plane. This implies that ΔT_o should be small or $\bar{\sigma}(T_o)$ should be small at those frequencies where Δ is significant, typically in the high-frequency range. This in turn implies that the loop gain $\bar{\sigma}(L_o)$ should be small for these frequencies. The conclusion is that the model errors impose an upper bound on the loop gains in the high-frequency range.

Note that more detailed analysis of closed-loop stability and performance in the presence of model errors is given in Sections 3.8 and 3.9.

The presentation given in this section shows that good performance requires to satisfy the conditions

$$\underline{\sigma}(GK) \gg 1, \quad \underline{\sigma}(KG) \gg 1, \quad \underline{\sigma}(K) \gg 1,$$

in some low-frequency range $(0, \omega_l)$ and good robustness and efficient suppression of the measurement noises requires to satisfy the conditions

$$\bar{\sigma}(GK) \ll 1, \quad \bar{\sigma}(KG) \ll 1, \quad \bar{\sigma}(K) \leq M$$

in some high-frequency range (ω_h, ∞) where M is bounded. These design requirements are shown in Figure 3.24 and can be interpreted in respect to the input loop (using the subscript i of the corresponding transfer function matrices) as well as in respect to the output loop (using the subscript o).

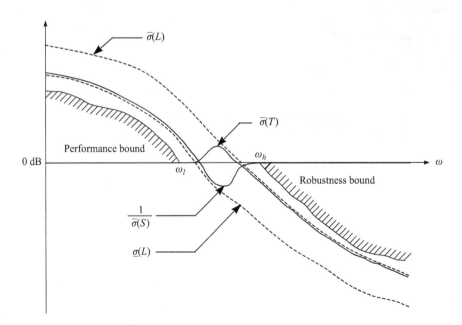

Figure 3.24 Specifying the desired singular values of L, S, and T

3.7   Uncertain systems

MATLAB functions used in this section

Function	Description
lft	Linear fractional transformations (LFT)
lftdata	M-Δ decomposition of uncertain system

The presence of uncertain dynamics in the system model may lead to significant differences between the dynamics of nominal and perturbed system. The increasing of model errors gradually worsen the performance and ultimately may cause instability of the closed-loop system. For this reason, it is important to take into account the possible uncertainties during the controller design and to investigate their effect on the closed-loop behavior. What is needed first is to obtain the mathematical description of a closed-loop system with uncertain plant.

As shown in Chapter 2, a plant with uncertainty (perturbed plant) G_p can be represented by a linear fractional transformation (LFT) given in Figure 3.25. The vector u_Δ is representing the inputs from uncertainty Δ and the vector y_Δ is representing the outputs to uncertainty. The perturbed plant is described by

$$G_p = F_U(G_o, \Delta),\tag{3.43}$$

where F_U stands for *upper* LFT. An equivalent representation with lower LFT F_L is also possible. It is assumed that the block Δ is stable and its elements are scaled so that

$$\max_\omega \bar{\sigma}(\Delta(j\omega)) \le 1.$$

Note that the latter condition is equivalent to

$$\|\Delta\|_\infty \le 1.$$

In case of unstructured uncertainty, the matrix Δ has no a special structure and represents a single "full" block.

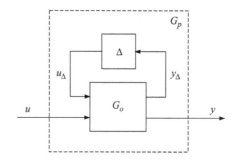

Figure 3.25   LFT representation of uncertain plant

The block diagram of a closed-loop system with uncertain plant G_p and controller K is shown in Figure 3.26.

To unify the representation of closed-loop systems with different structures, we shall make use of the standard block diagram given in Figure 3.27. The block denoted by P is the open-loop system and contains all known elements including the nominal plant model and the weighting functions. This block has three input sets: inputs u_Δ from uncertainty, the reference, disturbance, and noise signals collected in the vector w and control actions u. Three output sets are generated: outputs y_Δ to uncertainty, controlled outputs (errors) z, and measurements y. In this way, the uncertain closed-loop system is described by

$$z = F_U(F_L(P, K), \Delta)w = F_L(F_U(P, \Delta), K)w.$$

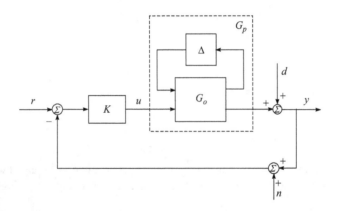

Figure 3.26   *Block diagram of uncertain closed-loop system*

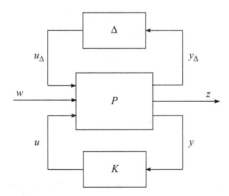

Figure 3.27   *Standard representation of uncertain closed-loop system*

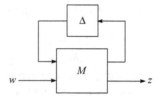

Figure 3.28   Block diagram for robustness analysis

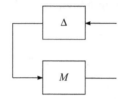

Figure 3.29   M-Δ loop

In this section, our attention will be focused on robustness analysis. That is why the controller K may be considered as a known system element and absorbed in the structure of open-loop system. Denote by

$$M(s) = F_L(P(s), K(s)) = \begin{bmatrix} M_{11}(s) & M_{12}(s) \\ M_{21}(s) & M_{22}(s) \end{bmatrix},$$

where the dimensions of M_{11} conform with those of Δ. Then, the general block diagram shown in Figure 3.27 is reduced to the block diagram given in Figure 3.28, where

$$z = F_U(M, \Delta)w = \left[M_{22} + M_{21} \Delta (I - M_{11} \Delta)^{-1} M_{12} \right] w.$$

When considering the robustness, we should take into account that the exogenous signals w do not affect the system stability. That is why in robust-stability analysis, we use the block diagram shown in Figure 3.29 neglecting the subscripts of M. The closed-loop system depicted in the Figure 3.29 is referred to as *M-Δ loop*.

In MATLAB, the construction of LFT interconnections is done using the function `lft` and the *M-Δ* decomposition is obtained by the function `lftdata`.

3.8   Robust-stability analysis

Robust Control Toolbox functions used in this section

Function	Description
`norm(T,inf)`	$\mathcal{H}_\infty$-norm of transfer function T
`robuststab`	Robust-stability analysis

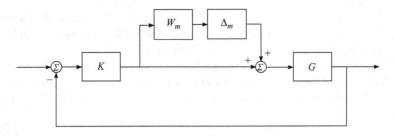

Figure 3.30   Closed-loop system with input multiplicative uncertainty

3.8.1   Unstructured uncertainty

The robust-stability analysis in case of unstructured uncertainty ("full" block Δ) can be done by using the so-called small gain theorem (see for instance [82, Section 9.2]). According to this theorem, the system shown in Figure 3.29 with stable M, is stable for all Δ with $\|\Delta\|_\infty \leq 1$, if and only if $\|M\|_\infty < 1$. This gives a simple robust-stability test if Δ corresponds to additive or multiplicative plant perturbation.

Example 3.5. Robust stability in case of input multiplicative uncertainty
Consider a closed-loop system with input multiplicative uncertainty shown in Figure 3.30 where the matrix W_m is frequency dependent weighting function and Δ_m is a "full" block Δ with $\|\Delta_m\|_\infty = 1$. In this case, the uncertain plant is given by

$$G_p = G(I + \Delta_m W_m)$$

and it is possible to show that

$$M = -W_m(I + KG)^{-1}KG = -W_m T_i,$$

where T_i is the input sensitivity function. As a result, the robust-stability condition is

$$\|W_m T_i\|_\infty < 1.$$

This condition is easily checked by using the MATLAB function `norm`. ❑

3.8.2   Structured singular value

The robust-stability test in presence of structured uncertainties is much more complicated and relies on the using of the so-called SSV denoted by $\mu(.)$. Further on, we shall assume the general case when the plant uncertainties include unstructured uncertainties, such as unmodeled dynamics, as well as parameter variations. According to the mixed uncertainty model presented in Chapter 2, all uncertain parts can be combined in a block-diagonal matrix and the closed-loop system can be rearranged in the standard configuration shown in Figure 3.31. There exist two type of blocks: *repeated scalar* blocks and *full* blocks. The block dimensions are denoted by $r_1, \ldots, r_s; m_1, \ldots, m_f$. Define the uncertain block Δ as

$$\Delta = \left\{ \text{diag}\left[\delta_1 I_{r_1}, \ldots, \delta_s I_{r_s}, \Delta_1, \ldots, \Delta_f\right] : \ \delta_i \in \mathbb{C}, \ \Delta_j \in \mathbb{C}^{m_j \times m_j} \right\}, \tag{3.44}$$

where $\sum_{i=1}^{s} r_i + \sum_{j=1}^{f} m_j = n$ and n is the dimension of Δ. The set of all matrices Δ is defined as $\mathbf{\Delta} \subset \mathbb{C}^{n \times n}$. The parameters δ_i of the repeated scalar blocks can be real numbers only, if the corresponding uncertainties are real.

The SSV is a function of the complex matrix M and the structure $\mathbf{\Delta}$. The definition of μ is done using the condition for stability of the loop shown in Figure 3.31. If the matrices M and Δ are stable, then according to Nyquist criterion (3.23), it is necessary to fulfill

$$\det(I - M(j\omega)\Delta(j\omega)) \neq 0 \text{ for each } \omega. \tag{3.45}$$

This condition motivates to find the smallest structured perturbation Δ (measured in terms of $\bar{\sigma}(\Delta)$), which leads to

$$\det(I - M(j\omega)\Delta(j\omega)) = 0$$

at some frequency ω, i.e., to closed-loop instability.

For $M \in \mathbb{C}^{n \times n}$, the SSV $\mu_{\mathbf{\Delta}}(M)$ is defined as

$$\mu_{\mathbf{\Delta}}(M) \overset{\text{def}}{=} \frac{1}{\min\{\bar{\sigma}(\Delta) : \Delta \in \mathbf{\Delta}, \ \det(I - M\Delta) = 0\}}. \tag{3.46}$$

If there is no $\Delta \in \mathbf{\Delta}$ such that $\det(I - M\Delta) = 0$, then $\mu_{\mathbf{\Delta}}(M) \overset{\text{def}}{=} 0$.

In accordance with this definition, *$\mu_{\mathbf{\Delta}}(M)$ is reciprocal of the size (measured in 2-norm) of smallest Δ from the set $\mathbf{\Delta}$, which makes the matrix $(I - M\Delta)$ singular.* The SSV μ may be considered as a strict estimate of robust stability in respect to the structured $\mathbf{\Delta}$ with multiple perturbation blocks.

Note that $\mu_{\mathbf{\Delta}}(M) = \bar{\sigma}(M)$ when $\Delta = \mathbb{C}^{n \times n}$ (the case of unstructured uncertainty). Thus $\mu_{\mathbf{\Delta}}(M)$ extends the notion of largest singular value to the case of structured uncertainties.

For systems described by transfer functions, μ is a frequency dependent function. It may be computed at each frequency substituting $s = j\omega$ in the transfer function matrix and finding μ for the constant complex matrix $M(j\omega)$.

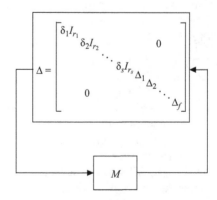

Figure 3.31   Closed-loop system used to define μ

From the definition of μ, it is possible to make the following conclusions:

- The SSV is defined not in respect to a particular perturbation Δ, but in respect to a whole set Δ of perturbations with given structure. This means that μ does not depend on the specific values of δ_i and Δ_j, but on the block numbers s, f and their dimensions r_i, m_j, respectively.
- μ is a function of two variables: the complex matrix M and the structure defined by Δ. For a fixed matrix M, it is possible to obtain different values of μ depending on the different structures of Δ. Further on, the symbol Δ in the notation $\mu_\Delta(M)$ sometimes will be skipped.
- Smaller values of μ mean better robustness.

3.8.3 Robust-stability analysis with μ

The SSV μ is used as a tool for robustness analysis in the frequency domain. Let $M(s)$ is stable transfer function and assume that Δ is a block-structure as in (3.44). Then, it is possible to prove the following result [82, Section 11.3].

Let $\beta > 0$. The loop shown in Figure 3.32 is stable for all stable transfer function matrices $\Delta \in \Delta$ with $\max_\omega \bar{\sigma}(\Delta(j\omega)) < 1/\beta$ if and only if

$$\max_\omega \mu_\Delta(M(j\omega)) \leq \beta.$$

According to this result, the peak value of the frequency response of μ determines the size of perturbations that the loop remains stable against.

If the uncertainty is normalized so that $\bar{\sigma}(\Delta) \leq 1$, the robust-stability condition is reduced to

$$\mu_\Delta(M(j\omega)) < 1, \text{ for each } \omega. \tag{3.47}$$

The condition (3.47) may be rewritten as

$$\mu(M(j\omega))\bar{\sigma}(\Delta(j\omega)) < 1, \text{ for each } \omega,$$

which can be interpreted as a "generalized small gain theorem" that takes into account the structure of Δ.

Except some simple cases, the SSV μ, which is a frequency dependent function, cannot be computed exactly. However, there exist efficient algorithms to determine upper and lower bounds on μ. That is why the conclusions about system robust stability should be done in terms of these bounds. Specifically, let the maximums

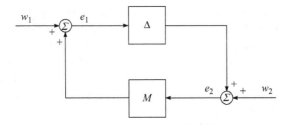

Figure 3.32   Robust stability

along frequency of the upper and lower bound on the SSVs be denoted, respectively, by β_u and β_l so that

$$\beta_l \leq \mu_\Delta(M(j\omega)) \leq \beta_u, \text{ for each } \omega.$$

Then

- The uncertain system under consideration is guaranteed stable for all structured uncertain matrices Δ with

$$\max_\omega \bar{\sigma}(\Delta(j\omega)) < 1/\beta_u.$$

- There exists a particular structured transfer matrix Δ with

$$\max_\omega \bar{\sigma}(\Delta(j\omega)) = \frac{1}{\beta_l}$$

that destabilizes the system.

In addition, for the case of normalized uncertainty

$$\max_\omega \bar{\sigma}(\Delta(j\omega)) \leq 1$$

it follows that

- If $\beta_u < 1$, the system is robustly stable in respect to the modeled uncertainty.
- If $\beta_l > 1$, the robust stability is not achieved.
- If $\beta_l < 1$ and $\beta_u > 1$, it is not possible to make sure conclusion about stability; it is possible that the system is not robustly stable.

The quantity

$$sm = \frac{1}{\max_\omega \mu_\Delta(M(j\omega))}$$

may be considered as a *stability robustness margin* in respect to the structured uncertainty which is acting on M. A stability robustness margin greater than one means that the uncertain system is stable for all possible values of the uncertainty Δ. A stability robustness margin less than one implies that there exists certain allowable Δ that leads to instability. Using the bounds on μ, the upper bound $sm_u = 1/\beta_l$ on stability margin is obtained as the reciprocal value of the lower bound of μ and the lower bound $sm_l = 1/\beta_u$ on stability margin is obtained in the same way from the upper bound on the SSV.

The main tool in Robust Control Toolbox for robust-stability analysis, based on computation of SSV, is the function robuststab. This function may be used for stability analysis of systems with unstructured, structured, or mixed uncertainty. The results of stability analysis are obtained in terms of upper and lower bounds on μ and on stability margin. The function robuststab produces also the structure destabunc containing a combination of uncertain parameter values closest to their nominal values that cause system instability along with the destabilizing frequency DestabilizingFrequency at which the instability occurs. Note that the closed-loop poles migrate across the stability boundary (imaginary axis in continuous-time systems, unit-disk in discrete-time systems) at the frequency given by `DestabilizingFrequency`.

Example 3.6. Robust-stability analysis using μ

Consider a SISO closed-loop system whose plant has uncertain parameters and unmodeled dynamics represented as input multiplicative uncertainty. The plant transfer function is

$$G = \frac{k}{s(T_1 s + 1)(T_2 s + 1)}(1 + W_m \delta),$$

where k, T_1, and T_2 are the uncertain parameters, δ is complex uncertain element with $|\delta| \leq 1$ and W_m is an uncertainty weighting function. The nominal values of the plant parameters are $k_{nom} = 0.4$, $T_{1_{nom}} = 0.2$ s, and $T_{2_{nom}} = 0.1$ s. The relative uncertainty of these parameters is 50 percent, i.e., $k \in [0.2\ 0.6]$, $T_1 \in [0.1\ 0.3]$, $T_2 \in [0.05\ 0.15]$. It is assumed that the maximum possible uncertainty due to the unmodeled dynamics is 2 percent in the low-frequency range gradually increasing to 100 percent at frequency 8 rad/s and reaches 400 percent in the high-frequency range. The corresponding uncertainty weighting function obtained by using the command `makeweight` is

$$W_m = \frac{4s + 0.6198}{s + 30.99}.$$

In this way, the 4×4 uncertainty matrix Δ contains three scalar real blocks corresponding to the parametric uncertainties in k, T_1, and T_2 and one scalar complex block δ corresponding to the unstructured plant uncertainty (the unmodeled dynamics).

The system is equipped with a controller with transfer function

$$K = 10\frac{s + 4}{s + 8},$$

which is determined so that to ensure stability and desired performance of the closed-loop system with nominal plant model.

The system robust stability is analyzed by aid of the function `robuststab` taking as an input argument the complementary sensitivity T. The computed upper and lower bounds on μ are shown in Figure 3.33. The function produces the stability robustness margin bounds

$$sm_u = 1.4348, \quad sm_l = 1.3946$$

that correspond to the μ bounds

$$\beta_l = 0.6970, \quad \beta_u = 0.7170.$$

It follows from the stability margin bounds that

- For the specified structure and uncertainty level, the closed-loop system is robustly stable (the lower bound sm_l of the stability margin is > 1).
- The system can tolerate uncertainties less than 139 percent of the specified uncertainties.
- There exists uncertainty of size 143 percent of the specified one, which destabilizes the system.

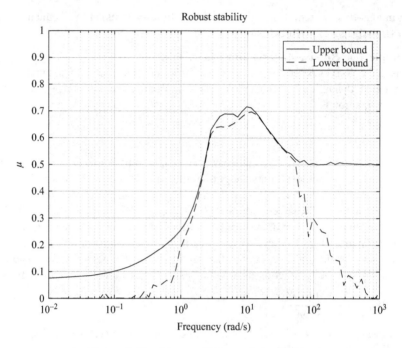

Figure 3.33   Upper and lower bounds of μ

The conclusions about system robust stability are contained in the variable `report`. This variable includes also information about the sensitivity of the stability robustness margin with respect to the uncertain elements. In the given case, the strongest effect on the stability margin has the uncertainty due to the unmodeled dynamics; the weakest effect has the uncertainty of the time constant T_2.

Substituting the destabilizing uncertainty `destabunc` in the complementary sensitivity function by using the function `usubs`, it is confirmed that the resulting closed-loop system is unstable with a pair of unstable poles equal to 0 ± 11.3258.   ❒

3.9   Robust performance analysis

Robust Control Toolbox functions used in this section

Function	Description
`robustperf`	Robust performance analysis
`wcgain`	Worst case gain
`wcmargin`	Worst case margin

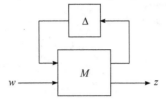

Figure 3.34   Robust performance definition

3.9.1   Using μ for robust performance analysis

In presence of parametric perturbations, the effect of exogenous disturbances and noises on the system dynamics may increase significantly. That is why apart from robust stability, in case of uncertainty it is necessary to ensure acceptable performance of the closed-loop system. This leads to the idea of *robust performance*, i.e., performance which remains acceptable in the presence of uncertain parameters and signals.

The performance of MIMO system can be characterized by the aid of the $\mathcal{H}_{\infty}$-norm. Specifically, it is assumed that *good performance* is equivalent to

$$\|T\|_{\infty} \stackrel{\text{def}}{=} \max_{\omega} \bar{\sigma}(T(j\omega)) \leq 1,$$

where T is some weighted closed-loop matrix transfer function, i.e., closed-loop transfer function in which participate certain weighting functions. In case of robust performance analysis of the uncertain system given in Figure 3.34, T is taken as the uncertain transfer function from w to z, so that $T = F_U(M, \Delta)$. The quantitative characterization of the performance is related to the determination of the "size" of the $F_U(M, \Delta)$, where Δ takes all admissible values. More precisely, it is said that the LFT in Figure 3.34 achieves *robust performance* if it is stable for all perturbations $\Delta \in \mathbf{\Delta}$ satisfying $\max_{\omega} \bar{\sigma}(\Delta(j\omega)) < 1$, and moreover, if $\|F_U(M, \Delta)\|_{\infty} \leq 1$ for all such perturbations.

The idea of robust performance analysis is to connect the size of the closed-loop transfer function matrix to a robust-stability test. Let T be a given stable system with input dimension n_w and output dimension n_z. According to Nyquist criterion (3.23) and small gain theorem, it follows that $\|T\|_{\infty} \leq 1$, if and only if the feedback loop in Figure 3.35 is stable for each stable $\Delta_F(s)$ (with dimension $n_w \times n_z$) satisfying $\|\Delta_F\| < 1$. Hence, a transfer matrix T is small ($\|T\|_{\infty} \leq \beta$), if and only if T can tolerate all admissible stable perturbations Δ_F (with $\|\Delta_F\| < 1/\beta$) avoiding instability. In this way, the size of the transfer matrix may be determined using a robust-stability test which allows to reduce the robust performance problem to a robust-stability problem.

According to the above considerations, it follows that $\|F_U(M, \Delta)\|_{\infty} \leq 1$ for all perturbations $\Delta \in \mathbf{\Delta}$ satisfying $\max_{\omega} \bar{\sigma}(\Delta(j\omega)) < 1$, if and only if the LFT, shown in Figure 3.36, is stable for all $\Delta \in \mathbf{\Delta}$ and for all stable Δ_F satisfying

$$\max_{\omega} \bar{\sigma}(\Delta(j\omega)) < 1 \text{ and } \max_{\omega} \bar{\sigma}(\Delta_F(j\omega)) < 1.$$

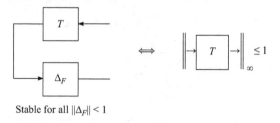

Stable for all $\|\Delta_F\| < 1$

Figure 3.35   Performance as robust stability

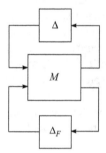

Figure 3.36   Robust stability with augmented uncertainty

But this is exactly a robust-stability problem for M subject to perturbation matrices in the form

$$\Delta_P = \begin{bmatrix} \Delta & 0 \\ 0 & \Delta_F \end{bmatrix}.$$

Hence, the robust-stability test was used on a larger problem computing $\mu_{\Delta_P}(M(j\omega))$ to determine bounds on the robust performance for the original problem. This leads to the use of additional (fictitious) uncertain block and determination of the robust stability of the augmented system deriving conclusions about the robust performance of the original system $T = F_U(M, \Delta)$.

Suppose that the peak value of the frequency response of μ is equal to β. This implies that for all perturbations $\Delta_P \in \mathbf{\Delta}_P$ satisfying $\max_\omega \bar\sigma(\Delta_P(j\omega)) < 1/\beta$, the perturbed system is stable and $\|F_U(M, \Delta_P)\|_\infty \le \beta$. Moreover, there exists a particular perturbation matrix $\Delta_P \in \mathbf{\Delta}_P$, satisfying $\max_\omega \bar\sigma(\Delta_P(j\omega)) = 1/\beta$, which causes $\|F_U(M, \Delta_P)\|_\infty = \beta$, or instability.

As in the case of robust-stability analysis, the algorithms for computation of SSV produce upper and lower bounds on μ. Let the maxima with respect to the frequency of the upper and lower bounds of the SSV $\mu_{\Delta_P}(M)$ be denoted by β_u and β_l, respectively. Then

- For all uncertainty matrices $\Delta_P \in \mathbf{\Delta}_P$ satisfying

$$\max_{\omega} \bar{\sigma}(\Delta_P(j\omega)) < \frac{1}{\beta_u}$$

the perturbed system is stable and $\|T_{zw}(s)\|_\infty \leq \beta_u$.
- There exists a particular perturbation matrix $\Delta_P \in \mathbf{\Delta}_P$ satisfying

$$\max_{\omega} \bar{\sigma}(\Delta_P(j\omega)) = \frac{1}{\beta_l}$$

which causes either $\|T_{zw}(s)\|_\infty \geq \beta_l$ or instability.

For a performance requirement $\|T_{zw}(s)\|_\infty \leq 1$ and normalized uncertainty

$$\|\Delta\|_\infty \leq 1,$$

the following conclusions hold.

- If $\beta_u < 1$, the system achieves robust performance for the modeled uncertainty (this includes also robust stability).
- If $\beta_l > 1$, the robust performance is not achieved.

Clearly, if $\beta_l < 1$ and $\beta_u > 1$, it is not possible to make a definite conclusion whether the system achieves robust performance.

In the analysis of robust performance, similarly to stability robustness margin, one may introduce the notion of *robust performance margin*. This quantity shows the uncertainty level, up to which the system possesses specified performance. The performance margin pm is equal to the reciprocal value of the maximum in respect to frequency of the SSV $\mu_{\Delta_P}(M)$, i.e., $pm = 1/\beta$. The upper bound pm_u and lower bound pm_l on the stability margin are obtained by β_l and β_u, respectively, as

$$pm_u = \frac{1}{\beta_l}, \quad pm_l = \frac{1}{\beta_u}.$$

The performance margin bounds are interpreted in the following ways: if $pm_u > 1$, the system achieves robust performance for the modeled uncertainty; if $pm_u < 1$, the robust performance is not achieved; if $pm_l < 1$ and $pm_u > 1$, it is not possible to make sure conclusion.

Note that the robust performance margin is less than the robust-stability margin.

In Robust Control Toolbox, the robust performance analysis is done by the function `robustperf`. This function produces upper and lower bounds on the SSV $\mu_{\Delta_P}(M)$ defined for the case of robust performance problem along with upper and lower bounds on the robust performance margin. The function `robustperf` produces also the structure `perfmargunc` of uncertain element values leading to worst performance degradation.

Example 3.7. Robust performance analysis using μ

Consider again the control system described in Example 3.4. Let the main requirements to system performance are to ensure reference (r) tracking with acceptable error (e) and limited the control action (u) magnitude. According to the first requirement, the sensitivity function S, connecting e with r, should be sufficiently small in the low-frequency range (r is a low-frequency signal). According to the second requirement, the transfer function KS connecting e with u should be sufficiently small. To account for both requirements, it is appropriate to use as a system performance index the quantity

$$\left\| \begin{bmatrix} W_pS \\ W_uKS \end{bmatrix} \right\|_\infty, \tag{3.48}$$

where W_p are W_u are weighting transfer functions. Let

$$W_p = 0.8\frac{s^2 + 18s + 50}{s^2 + 24s + 0.04} \qquad W_u = 0.02$$

are chosen so that when the condition

$$\left\| \begin{bmatrix} W_pS \\ W_uKS \end{bmatrix} \right\|_\infty < 1 \tag{3.49}$$

is fulfilled, then the control system has desired performance, i.e., desired accuracy of reference tracking and limited magnitude of control action are ensured.

The closed-loop system block diagram, used in the robust performance analysis, which involves the weighting functions W_p and W_u is shown in Figure 3.37. In the given case

$$T_{zw} = \begin{bmatrix} W_pS \\ W_uKS \end{bmatrix},$$

the output vector z contains the weighted "errors" and the input vector w being the reference r.

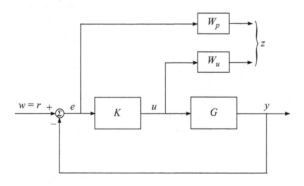

Figure 3.37   Closed-loop system with performance requirements

The system model is composed using the function `sysic`. The nominal closed-loop system has $\mathcal{H}_\infty$-norm equal to 0.8774, so that the condition (3.49) is satisfied and the closed-loop system achieves nominal performance.

The robust performance is analyzed by means of the function `robustperf` in the frequency range $[10^{-3} \ 10^3]$ rad/s.

The computed upper and lower bounds on $\mu_{\Delta_P}(M)$ are shown in Figure 3.38. The function produces the performance margin bounds

$$pm_u = 0.6691, \quad pm_l = 0.6656$$

that correspond to the μ bounds

$$\beta_l = 1.4944, \quad \beta_u = 1.5023.$$

Since $pm_u < 1$, the performance requirement (3.49) is not satisfied for the given uncertainty, i.e., the system does not achieve robust performance. It follows from the results obtained that for uncertainty level less than 66.56 percent of the given one, the value of the performance index (3.48) is less than or equal to $1/0.6656 = 1.5023$. Also, the analysis produces a combination of uncertain elements of level 66.91 percent of the given one, for which the value of (3.48) is greater than or equal to $1/0.6691 = 1.4945$.

Apart from the conclusions concerning system performance, the variable `report`, produced by `robustperf`, contains also information about the influence of uncertain elements on the stability margin. The strongest influence in the given case has the uncertainty in coefficient k.

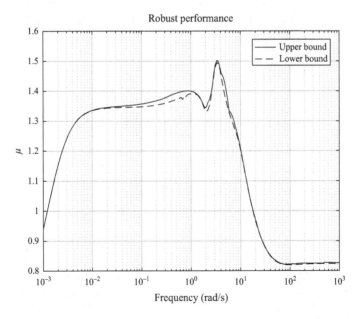

Figure 3.38 *Upper and lower bounds of* $\mu_{\Delta_P}(M)$

The structure `perfmargunc`, also produced by `robustperf`, contains uncertainty element values, which correspond to worst performance degradation. Substituting these elements in the closed-loop transfer function by using the function `usubs` it is obtained that $\|T_{zw}\|_\infty = 1.4944$ which is less than β_u as expected.   ❐

3.9.2 Worst case gain

The system performance may be assessed approximately by the maximum of the frequency response of the largest singular value ($\mathcal{H}_\infty$-norm) of the sensitivity or complementary sensitivity transfer function matrix. For uncertain systems, it is of interest to determine the largest value of this maximum for the allowed uncertainty. This value, representing the largest possible gain in the frequency domain, is defined as the "worst case" gain. Determining the maximum gain over all allowable values of the uncertain elements is referred to as a *worst case gain* analysis.

In Robust Control Toolbox, the worst case gain of uncertain systems may be determined by the function `wcgain`. It calculates upper and lower bounds on the worst gain and determines a structure `wcunc` containing a combination of uncertain element values which maximize the system gain. The function `wcgain` also produces information about the sensitivity of the worst case gain in respect to the uncertain elements.

Example 3.8. Worst case gain
Consider the uncertain control system described in Examples 3.6 and 3.7. To determine the worst case gain, we shall make use of the function `wcgain` with input argument the complementary sensitivity T. As a result, one obtains the following bounds on the worst case gain:

   Lower bound: 2.69342

   Upper bound: 2.69345

The strongest influence on the worst case gain in the given case has the uncertainty due to unmodeled dynamics.

In Figure 3.39, we show the magnitude responses corresponding to 30 random combinations of uncertainty values and compare them with the responses of system with worst case gain.

The step responses for 30 random uncertainty values and for the worst case gain are given in Figure 3.40.

It follows from the results obtained that the plant uncertainty can lead to significant degradation of closed-loop system performance. In the worst case, the transient response is very oscillatory with overshoot of about 43 percent and settling time which is almost seven times larger than this of the nominal closed-loop system.

Similar results about the worst case gain are obtained if instead of the complementary sensitivity T one is using the sensitivity function S.   ❐

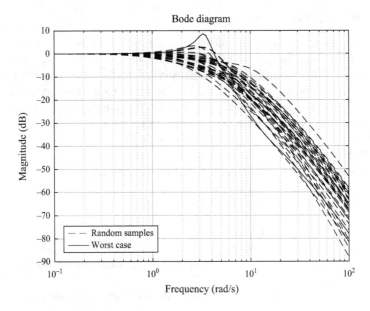

Figure 3.39   Random and worst case magnitude responses

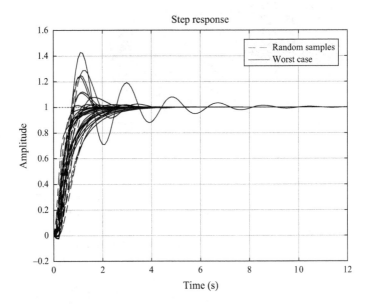

Figure 3.40   Random and worst case step responses

3.9.3 Worst case margin

Still another element of robust performance is the so-called worst case margin. Worst case margin defines the largest disk margin such that for all possible values of the uncertainty and all gain and phase variations inside the disk, the closed-loop system is stable (for a definition of the disk margin see Section 3.2). Thus, results from the worst case margin determination imply that the closed-loop system is stable for a given uncertainty set and would remain stable in the presence of an additional GM and PM variations in the specified input/output loop.

In Robust Control Toolbox, the worst case margin is computed using the function wcmargin. This function calculates the combined worst case input and output loop-at-a-time GM/PM of the feedback loop consisting of the loop transfer matrix $L(s)$ in negative feedback with an identity matrix.

Example 3.9. Worst case margin
It is instructive to compare the classical, disk, and worst case stability margins for the uncertain system considered in Examples 3.4–3.6. Note that the system plant has four uncertain parameters k, T_1, T_2, and δ.

The classical *cm* and disk margins *dm* are computed using the function loopmargin. As a result one obtains

```
cm =

      GainMargin:  8.9096
     GMFrequency:  9.7399
     PhaseMargin:  69.6771
     PMFrequency:  1.9757
     DelayMargin:  0.6155
     DMFrequency:  1.9757
          Stable:  1

dm =

      GainMargin:  [0.2725 3.6698]
     PhaseMargin:  [-59.5149 59.5149]
       Frequency:  4.0185
```

While the classical GM and PM (computed for the nominal system) are $8.9096 = 18.99$ dB and $\pm69.6771°$, respectively, the disk margin shows that the closed-loop will remain stable for independent, simultaneous, GM variation up to 0.2775 and $3.6698 = \pm11.29$ dB and PM variations of $\pm59.51°$.

The computation of the worst case margin by the function wcmargin produces

```
wcmarg =

      GainMargin:  [0.7092 1.4099]
     PhaseMargin:  [-19.3078 19.3078]
       Frequency:  3.3001
           WCUnc:  [1x1 struct]
     Sensitivity:  [1x1 struct]
```

The maximum allowable GM for all possible defined uncertainty ranges is up to 0.7092 and $1.4099 = \pm 2.98$ dB and PM variations of $\pm 19.3078°$. In this way, the variations of the parameters k, T_1, T_2, and δ may lead to significant reduction of the closed-loop stability margins.   ❏

3.10   Numerical issues in robustness analysis

Robust Control Toolbox function used in this section

Function	Description
mussv	Computation of the structured singular value

As demonstrated in the previous sections of this chapter, the robustness of an uncertain closed-loop system is characterized by the peak magnitude of the corresponding SSV μ. Since the exact value of μ may be computed only in some special cases, guaranteed results about the robustness properties may be found by using the upper bound on the SSV. That is why the reliability of stability robustness analysis and performance robustness analysis depends on the accuracy achieved in μ computation. Example 3.10 demonstrates that for some systems the error in computed upper bound on μ may be very large.

Example 3.10. Numerical issues in computing μ
Consider a second-order system described by the differential equation

$$\frac{d^2y}{dt^2} + a_1 \frac{dy}{dt} + a_2 y = b_1 u. \tag{3.50}$$

Let us assume that the coefficients a_1 and a_2 have relative uncertainties p_1 and p_2, respectively, so that these coefficients may be represented as

$$a_1 = \bar{a}_1(1 + p_1\delta_1), \quad a_2 = \bar{a}_2(1 + \delta_1), \quad |\delta_1| \le 1, \quad |\delta_2| \le 1,$$

where a_1, a_2 are the nominal values of the corresponding coefficients.
The block diagram of the uncertain parameter system is shown in Figure 3.41 where

$$M_1 = \begin{bmatrix} 0 & \bar{a}_1 \\ p_1 & \bar{a}_1 \end{bmatrix}, \quad M_2 = \begin{bmatrix} 0 & \bar{a}_2 \\ p_2 & \bar{a}_2 \end{bmatrix}.$$

Note that the matrices M_1 and M_2 are not unique and other representations of the uncertain coefficients are also possible.
Introducing the variables $x_1 = y$, $x_2 = \dot{y}$, the system (3.50) may be described in the state space form

$$\begin{aligned}
\dot{x}_1 &= x_2, \\
\dot{x}_2 &= -\bar{a}_2 x_1 - \bar{a}_1 x_2 - p_1 u_{\delta_1} - p_2 u_{\delta_2} + b_1 u, \\
y_{\delta_1} &= \bar{a}_1 x_2, \\
y_{\delta_2} &= \bar{a}_2 x_1
\end{aligned} \tag{3.51}$$

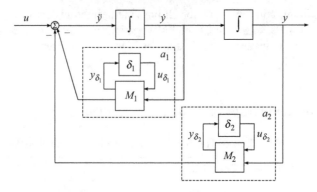

Figure 3.41   Block diagram of the uncertain second order system

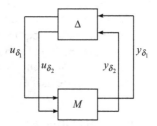

Figure 3.42   M-Δ loop corresponding to robustness stability analysis

For $u = 0$, the Laplace transform of (3.51) yields

$$\begin{bmatrix} y_{\delta_1}(s) \\ y_{\delta_2}(s) \end{bmatrix} = M(s) \begin{bmatrix} u_{\delta_1}(s) \\ u_{\delta_2}(s) \end{bmatrix}, \tag{3.52}$$

where

$$M(s) = \begin{bmatrix} \dfrac{\bar{a}_1 p_1 s}{s^2 + \bar{a}_1 s + \bar{a}_2} & \dfrac{\bar{a}_1 p_2 s}{s^2 + \bar{a}_1 s + \bar{a}_2} \\[3mm] \dfrac{-\bar{a}_2 p_1}{s^2 + \bar{a}_1 s + \bar{a}_2} & \dfrac{\bar{a}_2 p_2}{s^2 + \bar{a}_1 s + \bar{a}_2} \end{bmatrix}.$$

The system (3.52) may be represented as the M-Δ loop shown in Figure 3.42, where

$$\Delta = \begin{bmatrix} \delta_1 & 0 \\ 0 & \delta_2 \end{bmatrix}.$$

In the given case the set Δ consists of all diagonal 2×2 matrices whose diagonal elements have values between -1 and 1.

To derive an analytical expression for the SSV μ corresponding to stability robustness analysis, it is necessary to find an expression for $\det(I - M(j\omega)\Delta(j\omega))$. As a result, we obtain

$$\det(I - M(j\omega)\Delta(j\omega)) = (\bar{a}_2 - \omega^2 + j\bar{a}_1\omega + j\delta_1\bar{a}_1p_1\omega)$$
$$\times \frac{(\bar{a}_2 - \omega^2 + j\bar{a}_1\omega + \delta_2\bar{a}_2p_2)}{(\bar{a}_2 - \omega^2 + j\bar{a}_1\omega)^2} + \frac{j\delta_1\delta_2\bar{a}_1\bar{a}_2p_1p_2\omega}{(\bar{a}_2 - \omega^2 + j\bar{a}_1\omega)^2} \qquad (3.53)$$

Setting $\det(I - M(j\omega)\Delta(j\omega)) = 0$, we find that the M-Δ loop is unstable if the uncertainties δ_1 and δ_2 satisfy

$$\delta_1^* = -\frac{1}{p_1}, \quad \delta_2^* = \frac{\bar{a}_2 - \omega^2}{\bar{a}_2p_2}. \qquad (3.54)$$

Hence, the minimum size uncertainty which leads to instability of the system (3.50) is

$$\Delta_{\min} = \begin{bmatrix} \delta_1^* & 0 \\ 0 & \delta_2^* \end{bmatrix}.$$

The maximum singular value of $\Delta_{\min}$ is equal to $\max\{|\delta_1^*|, |\delta_2^*|\}$ so that the exact value of μ as a function of the frequency ω is given by

$$\mu_\Delta(M) = \frac{1}{\max\{|\delta_1^*|, |\delta_2^*|\}}. \qquad (3.55)$$

The comparison of the exact value of μ determined by (3.55) and its upper bound, computed by the function mussv from Robust Control Toolbox, is done for three second-order systems with the following parameters:

System 1 $\bar{a}_1 = 2$, $\bar{a}_2 = 20$, $b_1 = 1$
System 2 $\bar{a}_1 = 0.3$, $\bar{a}_2 = 100$, $b_1 = 1$
System 3 $\bar{a}_1 = 0.1$, $\bar{a}_2 = 190$, $b_1 = 1$

For all systems, we assume relative uncertainties $p_1 = 0.5$, $p_2 = 0.5$.

The exact values of μ for the three systems, computed by (3.55), are shown in Figure 3.43. The values of μ do not exceed 0.5 so that these systems remain stable for all perturbations satisfying $|\delta_1| < 2.0$, $|\delta_2| < 2.0$. This is in accordance with the stability analysis of second order systems which are always stable for positive coefficients a_1 and a_2.

The upper bounds of μ for the three systems computed by mussv are shown in Figure 3.44. The accuracy of the bound for Systems 2 and 3 is very low, and according to the computed values, these systems are not robustly stable for the given perturbations. Of course, this is not true for the prescribed relative uncertainties. The large errors in the computed μ upper bound arise because the corresponding

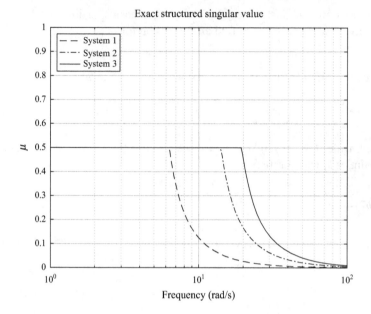

Figure 3.43   Exact values of the structured singular value

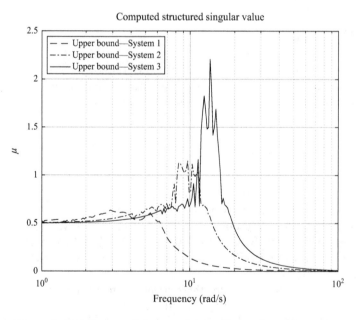

Figure 3.44   Computed values of the structured singular value

systems are lightly damped. The damping ratios ζ of the three systems are given in the table below.

	ζ
System 1	0.2236
System 2	0.0150
System 3	0.0036

It is seen from the table that the error in computation of μ upper bound increases with the decreasing of the damping ratio which leads to wrong conclusions about the system robustness. ☐

The numerical difficulties associated with the computation of μ may be removed if the optimization problem related to the determination of the upper bound of μ is solved by using a linear matrix inequality (LMI) solver. This allows to employ the methods of semidefinite programming to find more accurate upper bound of μ. The implementation of LMI solver is done by using the function `mussv` with the option `'a'`. With such option, the computed upper bound on μ practically coincides with the exact value of μ.

3.11 Notes and references

The material presented in this chapter is available in lots of sources. Thorough exposition of the classical control theory for SISO systems is done in several excellent textbooks such as Franklin, Powell, and Emami-Naeini [69]; Goodwin, Graebe, and Salgado [43]; Golnaraghi and Kuo [62]; Ogata [68]; Dorf and Bishop [90]. An accessible introduction to feedback system theory is given in Åström and Murray [89]. The modern approach to the theory of SISO systems based on $\mathcal{H}_\infty$ methods is presented in Doyle, Francis, and Tannenbaum [91]; Skogestad and Postlethwaite [87]; Helton and Merino [92]. The limitations and trade-offs in the design of SISO systems are discussed in depth in Freudenberg and Looze [88,93], Goodwin, Graebe, and Salgado [43]; Åström and Murray [89]; Leong and Doyle [94]; Skogestad and Postlethwaite [87]; Lurie and Enright [95]; see also the paper of Åström [96] and the discussion therein. The book of Seron, Braslavsky, and Goodwin [97] is entirely devoted to the fundamental limitations in filtering and control problems. The multivariable system theory based on frequency-domain methods is presented in several books, see, e.g., Callier and Desoer [98], Green and Limebeer [99], Maciejowski [100], and Skogestad and Postlethwaite [87]. The limitations in the design of MIMO systems are discussed in Zhou, Doyle, and Glover [82]; Goodwin, Graebe, and Salgado [43]; Morari and Zafiriou [101]. The modern theory of robust control may be found in the books of Zhou, Doyle, and Glover [82] and its textbook version [102], as well as in the books of Green and Limebeer [99], Skogestad and Postlethwaite [87], Sánchez-Peña and Sznaier [103], Dullerud and Paganini [104]. The practical stability robustness and performance robustness analysis is based on the computation of SSV μ. As it was shown by Braatz, Young, Doyle, and Morrari [105], the computational complexity of μ calculation in case of mixed real and complex uncertainties is nonpolynomial hard. This suggests that in the general case, the volume of computations in determining μ depends exponentially on the problem size. Therefore, it is pointless to pursue exact methods for the calculation of μ even for medium-size problems (less than 100 real parameters). Efficient computational methods to determine upper and lower bounds on μ are proposed by Fan, Tits, and Doyle [106] and Young, Newlin, and Doyle [107,108]. A comparison of the available algorithms to compute lower bounds on the SSV is done by Roos and Bianic [109].

Chapter 4

Controller design

The controller synthesis is probably the most difficult and time consuming stage of embedded control system design. In this chapter, we present the design and analysis of five different discrete-time controllers which may be implemented successfully in embedded systems. To compare the controller properties, they are applied in single precision to steer one and the same system, namely, the cart–pendulum system presented in Chapter 2. This system has some peculiarities which lead to difficulties in the implementation of design methods. One of our primary goals is to investigate the behavior of the corresponding closed-loop systems in presence of plant uncertainty.

As is discussed in Chapter 1, the design of discrete-time controllers can be carried out in two distinctly different ways. One is to design the controller in continuous-time and then to derive a discrete version of it. The alternative method is to discretize the continuous-time plant model and perform the design in discrete time. In this chapter, we adopt the second approach. The performance specifications are usually formulated in continuous-time due to the clear physical interpretation in this case. After plant discretization, a discrete-time controller is computed which may be implemented directly for simulation or real-time control. Using such approach avoids the errors which might be introduced in converting a continuous-time controller to discrete time.

The controllers considered in this chapter are chosen somewhat subjectively, but we tried to include regulators which are proven in practice. They are described in order of their increasing complexity and robustness. First, we discuss the properties of the classical proportional-integral-derivative (PID) controller which is widely used due to its simple implementation although it is not sufficiently robust to plant variations. Next, we present a linear quadratic Gaussian (LQG) controller that involves linear quadratic regulator (LQR) and uses state estimate obtained by Kalman filter. When applied to the case of colored noises, the LQG controller design may be associated with some numerical difficulties which are briefly discussed. Similar performance has a controller comprising LQR and $\mathcal{H}_\infty$ filter. These three controllers are designed based on performance requirements formulated in the time domain. It is shown then that better results in respect to the robustness of the closed-loop system are obtained by applying $\mathcal{H}_\infty$ controller or μ regulator which are designed by using performance specifications in the frequency domain. Finally, we present the results from hardware-in-the-loop (HIL) simulations of the five controllers.

In our presentation, we do not discuss the implementation of state regulators and state observers designed by pole assignment since, in authors' opinion, they cannot ensure always the necessary closed-loop performance and robustness. Also, we do not include description of robust loop shaping procedures which are considered in detail in [50,110]. Other design methods are described in the notes and references given at the end of the chapter.

4.1 PID controller

MATLAB® files used in this section

File	Description
PID_controller_design.m	Design of PID controller
err_fun.m	Evaluates optimization performance index
discrete_PID_pendulum.slx	Generates outputs and control for optimization
unc_lin_model.m	Generates the uncertainty cart–pendulum model
sim_PID_pendulum_nominal.m	Simulation of nominal system
clp_PID_pendulum.slx	Simulink® model of closed-loop system
PID_robstab.m	Robust-stability analysis
PID_wcp.m	Frequency responses and worst gain analysis
sim_MC_PID.m	Monte-Carlo simulation
PID_controller_open_loop.slx	Simulink model of PID controller for open loop connection

Due to its simplicity and clear physical interpretation, the PID controller and its modifications are the most widely used in practice control algorithms. Usually, they are implemented in single embedded devices and/or in industrial factories with dozens of controllers. There is a large amount of literature dedicated to methods for design of PID controllers for both single-input-single-output (SISO) and multiple-input-multiple-output (MIMO) plants. In case of MIMO plant, the PID controller tuning becomes a difficult problem because of interactions between different inputs and outputs. For such plants, the most common used control scheme is the decentralized PID control. Unfortunately, it has some disadvantages when one should control non-minimum phase or unstable plant with more outputs then inputs (plant with rectangular transfer matrix). In this case, optimization approach can be used for PID controller tuning. The main advantages of this approach are that it may be used with linear or nonlinear MIMO models, and the numerical optimization can be done with many different combinations between optimization methods with performance indices. The main disadvantage is that the robust stability and robust performance of control system are not guaranteed.

In this section, we present the optimization approach for PID controller tuning. The controller aim is to ensure accurate reference tracking and regulation in presence of disturbances and noises. It is shown that the PID controller ensures nominal control system performance but cannot guarantee good performance when plant model varies.

The basic configuration of closed-loop system with PID controller is shown in Figure 4.1. The PID controller design may be presented briefly as follows. Assume

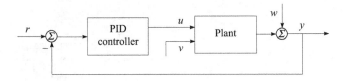

Figure 4.1 Closed-loop system with PID controller

that the time-invariant plant is described by the differential state and algebraic output equations

$$\dot{x}(t) = f(x, u, v, t),$$
$$y(t) = h(x, u, w, t),$$
(4.1)

where $x(t)$ is the state vector, $y(t)$ is measurement vector, and $v(t), w(t)$ are the process and measurement noises, respectively; $f(\cdot), h(\cdot)$ are smooth nonlinear functions. In design procedure, both the deterministic model, obtained from (4.1) for $v(t) = 0, w(t) = 0$ and the model with noises (4.1) can be used. Note that, the linear stochastic or deterministic state space model can be used instead of the model (4.1),

$$\dot{x}(t) = Ax(t) + Bu(t) + v(t),$$
$$y(t) = Cx(t) + w(t),$$
(4.2)

where $A, B, C,$ and D are constant matrices with appropriate dimensions. The model (4.2) can be obtained by linearization of model (4.1) or by the identification methods described in Appendix D and Chapter 2.

The PID controller is described by the equation

$$u(t) = K_p \left(e(t) + \frac{1}{T_i} \int_0^t e(\tau) d\tau + T_d \frac{de(t)}{dt} \right),$$
(4.3)

where $e(t) = r(t) - y(t)$ is the error; $K_p, T_i,$ and T_d are the proportional gain, integral time constant and derivative time constant, respectively. After Laplace transform, (4.3) takes the form

$$u(s) = K_p \left(e(s) + \frac{1}{T_i s} e(s) + T_d s e(s) \right).$$
(4.4)

The controller (4.4) is idealized one and cannot be implemented in practice. Some modifications should be performed to obtain useful for practical implementation controller. A drawback of the derivative term of (4.3) is that it has high gain in high frequencies. This means that measurement noise will generate large variations of control signal, which is undesirable for the actuators. This effect can be reduced by implementation of first-order low-pass filter derivative term $NT_d s/(s+N)$ instead of the ideal term $T_d s$. Large value of the filter pole N means high cut-off frequency. Ideal derivative term is obtained for $N \to \infty$. Typically, N takes the value in range 2–100. Other significant drawback in practical implementation of ideal PID controller is the so-called integrator windup. This situation arises when the control signal reaches the actuator limits (note that we always have actuator limits, e.g., the valve cannot

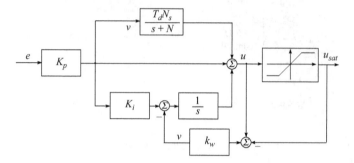

Figure 4.2   PID controller with low pass filter in derivative and antiwindup

be more than fully opened or fully closed and the electrical drive has limited speed). Then, the control signal remains at its limit independently of plant output as long as the actuator remains saturated. The integral term is also build up since the error is nonzero. The controller output and the integral term may become very large. The control signal then will remain saturated even when the error changes, and it takes a long time before the controller output fall back in saturation range. The consequence is that a large deviation of plant output from reference can occurred and the settling time becomes very large. This situation is known as integrator windup. There are many methods to reduce the windup effect. One useful method is illustrated in Figure 4.2, where the PID controller has an extra condition feedback. The feedback forms the signal which is the weighted difference between the PID controller output $u(t)$ and output of actuator saturation mathematical model $u_{sat}(t)$.

Thus, the useful in practice PID controller with first-order low-pass filter in derivative term and antiwindup algorithm is described by

$$u(s) = K_p\left(e(s) + \frac{T_d N s}{s + N}e(s)\right) + (K_p K_i e(s) - k_w v(s))\frac{1}{s}, \tag{4.5}$$

where $K_i = 1/T_i$ is the integral gain, k_w is the condition feedback gain, and $v(s) = u(s) - u_{sat}(s)$. The signal v is zero when there is no saturation and then condition feedback has not effect on the control system. When the control signal saturates, the signal $v(s)$ is subtracted from the error $e(s)$ before its integration. In this manner, the integral term value and controller output are decreased, and the windup effect is reduced. The rate at which controller output is reset depends of condition feedback gain k_w. The large value of k_w ensures fast reset of integrator, but k_w cannot be too large because then the measurement noise can cause an undesirable reset or oscillations with constant magnitude equal to the difference between the upper and the lower limits of actuator saturation mathematical model. A usual choice of k_w is as a fraction of $1/T_i$.

In embedded control systems, PID controller algorithm is implemented in digital microcontroller. For this reason, algorithm (4.5) must be converted to discrete-time form. There are many technics for discretization of PID controller terms. Here, we use backward Euler method for discretization of filtered derivative term and forward Euler method for discretization of integration term. Thus, the description of discrete-time PID controller is given by

$$u(z) = K_p\left(e(z) + \frac{T_d N(z-1)}{(1+NT_s)-1}e(z)\right) + (K_p K_i e(s) - k_w v(z))\frac{T_s}{z-1}, \qquad (4.6)$$

where $v(z) = u(z) - u_{sat}(z)$.

The PID controller design problem for system (4.1) or (4.2) consists of finding appropriate values of parameters $\phi = [K_p, K_i, T_d, N, k_w]^T$ which ensure the desired control system performance. In case of MIMO plant, we should find the values of

$$\phi = [K_{p11}, K_{i11}, T_{d11}, N_{11}, k_{w11}, K_{p21}, K_{i21}, T_{d21}, N_{21}, k_{w21}, \dots,$$
$$K_{prm}, K_{irm}, T_{drm}, N_{rm}, k_{wrm}]^T,$$

where $K_{pij}, K_{ij}, T_{dij}, N_{ij}, k_{wij}$ are the parameters of PID controller between ith plant output and jth plant input.

After the optimization of PID regulator parameters, we should test robust stability and performance of control system with the obtained optimal parameters. The most frequently used performance index is

$$J(\phi) = \sum_{i=0}^{t} e(i,\phi)^T Q e(i,\phi) + \sum_{i=0}^{t} u(i,\phi)^T R u(i,\phi), \qquad (4.7)$$

where Q and R are symmetric positive semidefinite matrices with appropriate dimensions. The diagonal elements of matrices Q and R may be chosen in accordance to the relative significance of components of error signal and control signal. For example, if the designer knows that the error between reference signal and first output is more significant than other errors, then the element $Q(1,1)$ should be chosen to be large relative to other elements of Q. Often, the matrix R is chosen as $R = \rho I, \rho > 0$. Then, large value of coefficient ρ means large relative weight of the control signal in performance index (4.7), which ensures control signals with small magnitudes and slow transient responses. The optimal PID controller parameters are obtained as

$$\phi_{opt} = \min_{\phi} J(\phi). \qquad (4.8)$$

The performance index can be minimized by wide variety of classical and contemporary optimization methods. When a nonlinear plant model is available and/or the objective function has several local minima, methods for global optimization are used. For example, methods as simulated annealing, pattern search, simulated annealing, and genetic algorithms (GAs) are especially useful. The GA is a method for solving optimization problems that is based on natural selection, the process that drives biological evolution. The GA repeatedly modifies a population of individual solutions. At each step, it selects individuals from the current population to be "parents" and uses them to produce the "children" for the next generation. Over successive generations, the population "evolves" toward an optimal solution. The GA mainly differs from classical optimization algorithms in two ways: it generates a population of points at each iteration. The best point in the population approaches an optimal solution; it selects the next population by computation which uses random number generators. Other advantage of GA is that the initial population is created by uniform random generator. Thus, the designer is not obliged to set the initial PID controller parameters. Due to random generation of initial population, it is recommended to run several times the optimization procedure. Then one can use the best solution obtained.

Example 4.1. Design of PID controller for the cart–pendulum system

Consider the design of a PID controller for the cart–pendulum system presented in Chapter 2. The block diagram of the closed-loop system comprising the plant and the PID controller is shown in Figure 4.3. The cart position and the pendulum angle are measured by 12-bit encoders which generate measurement noises w_1 and w_2.

The state-space equations describing the fourth-order continuous-time plant are

$$\dot{x}(t) = f(x, u, t),$$
$$y(t) = h(x, w, t), \tag{4.9}$$

where $x(t) = \begin{bmatrix} p, & \theta, & \dot{p}, & \dot{\theta} \end{bmatrix}^T$, $y(t) = [p, \ \theta]^T$, $w(t) = [w_1(t), \ w_2(t)]^T$, and $p(t), \dot{p}(t)$, $\theta(t), \dot{\theta}(t)$ are the cart position, cart velocity, pendulum angle, and pendulum angular velocity, respectively. The functions in model (4.9) are given by

$$f(x, u, t) = \begin{bmatrix} \dot{p} \\ \dot{\theta} \\ \dfrac{m^2 l^2 g \sin(\theta)\cos(\theta) - ml I_t \dot{\theta}^2 \sin(\theta) - I_t f_c \dot{p} - f_p lm \cos(\theta)\dot{\theta} + I_t k_F u}{den} \\ \dfrac{M_t mlg \sin(\theta) - m^2 l^2 \sin(\theta)\cos(\theta)\dot{\theta}^2 - f_c ml \cos(\theta)\dot{p}}{den} \\ -\dfrac{f_p M_t \dot{\theta} + ml \cos(\theta) k_F u}{den} \end{bmatrix},$$

$$h(x, w, t) = \begin{bmatrix} 1 & 0 & 0 & 0 \\ 0 & 1 & 0 & 0 \end{bmatrix} \begin{bmatrix} p \\ \theta \\ \dot{p} \\ \dot{\theta} \end{bmatrix} + \begin{bmatrix} w_1(t) \\ w_2(t) \end{bmatrix},$$

where $I_t = I + ml^2$, $M_t = M + m$, $den = M_t I_t - ml^2$, and the plant parameters with their tolerances are given in Table 4.1.

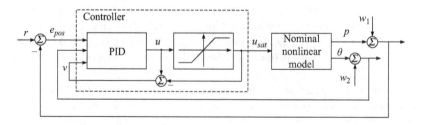

Figure 4.3   PID controller of cart–pendulum system

Table 4.1 Cart–pendulum model parameters and tolerances

Parameter	Description	Value	Units	Tolerance (%)
m	Equivalent pendulum mass	0.104	kg	
M	Equivalent cart mass	0.768	kg	±10.0
l	Distance from the base to the system center of mass	0.174	m	
I	Pendulum moment of inertia	2.83×10^{-3}	kg m^2	±20.0
f_c	Dynamic cart friction coefficient	0.5	N s/m	±20.0
f_p	Rotational friction coefficient	6.65×10^{-5}	N ms/rad	±20.0
k_F	Control force to PWM signal ratio	9.4	N	

The designed PID controller has three inputs and one output. The first input is cart-position error, the second one is the pendulum angle, and the third is the difference between the controller output u and output of actuator saturation mathematical model u_{sat}. The control signal is set up by the following equation:

$$u(z) = \begin{bmatrix} K_{Ppos} & -K_{p\theta} & 0 \\ 0 & -K_{p\theta}K_{i\theta}\dfrac{T_s}{z-1} & \dfrac{T_s}{z-1} \\ \dfrac{K_{Ppos}T_{dpos}N_{pos}(z-1)}{(1+N_{pos}T_s)z-1} & -\dfrac{K_{p\theta}T_{d\theta}N_{\theta}(z-1)}{(1+N_{pos}T_s)z-1} & 0 \end{bmatrix} e(z), \qquad (4.10)$$

where $e(z) = [e_{pos}(z)\theta(z)v(z)]^T$, $e_{pos}(z) = r_{pos}(z) - p(z)$ is the cart position error, $T_s = 0.01$ s is the sampling time, $K_{p\theta}, K_{i\theta}, T_{d\theta}, N_{\theta}, K_{Ppos}, T_{dpos}, N_{pos}$ are the coefficients of multiple-input–single-output PID controller. As can be seen from (4.10), for stabilization of pendulum angle a PID controller with antiwindup mechanism is used, whereas for control of cart position, a PD controller is implemented. The vector with controller parameters is set as

$$\phi = \begin{bmatrix} K_{p\theta}, K_{i\theta}, T_{d\theta}, N_{\theta}, K_{Ppos}, T_{dpos}, N_{pos} \end{bmatrix}. \qquad (4.11)$$

The values of parameters contained in the vector ϕ are obtained by minimizing the performance index (4.7) for

$$Q = \begin{bmatrix} 1 & 0 & 0 \\ 0 & 1 & 0 \\ 0 & 0 & 0 \end{bmatrix}, \qquad R = 0.$$

Due to the fact that we cannot set "good" initial value of vector ϕ and due to the nonlinearity of model (4.9), the minimization of performance index (4.7) is done by GA. Such algorithm is implemented in function ga of MATLAB Global Optimization Toolbox. Initial population of GA is created by random generator. Due to this fact, sometimes the algorithm can return local minima. This possibility decreases significantly if greater number of individuals is used and/or several runs of procedure are done.

The PID controller design for cart–pendulum system is done by the M-file `PID_controller_design` which uses the MATLAB function `ga`. The fitness function (4.7) is calculated by user-defined MATLAB function `err_fun.m`, whereas the cart-position error and pendulum angle are evaluated by the Simulink model `discrete_PID_pendulum.slx`. Note that GA computes successively a fitness function for each individual in the current population. However, if the so-called vectorized fitness function is used, than the algorithm runs very quickly owing to that the fitness function computation is done once simultaneously for all individuals in the current population. Fifteen runs of optimization procedure with initial population of 300 individuals and option "vectorized" for fitness function are performed.

The values of PID controller parameters, performance index (4.7) and number of performed generations for each run of optimization are presented in Table 4.2. As can be seen in runs 1, 3, 8, 11, and 12 almost the same small values of performance index (4.7) are achieved, which is indication for successful minimization. For other runs, the fitness function values are large and optimization procedure does not reach global minima. Moreover, some of these solutions do not stabilize control system. In all runs optimization stops due to that the average change in the fitness value is less than prescribed tolerance, which means finding of global or local minima. The closed-loop system with parameters obtained in 1st, 11th, and 12th runs has real pole whose value exceeds 1 in the last decimal digit, for example, the closed-loop pole for system parameters obtained in 11 run is 1.000000000000003. Obviously, this is a numerical problem that can cause instability of closed-loop system if the controller will be implemented on a microcontroller. Thus, the parameters obtained in the 14th run are chosen as a solution of controller design task. The closed-loop system with PID controller is simulated by the Simulink file `clp_PID_pendulum.slx`.

The PID controller model is shown in Figure 4.4. Its implementation is done by single precision arithmetic taking into account that the controller will be implemented on a processor having floating-point unit.

Table 4.2   PID controller design iterations

Run	n_{gen}	$J(\phi)$	K_{p_θ}	K_{i_θ}	T_{d_θ}	N_θ	$K_{p_{pos}}$	$T_{d_{pos}}$	N_{pos}
1	171	8.73×10^1	2.77	0.50	0.25	30.00	−0.03	0.35	13.35
2	68	1.61×10^8	−7.87	0.13	7.07	2.67	10.63	10.28	28.70
3	83	8.75×10^1	4.85	0.56	0.13	29.98	−0.07	0.09	8.78
4	76	1.61×10^8	−9.70	0.06	3.93	2.02	8.75	9.08	29.14
5	138	1.61×10^8	3.76	−0.12	−7.44	2.43	4.70	9.34	29.19
6	70	1.61×10^8	−3.81	0.12	8.29	2.50	7.81	7.50	29.46
7	84	1.61×10^8	−7.98	0.05	2.82	2.43	8.37	6.73	28.80
8	126	8.77×10^1	6.08	0.59	0.09	30.00	−0.09	0.00	10.70
9	245	1.19×10^8	−8.72	0.07	−0.29	18.03	2.66	3.11	25.30
10	85	1.61×10^8	−5.02	0.16	8.39	3.19	9.02	10.07	29.35
11	166	8.75×10^1	5.28	0.56	0.12	30.00	−0.08	0.07	7.23
12	85	8.76×10^1	5.48	0.58	0.10	30.00	−0.08	0.02	30.00
13	109	1.61×10^8	−3.60	0.17	8.78	3.19	7.54	9.06	29.07
14	112	6.30×10^3	7.30	9.63	1.24	21.49	0.23	−7.28	14.66
15	112	1.38×10^3	−8.13	0.95	−0.44	13.42	8.02	2.07	9.82

Single precision PID controller with antiwindup mechanism

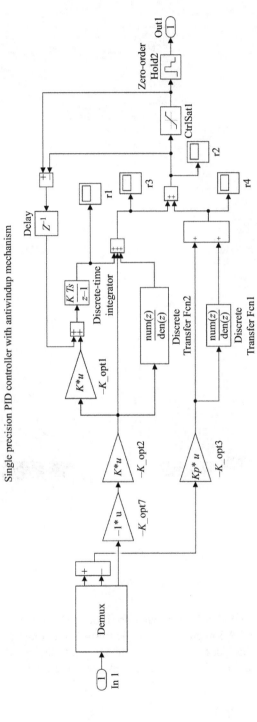

Figure 4.4   Simulink model of PID controller with antiwindup

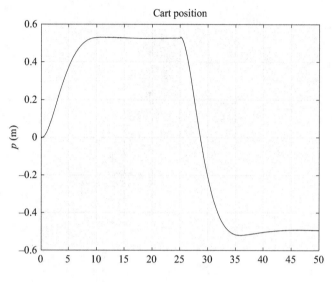

Figure 4.5   Cart position

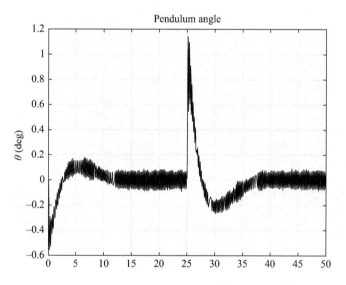

Figure 4.6   Pendulum angle

The results from simulation of nominal closed-loop system for step references of magnitude 0.5 m and −0.5 m are shown in Figures 4.5–4.7. Regarding cart position, the control system has good performance, the settling time is approximately 9 s and the overshoot is negligible. The deviation of pendulum angle is about 1 deg which

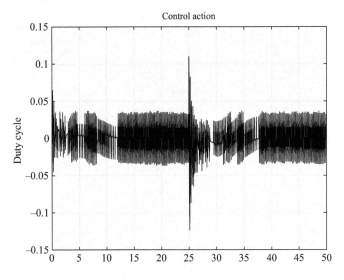

Figure 4.7   Control action

is sufficiently small, but it should be noted that the pendulum angle is affected by the noise due to quantization error. This noise can be seen in control action too, which is undesirable for the actuator. The noise influence of control action may be reduced by using a higher bit analog-to-digital converter (ADC) or by limitation of closed-loop bandwidth, or by filtration with optimal filters like Kalman filter and $\mathcal{H}_\infty$ filter.

The nominal control system has good performance, but the robust stability and robust performance are not guaranteed due to the used approach for PID controller tuning. A robust-stability analysis should be performed. The classical and disk stability margins are obtained as

```
cm =
    GainMargin: [0.131614125528137 1.298114244837111]
    GMFrequency: [1.086189627370737 31.275983301257536]
    PhaseMargin: 6.806361997034553
    PMFrequency: 26.389644674786840
    DelayMargin: [0.450151672729292 1]
    DMFrequency: [26.389644674786840 3.141592653589793e+02]
        Stable: 1

dm =

    GainMargin: [0.896753236434440 1.115133973729581]
    PhaseMargin: [-6.231460596250620 6.231460596250620]
    Frequency: 27.204496764576167
```

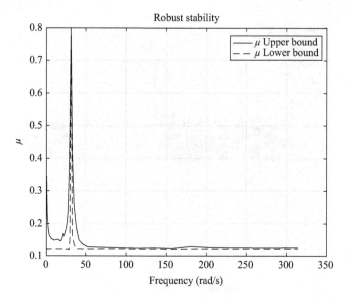

Figure 4.8   Robust stability

The gain margin of 1.298 deg and phase margin of 6.806 deg are sufficiently large. Additional robust-stability analysis is done by M-file `PID_robstab.m`, which produces the report

```
REPORT =

Uncertain system is robustly stable to modeled uncertainty.
 -- It can tolerate up to 125% of the modeled uncertainty.
 -- A destabilizing combination of 128% of the modeled uncertainty
                                                was found.
 -- This combination causes an instability at 31.3 rad/seconds.
 -- Sensitivity with respect to the uncertain elements are:
    'I' is 9%.   Increasing 'I' by 25% leads to a 2% decrease
                                             in the margin.
    'M' is 0%.   Increasing 'M' by 25% leads to a 0% decrease
                                             in the margin.
    'f_c' is 4%.  Increasing 'f_c' by 25% leads to a 1% decrease
                                             in the margin.
    'f_p' is 0%.  Increasing 'f_p' by 25% leads to a 0% decrease
                                             in the margin.
```

The report shows that the control system is robustly stable. The peak value of structured singular value μ which corresponds to the robust stability, is 0.798 (Figure 4.8). Thus, the stability robustness margin is equal to $sm = 1/0.798 = 1.25$ so that the system can tolerate up to 125 percent of the modeled uncertainty. The robust

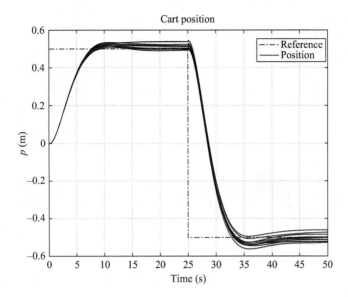

Figure 4.9   Cart position obtained by Monte-Carlo simulation

stability of closed-loop system with nonlinear plant model is illustrated by using the Monte-Carlo analysis. It is done by the program sim_MC_PID.m which obtains the transient responses for ten random combinations of the uncertain parameters given in Table 4.1.

The results from Monte-Carlo simulation are shown in Figures 4.9 and 4.10. It is seen that the closed-loop system with nonlinear model is robustly stable but does not achieve robust performance. For some simulations, the transient response cannot reach steady-state value for 25 s, because of significant influence of noises on control action (Figure 4.11). The worst performance of the cart–pendulum closed-loop system is assessed by the largest possible gain in the frequency domain ("worst case" gain). The worst case analysis is done by the M-file PID_wcp.m which utilizes the function wcgain from Robust Control Toolbox. The structure wcunc containing a combination of uncertain element values which maximize the system gain is substituted in the complementary sensitivity function. The magnitude plot of the closed-loop system in respect to the cart position for 30 random samples of the uncertain parameters and for the worst case gain is shown in Figure 4.12. The obtained closed-loop system bandwidth of 0.42 rad/s for worst case uncertainty is comparatively small. The closed-loop transient responses in respect to the cart position and pendulum angle are given in Figures 4.13 and 4.14, respectively. It is seen that the system has almost the same aperiodic transient responses. The pendulum angle deviations from zero are sufficiently small. The M-file PID_wcp computes also the worst case gain

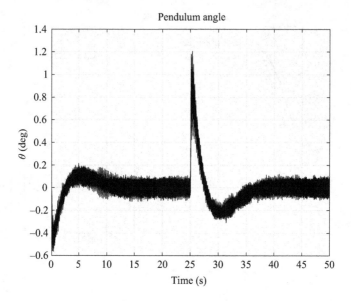

Figure 4.10   Pendulum angle obtained by Monte-Carlo simulation

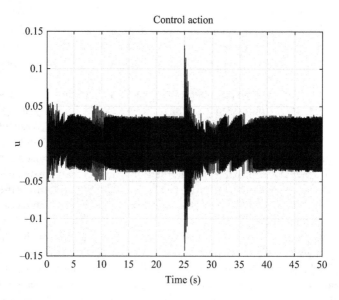

Figure 4.11   Control action obtained by Monte-Carlo simulation

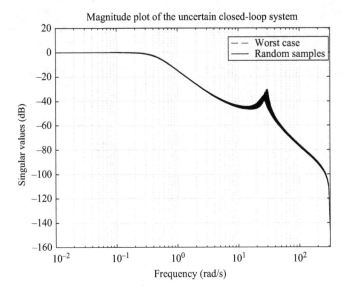

Figure 4.12 Worst case closed-loop system gain

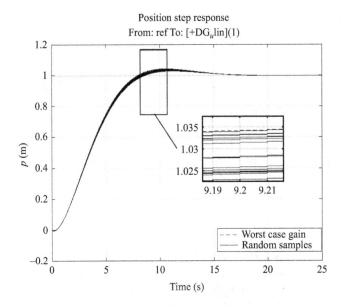

Figure 4.13 Worst case cart position

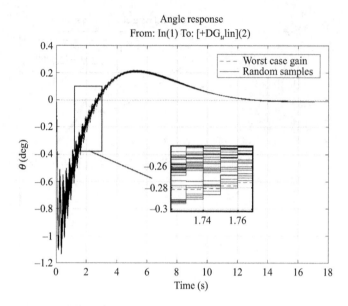

Figure 4.14   Worst case pendulum angle

margin of the closed-loop system using the open-loop transfer matrix $L(s)$. As a result, one obtains

```
wcmarg =

    GainMargin: [0.896631324710085 1.115285594470323]
   PhaseMargin: [-6.239204284065615 6.239204284065615]
     Frequency: 27.085636561297559
         WCUnc: [1x1 struct]
   Sensitivity: [1x1 struct]
```

This means that the maximum allowable gain margin for all possible defined uncertainty ranges is up to 0.8966 and $1.1153 = \pm 0.947$ dB and phase margin variations of ± 6.239 deg. This result is in agreement with the results from robust-stability analysis.

In Figure 4.15, we show the magnitude plot of closed-loop sensitivity of control action to the reference and encoder noises. It is seen that the noise in measuring the pendulum angle θ is more significant than the noise in measuring the car position p. The sensitivity of control action to noise in pendulum angle is significant in wide frequency range especially in high frequencies. The peak value of 7.4 dB means that in this frequency range, the controller will increase a noise in pendulum angle more than two times. Note that this peak is in the range where the noise in θ is significant. This is in agreement with results from Monte-Carlo simulation of closed loop with nonlinear model and leads to undesirable variations of control action (Figure 4.11). ❐

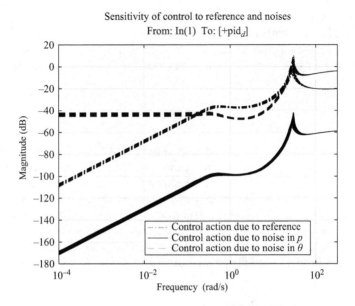

Figure 4.15   Control sensitivity to reference and noises

4.2   LQG controller with integral action

MATLAB files used in this section

Files for basic LQG controller	Description
LQG_design	Design of LQG controller
LQG_robstab	Robust-stability analysis
dfrs_LQG	Frequency responses
sim_LQG	Simulation
sim_MC_LQG	Monte-Carlo simulation

Files for LQG controller with bias compensation	Description
LQG_bias_design	Design of LQG controller
LQG_bias_robstab	Robust-stability analysis
dfrs_LQG_bias	Frequency responses
sim_LQG_bias	Simulation
sim_MC_LQG_bias	Monte-Carlo simulation
LQG_bias_wcp	Worst case gain

In this section, we present two versions of LQG controller with Kalman filter state estimation and integral action. The purpose of these controllers is to ensure accurate tracking and regulation in presence of disturbances and noises. It is shown that the basic version of such controller produce satisfactory results in case of nominal plant model but may have a bad performance when the plant model vary. That is why, a bias compensation is done in case of uncertain plant by adding additional states to the Kalman filter state vector. This may increase significantly the closed-loop system robustness, as demonstrated by examples.

4.2.1　Discrete-time LQG controller

The basic configuration of a closed-loop system with LQG controller comprising LQR and a Kalman filter is shown in Figure 4.16. The LQR is designed to minimize a quadratic performance index, and the Kalman filter produces optimal system state estimate in presence of process noise v and measurement noise w. The system error e is integrated in order to reject constant disturbances and/or achieving of set point regulation properties (zero steady-state error in case of step reference r).

The discrete-time LQG controller design may be presented briefly as follows. Assume that the time-invariant system is described by the difference state and output equations

$$
\begin{aligned}
x(k+1) &= Ax(k) + Bu(k) + v(k), \\
y(k) &= Cx(k) + w(k),
\end{aligned}
\tag{4.12}
$$

where $x(k)$ is the state vector, $y(k)$ is the measurement vector, and $v(k), w(k)$ are the process noise and measurement noise, respectively; $A, B,$ and C are constant matrices with appropriate dimensions. It is assumed that $v(k)$ and $w(k)$ are stationary independent zero-mean Gaussian white noises with covariances

$$
E\left\{v(k)v(j)^T\right\} = V\delta(j-k), \quad E\left\{w(k)w(j)^T\right\} = W\delta(j-k),
$$

where $E\{.\}$ denotes mathematical expectation, $\delta(\cdot)$ is the unit impulse function, and V, W are known covariance matrices.

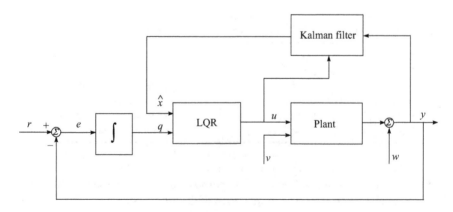

Figure 4.16　LQG control with integral action

The discrete-time LQG problem for the system (4.12) consists of finding the optimal control $u = u_{opt}$ which minimizes the quadratic performance index

$$J = E\left\{\sum_0^\infty x(k)^T Qx(k) + u(k)^T Ru(k)\right\}, \tag{4.13}$$

where Q is a symmetric positive semidefinite matrix and R is a symmetric positive definite matrix.

It is well known that according to the separation principle [44,73,111], the LQG problem is decomposed into two subproblems, namely, LQR design of optimal control law

$$u(k) = -K_{opt}x(k) \tag{4.14}$$

of the deterministic system

$$\begin{aligned} x(k+1) &= Ax(k) + Bu(k), \\ y(k) &= Cx(k), \end{aligned} \tag{4.15}$$

and design of optimal state estimator (*Kalman filter*) (see Section 2.4)

$$\hat{x}(k+1) = A\hat{x}(k) + Bu(k) + L(y(k+1) - CA\hat{x}(k) - CBu(k)) \tag{4.16}$$

of the stochastic system (4.12). The optimal control of the system (4.12) which minimizes (4.13) is then given by the state estimate feedback control law

$$u(k) = -K_{opt}\hat{x}(k), \tag{4.17}$$

which stabilizes the closed-loop system.

The gain matrices K_{opt} and L arising in the solution of the LQG problem are determined by solving two matrix algebraic Riccati equations as follows.

If the matrix pair (A, B) is stabilizable and the matrix pair $(Q^{1/2}, A)$ is detectable, the optimal gain matrix K_{opt} in (4.14) is determined as

$$K_{opt} = (R + B^T SB)^{-1} B^T SA,$$

where S is the positive semidefinite solution of the discrete-time matrix Riccati equation

$$A^T SA - S + Q - A^T SB(R + B^T SB)^{-1} B^T SA = 0.$$

In such case, the resulting closed-loop system is asymptotically stable (the eigenvalues of $A - BK_{opt}$ are inside the unit circle in the complex plane). Moreover, the closed-loop system has guaranteed gain and phase margins for all inputs.

Provided the matrix pair $(A, V^{1/2})$ is stabilizable and the matrix pair (C, A) is detectable, the optimal gain matrix L of the Kalman filter (4.16) is determined as

$$L = PC^T(CPC^T + W)^{-1},$$

where P is the positive semidefinite solution of the matrix Riccati equation

$$APA^T - P + V - APC^T(CPC^T + W)^{-1}CPA^T = 0.$$

In such case, the Kalman filter is stable (the eigenvalues of $(I - LC)A$ are inside the unit circle), i.e., $E\{\hat{x}(k)\} \to E\{x(k)\}$ as far as $k \to \infty$ and the error covariance is given by

$$E\{(x(k) - \hat{x}(k))(x(k) - \hat{x}(k))^T\} = P.$$

Hence, the elements of the matrix P can be used to assess the error covariance of the state estimate $\hat{x}(k)$.

As opposite to the case of LQR control used alone, there are no guaranteed stability margins of the closed-loop system involving LQ regulator and Kalman filter.

The optimal state regulator (4.14) is computed in MATLAB using the function dlqr and the discrete-time Kalman filter is designed using the function kalman. The function kalman designs a Kalman filter of a system with the general description

$$x(k + 1) = Ax(k) + Bu(k) + Gv(k),$$
$$y(k) = Cx(k) + Hv(k) + w(k)$$

where the process noise $v(k)$ and the measurement noise $w(k)$ can be correlated. An alternative to the functions dlqr and kalman is the function lqg which determines the whole LQG controller. The function lqg also has an option to include integral action in the controller.

Consider now how to include integral action in the controller. It is well known that the conventional LQG controller can ensure zero steady-state control system error if the plant gain is exactly known or/and the plant operates without output disturbances. In practice, we always have output disturbances and we often have uncertainty in plant gain. That is why, it is important to implement integral action in the LQG controller.

An approximation of the discrete-time integral of the system error $e(k)$ may be computed by the difference equation

$$q(k + 1) = q(k) + T_s e(k) = q(k) + T_s(r(k) - y(k)), \qquad (4.18)$$

where $q(k)$ has the dimension of $y(k)$ and T_s is the sampling interval. Combining this equation with the system equations (4.15), one obtains the augmented system

$$\begin{aligned}\bar{x}(k + 1) &= \bar{A}\bar{x}(k) + \bar{B}u(k) + \bar{G}r(k) \\ y(k) &= \bar{C}\bar{x}(k)\end{aligned} \qquad (4.19)$$

with state vector

$$\bar{x}(k) = \begin{bmatrix} x(k) \\ q(k) \end{bmatrix}$$

and matrices

$$\bar{A} = \begin{bmatrix} A & 0 \\ -T_s C & I \end{bmatrix}, \quad \bar{B} = \begin{bmatrix} B \\ 0 \end{bmatrix}, \quad \bar{C} = \begin{bmatrix} C & 0 \end{bmatrix}, \quad \bar{G} = \begin{bmatrix} 0 \\ T_s I \end{bmatrix}. \qquad (4.20)$$

It may be proved that the system (4.19) is stabilizable provided the original system (4.12) is stabilizable and if there is no zero of the original system on the unit circle. Hence, the system (4.19) can be optimized in respect to a quadratic performance index considering as a state vector the vector $\bar{x}(k)$. As a result, the optimal control law is obtained in the form

$$u(k) = -K_o\bar{x}(k) = -K_x x(k) - K_q q(k),$$

where the optimal state feedback matrix

$$K_o = \begin{bmatrix} K_x & K_q \end{bmatrix}$$

is partitioned according to the dimensions of the vectors $x(k)$ and $q(k)$. In this way, the optimal control law is obtained as a feedback in the plant and integrator states.

The integral action LQR controller may be designed again by the function dlqr, but instead of matrices A, B one has to use the matrices $\bar{A}, \bar{B}$.

4.2.2 Colored measurement noise

If the process and measurement noises are colored, then the original system dynamics has to be augmented with the dynamics of the corresponding shaping filters. As it is shown below, this may lead to numerical difficulties.

Consider the case of colored measurement noise w. This noise may be represented as an output of the shaping filter

$$\begin{aligned} \psi(k+1) &= A_w \psi(k) + B_w \eta(k), \\ w(k) &= C_w \psi(k) + D_w \eta(k), \end{aligned} \tag{4.21}$$

whose input is a white noise $\eta(k)$ with covariance

$$E\left\{\eta(k)\eta(j)^T\right\} = I\delta(j-k),$$

where the covariance matrix I is an unit matrix of dimension equal to the dimension of $\eta(k)$. Combining (4.12) and (4.21), one obtains

$$\begin{aligned} \begin{bmatrix} x(k+1) \\ \psi(k+1) \end{bmatrix} &= \begin{bmatrix} A & 0 \\ 0 & A_w \end{bmatrix}\begin{bmatrix} x(k) \\ \psi(k) \end{bmatrix} + \begin{bmatrix} B \\ 0 \end{bmatrix}u(k) + \begin{bmatrix} I & 0 \\ 0 & B_w \end{bmatrix}\begin{bmatrix} v(k) \\ \eta(k) \end{bmatrix}, \\ y(k) &= \begin{bmatrix} C & C_w \end{bmatrix}\begin{bmatrix} x(k) \\ \psi(k) \end{bmatrix} + \begin{bmatrix} 0 & 0 \\ 0 & D_w \end{bmatrix}\begin{bmatrix} v(k) \\ \eta(k) \end{bmatrix}. \end{aligned} \tag{4.22}$$

Denoting

$$\tilde{x}(k) = \begin{bmatrix} x(k) \\ \psi(k) \end{bmatrix}, \quad \tilde{v}(k) = \begin{bmatrix} v(k) \\ \eta(k) \end{bmatrix}$$

and

$$\tilde{A} = \begin{bmatrix} A & 0 \\ 0 & A_w \end{bmatrix}, \quad \tilde{B} = \begin{bmatrix} B \\ 0 \end{bmatrix}, \quad \tilde{C} = \begin{bmatrix} C & C_w \end{bmatrix},$$

$$\tilde{G} = \begin{bmatrix} I & 0 \\ 0 & B_w \end{bmatrix}, \quad \tilde{H} = \begin{bmatrix} 0 & 0 \\ 0 & D_w \end{bmatrix}, \tag{4.23}$$

(4.22) can be written as

$$\tilde{x}(k+1) = \tilde{A}\tilde{x}(k) + \tilde{B}u(k) + \tilde{G}\tilde{v}(k),$$
$$y(k) = \tilde{C}\tilde{x}(k) + \tilde{H}\tilde{v}(k),$$
(4.24)

where

$$E\left\{\tilde{v}(k)\tilde{v}(k)^T\right\} = E\left\{\begin{bmatrix} v(k) \\ \eta(k) \end{bmatrix}\begin{bmatrix} v(k)^T & \eta(k)^T \end{bmatrix}^T\right\} = \begin{bmatrix} V & 0 \\ 0 & I \end{bmatrix}.$$

In this way, we obtain the augmented system (4.24) which has no measurement noise. It is not possible to design Kalman filter for such systems using the solution of Riccati equation since the measurement noise variance matrix W should be nonsingular matrix. Alternative solution, which does not rely on the augmentation of the state vector, is described in [44, Section 2.7.3], [78, Section 7.2.3]. A simple solution may be obtained as follows. Instead of estimating the state of system (4.24), the Kalman filter is designed to estimate the state of the system

$$\tilde{x}(k+1) = \tilde{A}\tilde{x}(k) + \tilde{B}u(k) + \tilde{G}\tilde{v}(k),$$
$$y(k) = \tilde{C}\tilde{x}(k) + \tilde{H}\tilde{v}(k) + \tilde{w}(k),$$
(4.25)

where the matrix $W = E\left\{\tilde{w}(k)\tilde{w}(k)^T\right\}$ is chosen in the form $10^{-p}I$ for some sufficiently large positive p. Obviously, this is equivalent to introducing a small measurement noise in the system equations. Note that the matrix W chosen in this way is perfectly conditioned so that its inversion will not pose a problem. The price for this simple solution is that the estimate produced by the Kalman filter is no more optimal, since the filter should fight an additional fictitious noise. Fortunately, this noise may be chosen sufficiently small.

Example 4.2. Design of a LQG controller for the cart–pendulum system
Consider the design of a LQG controller with integral action for the cart–pendulum system presented in Chapter 2. The block diagram of the closed-loop system comprising the plant and the controller is shown in Figure 4.17. The plant input is the control signal to the cart motor actuator, and the plant outputs are cart position p and pendulum angle θ. The control system goal is to keep the cart position equal to the reference in presence of noises and disturbances stabilizing in the same time the pendulum in the inverted position. The variables p and θ are measured by 12-bit encoders which generate measurement noises, the cart velocity $\dot{p}$ and pendulum angular velocity $\dot{\theta}$ being not measured. This makes necessary to find an estimate of system state by using a Kalman filter in the controller. The cart position error is integrated by a discrete-time integrator and included as an additional system state in the LQR design.

The state-space equations describing the fourth-order continuous-time plant are

$$x_c(t) = A_c x_c(t) + B_c u(t),$$
$$y_c(t) = C_c x_c(t),$$

where

$$x_c(t) = [p \ \theta \ \dot{p} \ \dot{\theta}]^T, \quad y_c(t) = [p \ \theta]^T.$$

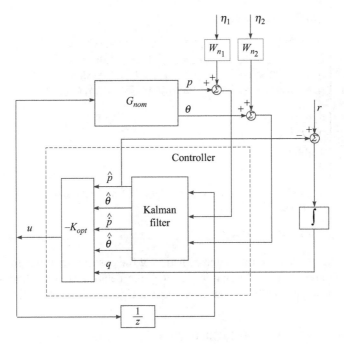

Figure 4.17   LQG controller of cart–pendulum system

The matrices A_c, B_c, C_c are given by

$$A_c = \begin{bmatrix} 0 & 0 & 1 & 0 \\ 0 & 0 & 0 & 1 \\ 0 & m^2 l^2 g/den & -f_c I_t/den & -f_p lm/den \\ 0 & M_t mgl/den & -f_c ml/den & -f_p M_t/den \end{bmatrix},$$

$$B_c = k_F \begin{bmatrix} 0 \\ 0 \\ I_t/den \\ lm/den \end{bmatrix},$$

$$C_c = \begin{bmatrix} 1 & 0 & 0 & 0 \\ 0 & 1 & 0 & 0 \end{bmatrix},$$

where the plant parameters with their tolerances are given in Table 4.1.

The LQG controller design for cart–pendulum system is done by the M-file `LQG_design`.

The nominal discrete-time plant model (4.15), obtained for sampling interval $T_s = 0.01$ s, has a state vector

$$x(k) = [p(k)\ \theta(k)\ \dot{p}(k)\ \dot{\theta}(k)]^T$$

and matrices (to four decimal digits)

$$
A = \begin{bmatrix}
1 & 3.2815 \times 10^{-4} & 9.9695 \times 10^{-3} & 9.7135 \times 10^{-8} \\
0 & 1.0016 \times 10^{0} & -9.2424 \times 10^{-5} & 1.0005 \times 10^{-2} \\
0 & 6.5578 \times 10^{-3} & 9.9390 \times 10^{-1} & 3.0358 \times 10^{-5} \\
0 & 3.1691 \times 10^{-1} & -1.8470 \times 10^{-2} & 1.0015 \times 10^{0}
\end{bmatrix},
$$

$$
B = \begin{bmatrix}
7.8521 \times 10^{-4} \\
2.3771 \times 10^{-3} \\
1.5689 \times 10^{-1} \\
4.7506 \times 10^{-1}
\end{bmatrix}, \quad
C = \begin{bmatrix}
1 & 0 & 0 & 0 \\
0 & 1 & 0 & 0
\end{bmatrix}.
$$

The augmented plant discrete-time model of fifth order used in the LQR design has the form (4.19) with state vector

$$
\bar{x}(k) = [p(k)\ \theta(k)\ \dot{p}(k)\ \dot{\theta}(k)\ q(k)]^{T},
$$

where $q(k)$ is the discrete-time approximation of the position error integral obtained by the difference (4.18). The matrices $\bar{A}, \bar{B}, \bar{C}$ are determined according to (4.20). Note that the augmented state matrix $\bar{A}$ has two eigenvalues on the unit circle. The weighting matrices

$$
Q = \mathrm{diag}(1, 1{,}000, 100, 1{,}000, 10{,}000), \quad R = 1{,}000
$$

are chosen so as to obtain satisfactory closed-loop transient responses with control action less than 0.5.

The LQR design, done for matrices $\bar{A}, \bar{B}, Q, R$, produces the optimal feedback matrix

$$
K_{o} = [-2.9779, 7.2230, -1.8493, 1.5156, 2.4131].
$$

The design of the Kalman filter takes into account the quantization noises n_1 and n_2 produced by the encoders which measure the cart position and the pendulum angle, respectively. As shown in Chapter 2, n_1 and n_2 may be considered as white noises with variances equal to

$$
V_1 = \frac{0.235}{2^{12}} \frac{1}{\sqrt{12}}, \quad V_2 = \frac{2\pi}{2^{12}} \frac{1}{\sqrt{12}},
$$

respectively. However, in the Kalman filter design, we assume that the noises in p and θ are generated by first-order shaping filters with unit variance input white noises η_1 and η_2, respectively. The shaping filters transfer functions are chosen as

$$
W_{n_1}(s) = V_1 \frac{5s + 50}{0.01s + 1}, \quad W_{n_2}(s) = V_2 \frac{0.5s + 10}{0.01s + 1}.
$$

In the low-frequency range, these shaping filters produce outputs which approximate the encoder noises n_1, n_2 (Figure 4.18). In the high-frequency range, the intensities of the filter outputs increase to take into account other noises in the physical system like the actuator noise.

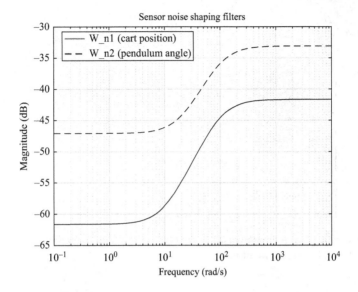

Figure 4.18   Magnitude plots of noise shaping filters

In the given case, we have that

$$\tilde{v} = \eta = \begin{bmatrix} \eta_1 \\ \eta_2 \end{bmatrix}$$

and from (4.24), we obtain the sixth-order system

$$\tilde{x}(k+1) = \tilde{A}\tilde{x}(k) + \tilde{B}u(k) + \tilde{G}\eta(k),$$
$$y(k) = \tilde{C}\tilde{x}(k) + \tilde{H}\eta(k),$$

where

$$\tilde{x}(k) = \begin{bmatrix} x(k) \\ \psi(k) \end{bmatrix},$$

$\psi(k)$ is a vector containing the shaping filters states and

$$E\left\{\eta(k)\eta(k)^T\right\} = I = \begin{bmatrix} 1 & 0 \\ 0 & 1 \end{bmatrix}.$$

According to (4.23), the system matrices are determined as

$$\tilde{A} = \begin{bmatrix} & & & & 0 & 0 \\ & & & & 0 & 0 \\ & & A & & 0 & 0 \\ & & & & 0 & 0 \\ 0 & 0 & 0 & 0 & 3.6788 \times 10^{-1} & 0 \\ 0 & 0 & 0 & 0 & 0 & 3.6788 \times 10^{-1} \end{bmatrix},$$

$$\tilde{B} = \begin{bmatrix} B \\ \hline 0 \\ 0 \end{bmatrix}, \quad \tilde{C} = \begin{bmatrix} C & \begin{matrix} -7.4530 \times 10^{-1} & 0 \\ 0 & -1.7713 \times 10^{0} \end{matrix} \end{bmatrix},$$

$$\tilde{G} = \begin{bmatrix} 0 & 0 \\ 0 & 0 \\ 0 & 0 \\ 0 & 0 \\ \hline 6.3212 \times 10^{-3} & 0 \\ 0 & 6.3212 \times 10^{-3} \end{bmatrix},$$

$$\tilde{H} = \begin{bmatrix} 8.2811 \times 10^{-3} & 0 \\ 0 & 2.2141 \times 10^{-2} \end{bmatrix}.$$

Since there is no formal measurement noise in the system equations, the variance matrix W of the measurement noise is chosen as

$$W = 10^{-3} \begin{bmatrix} 1 & 0 \\ 0 & 1 \end{bmatrix}.$$

The Kalman filter designed is of sixth order. Its optimal gain matrix is

$$L = \begin{bmatrix} 4.0253 \times 10^{-5} & 1.7184 \times 10^{-3} \\ 2.1381 \times 10^{-3} & 9.1278 \times 10^{-2} \\ 2.2563 \times 10^{-4} & 9.6325 \times 10^{-3} \\ 1.1985 \times 10^{-2} & 5.1165 \times 10^{-1} \\ 3.6716 \times 10^{-2} & -9.5254 \times 10^{-5} \\ -2.7569 \times 10^{-4} & 6.0634 \times 10^{-2} \end{bmatrix}.$$

The diagonal elements of the covariance matrix P are

$$\{4.47 \times 10^{-8}, 1.26 \times 10^{-4}, 1.40 \times 10^{-6}, 3.96 \times 10^{-3}, 4.45 \times 10^{-5}, 3.86 \times 10^{-5}\}$$

and can be used to determine error bounds on the estimate $\hat{x}(k)$. In particular, the estimate of $\dot{\theta}$ has the largest error bound corresponding to mean square error equal to 0.0629 rad/s.

The closed-loop system with LQG controller is simulated by the Simulink file clp_LQG.slx. The controller model is shown in Figure 4.19. The controller involves Kalman filter, optimal feedback matrix, and discrete-time position error integrator. The presence of unit sample delay element reflects the fact that in the estimation at the moment k, we may use the value of control action u at the moment $k-1$. (The lack of the delay element will lead to the appearing of algebraic loop in the simulation diagram.) The implementation of the Kalman filter is done in state-space using single precision taking into account that the controller will be implemented on a processor having floating-point unit (Figure 4.20).

The results from simulation of nominal closed-loop system for step references of magnitude 0.5 and −0.5 m are shown in Figures 4.21 and 4.22. The steady-state error of cart position is zero due to the using of integral action controller. It should be noted that although the cart position and pendulum angle measurements are discontinuous

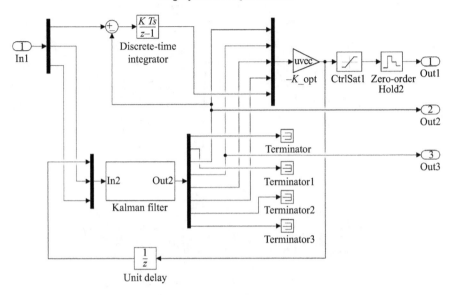

Figure 4.19 *Simulink model of the LQG controller*

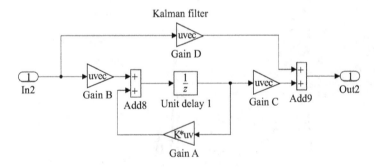

Figure 4.20 *Simulink model of the Kalman filter*

functions due to quantization errors, the corresponding estimates obtained by the Kalman filter are smooth. The control action in case of LQG regulator is shown in Figure 4.23).

Consider now the robustness properties of the closed-loop system. The classical and disk stability margins are obtained as

```
cm =

     GainMargin: [2.1494 9.0235e+03]
    GMFrequency: [2.2527 87.9130]
    PhaseMargin: 60.6182
    PMFrequency: 0.8368
    DelayMargin: 126.4363
    DMFrequency: 0.8368
         Stable: 1
```

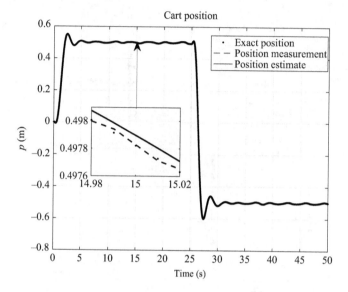

Figure 4.21   Cart position

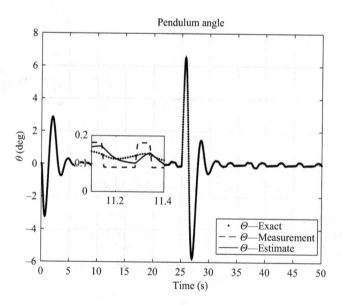

Figure 4.22   Pendulum angle

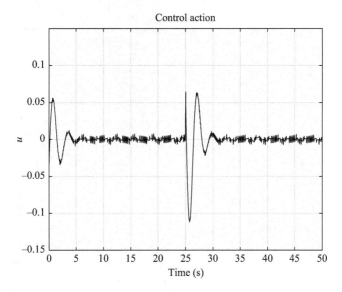

Figure 4.23   Control action

```
dm =

    GainMargin: [0.4802 2.0826]
   PhaseMargin: [-38.7026 38.7026]
     Frequency: 2.0360
```

Clearly, the gain and phase margins are satisfactory. However, the robust-stability analysis of uncertain system, done by the M-file LQG_robstab produces the report

```
report =

Assuming nominal UFRD system is stable...
Uncertain system is possibly not robustly stable to modeled
                                              uncertainty.
 -- It can tolerate up to 91% of the modeled uncertainty.
 -- A destabilizing combination of 104% of the modeled uncertainty
                                              was found.
 -- This combination causes an instability at 3.02 rad/seconds.
 -- Sensitivity with respect to the uncertain elements are:
    'I' is 0%.  Increasing 'I' by 25% leads to a 0% decrease
                                        in the margin.
    'M' is 3%.  Increasing 'M' by 25% leads to a 1% decrease
                                        in the margin.
    'f_c' is 7%.  Increasing 'f_c' by 25% leads to a 2% decrease
                                        in the margin.
    'f_p' is 0%.  Increasing 'f_p' by 25% leads to a 0% decrease
                                        in the margin.
```

The report shows that under the parameter tolerances given in Table 4.1 the closed-loop system is not robustly stable. In the given case, the peak value of the structured

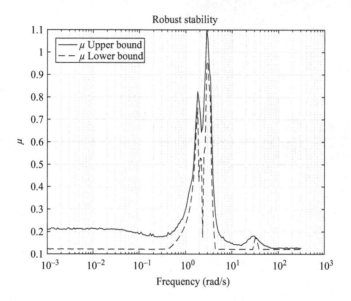

Figure 4.24   Robust stability

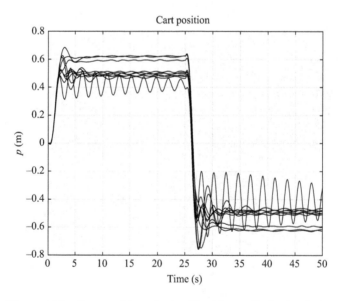

Figure 4.25   Cart position obtained by Monte-Carlo simulation

singular value of μ, which corresponds to robust-stability analysis, is greater than 1 (Figure 4.24).

The poor robustness properties of the closed-loop system are illustrated by the Monte-Carlo analysis shown in Figure 4.25. This analysis is done by the program sim_MC_LQG which obtains the transient responses for ten random combinations of the uncertain parameters given in Table 4.1. The large steady-state errors are due to the large errors of state estimates produced by the Kalman filter when the model varies (the Kalman filter is designed for the nominal plant model). Note that there exist some uncertain parameter combinations for which the closed-loop system is even unstable, as revealed by the robust-stability analysis. ☐

4.2.3 LQG controller with bias compensation

The steady-state errors of uncertain closed-loop systems with LQG controllers may be removed by adding bias components to the state vector which are estimated by the Kalman filter and used to compensate the estimation errors when the plant model varies. This approach also improves the overall robustness of the closed-loop system.

Assume that we desire to remove the steady-state error in the variable $x_i(k)$ for some i. We replace the ith component of the state vector by the sum $x_i(k) + \beta(k)$, where $\beta(k)$ accounts for the bias in $x_i(k)$. The variable β can be modeled as random walk process in the form

$$\beta(k+1) = \beta(k) + g_b \xi(k) \tag{4.26}$$

for a white noise input $\xi(k)$ with $E\{\xi(k)\xi(j)\} = 1\delta(k-j)$ and some small constant g_b. Once we obtain an estimate $x_i(k) + \beta(k)$, it may be used in the computation of optimal control. Therefore, the Kalman filter in the given case has to be designed for the system

$$\begin{bmatrix} x(k+1) \\ \beta(k+1) \end{bmatrix} = \begin{bmatrix} \tilde{A} & \tilde{a} \\ 0 & 1 \end{bmatrix} \begin{bmatrix} x(k) \\ \beta(k) \end{bmatrix} + \begin{bmatrix} \tilde{B} \\ 0 \end{bmatrix} u(k) + \begin{bmatrix} \tilde{G} & 0 \\ 0 & g_b \end{bmatrix} \begin{bmatrix} \eta(k) \\ \xi(k) \end{bmatrix},$$

$$y(k) = \begin{bmatrix} \tilde{C} & \begin{matrix} 0 \\ \vdots \\ 0 \end{matrix} \end{bmatrix} \begin{bmatrix} x(k) \\ \beta(k) \end{bmatrix} + \begin{bmatrix} \tilde{H} & \begin{matrix} 0 \\ \vdots \\ 0 \end{matrix} \end{bmatrix} \begin{bmatrix} \eta(k) \\ \xi(k) \end{bmatrix}, \tag{4.27}$$

where

$$\tilde{a} = \begin{bmatrix} 0 \\ \vdots \\ 1 \\ \vdots \\ 0 \end{bmatrix} \leftarrow i$$

and

$$E\begin{bmatrix} \eta(k) \\ \xi(k) \end{bmatrix} [\eta(j)^T \xi(j)^T] = \begin{bmatrix} I & 0 \\ 0 & 1 \end{bmatrix} \delta(k-j).$$

The steady-state errors of other state vector components may be removed in a similar way provided the corresponding augmented system is detectable.

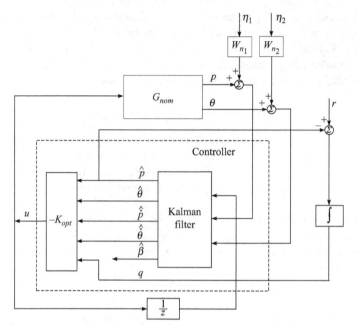

Figure 4.26   LQG controller of cart–pendulum system with bias compensation

Example 4.3. Design of LQG controller with bias compensation for the cart–pendulum system

Consider how to design LQG controller with integral action for the cart–pendulum system removing the steady-state error of cart position.

The LQG controller is designed by the M-file `LQG_bias_design`. The Kalman filter is designed using (4.27) for $i = 1$ and taking

$$V = \begin{bmatrix} \eta(k) \\ \xi(k) \end{bmatrix} [\eta(k)^T \quad \xi(k)] = I_3, \quad W = 10^{-3} \begin{bmatrix} 1 & 0 \\ 0 & 1 \end{bmatrix}.$$

The constant g_b is determined experimentally as 0.01. The filter obtained is of seventh order with optimal gain matrix

$$L = \begin{bmatrix} 7.4207 \times 10^{-1} & 2.2698 \times 10^{-5} \\ 1.2331 \times 10^{-5} & 9.1312 \times 10^{-2} \\ 1.3012 \times 10^{-6} & 9.6361 \times 10^{-3} \\ 6.9118 \times 10^{-5} & 5.1184 \times 10^{-1} \\ 1.0891 \times 10^{-2} & -3.6267 \times 10^{-7} \\ -1.5899 \times 10^{-6} & 6.0630 \times 10^{-2} \\ 2.0272 \times 10^{-1} & -8.0760 \times 10^{-5} \end{bmatrix}.$$

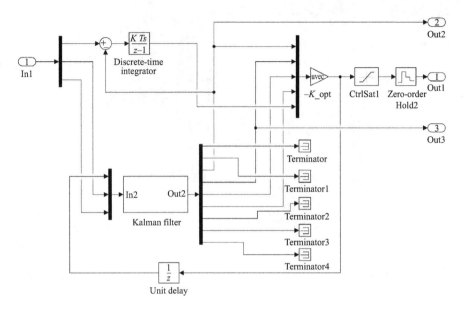

Figure 4.27   Simulink model of the LQG controller with bias compensation

The block diagram of the closed-loop system for the case of bias compensation is shown in Figure 4.26. The corresponding Simulink controller model is displayed in Figure 4.27.

The results from Monte-Carlo simulation of the closed-loop system, obtained by the M-file sim_MC_LQG_bias, are shown in Figures 4.28–4.30. It is seen that the cart position has not steady-state error.

The frequency response plot of the structured singular value μ in case of bias compensation is shown in Figure 4.31. The peak value of μ is equal to 0.892 which guarantees that the closed-loop system achieves robust stability, i.e., for each combination of uncertain plant parameters the system will remain stable. The stability robustness margin is equal to $sm = 1/0.892 = 1.12$ so that the system can tolerate up to 112 percent of the modeled uncertainty.

The worst performance of the cart–pendulum closed-loop system is assessed by the largest possible gain in the frequency domain ("worst case" gain). The worst case analysis is done by the M-file LQG_bias_wcp which utilizes the function wcgain from Robust Control Toolbox™. The structure wcunc containing a combination of uncertain element values which maximize the system gain is substituted in the complementary sensitivity function.

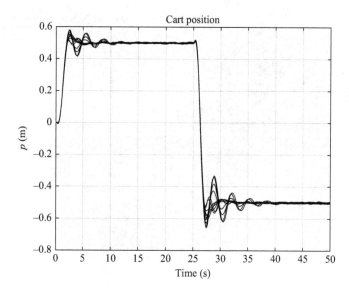

Figure 4.28   Monte-Carlo simulation in case of bias compensation. Cart position

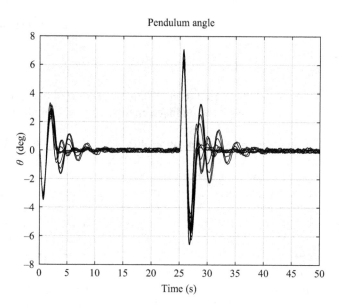

Figure 4.29   Monte-Carlo simulation in case of bias compensation.
Pendulum angle

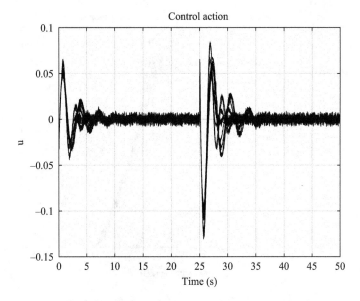

Figure 4.30   Monte-Carlo simulation in case of bias compensation. Control action

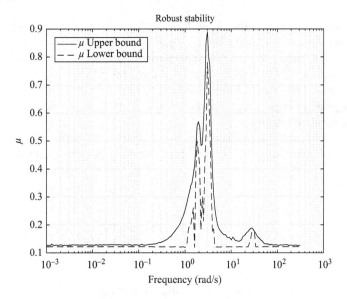

Figure 4.31   Robust stability in case of bias compensation

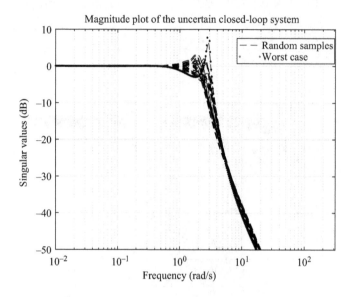

Figure 4.32   Worst case closed-loop system gain

The magnitude plot of the closed-loop system in respect to the cart position for 30 random samples of the uncertain parameters and for the worst case gain is shown in Figure 4.32. The peak of the magnitude response is 10.02 dB which leads to oscillatory transient responses. The worst case bandwidth is 1.99 rad/s.

The closed-loop transient responses in respect to the cart position and pendulum angle are given in Figures 4.33 and 4.34, respectively. It is seen that the closed-loop system is close to instability.

The M-file LQG_bias_wcp computes also the worst case gain margin of the closed-loop system using the loop transfer matrix $L(s)$. As a result, one obtains

```
wcmarg =

    GainMargin: [0.7823 1.2782]
   PhaseMargin: [-13.9246 13.9246]
     Frequency: 2.9152
         WCUnc: [1x1 struct]
   Sensitivity: [1x1 struct]
```

Hence, even in the worst case, the phase margin is about 13.9 deg which is in agreement with the results from robust-stability analysis.

In Figure 4.35, we show the magnitude plot of control action closed-loop sensitivity to reference and encoder noises. It is seen that the noise in measuring the

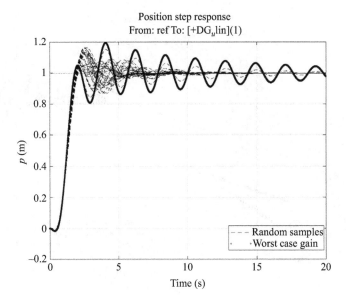

Figure 4.33   Worst case cart position

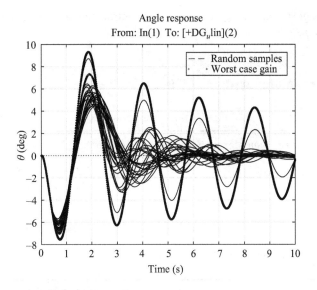

Figure 4.34   Worst case pendulum angle

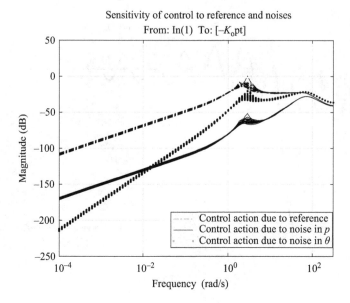

Figure 4.35   Control sensitivity to reference and noises

pendulum angle θ has a stronger influence on control than the noise in measuring the cart position p.   ❏

4.3   LQ regulator with $\mathscr{H}_\infty$ filter

MATLAB files used in this section

Files for basic LQ regulator with $\mathscr{H}_\infty$ filter	Description
LQR_Hinf_design	Design of LQ regulator and $\mathscr{H}_\infty$ filter
LQR_Hinf_robstab	Robust-stability analysis
dfrs_LQR_Hinf	Frequency responses
sim_LQR_Hinf	Simulation
sim_MC_LQR_Hinf	Monte-Carlo simulation
Files for LQ regulator with $\mathscr{H}_\infty$ filter and bias compensation	**Description**
LQR_Hinf_bias_design	Design of LQ regulator and $\mathscr{H}_\infty$ filter
LQR_Hinf_bias_robstab	Robust-stability analysis
dfrs_LQR_Hinf_bias	Frequency responses
sim_LQR_Hinf_bias	Simulation
sim_MC_LQR_Hinf_bias	Monte-Carlo simulation
LQR_Hinf_bias_wcp	Worst case gain

In this section, we present a LQR with integral action which uses a state estimate obtained by the so-called $\mathcal{H}_\infty$ filter. Since the plant uncertainty may lead to steady-state errors, additional filter states are used in order to remove the estimate bias, similarly to the case of using Kalman filter. It is demonstrated by the example of cart–pendulum system that implementing such controller makes possible to obtain better closed-loop robustness in comparison to the case of Kalman filter implementation.

4.3.1 Discrete-time $\mathcal{H}_\infty$ filter

Consider the time-invariant discrete-time system

$$
\begin{aligned}
x(k+1) &= Ax(k) + Bu(k) + v(k), \\
y(k) &= Cx(k) + w(k),
\end{aligned}
\tag{4.28}
$$

where $v(k)$ and $w(k)$ are noisy terms. These noise terms may be random with possibly unknown statistics and nonzero mean. A distinct feature of the $\mathcal{H}_\infty$ filter, used to obtain an estimate of the system (4.28), is that $v(k)$ and $w(k)$ may be deterministic disturbances. In this way, the $\mathcal{H}_\infty$ filter does not make any assumptions about the statistics of the process and measurement noise, although this information can be used in the design of this filter if it is available.

Further on by

$$
\|x\| = \sqrt{x^T x}
$$

and

$$
\|x\|_Q = \sqrt{x^T Q x},
$$

we denote the two-norm of a vector x and the Q-weighted norm of x, respectively, where Q is a positive definite matrix.

Our goal is to find an estimate $\hat{x}(k)$ of the state vector $x(k)$ such that

$$
J = \lim_{N \to \infty} \frac{\sum_{k=0}^{N-1} \|x(k) - \hat{x}(k)\|^2}{\sum_{k=0}^{N-1} \left(\|v(k)\|_{V^{-1}}^2 + \|w(k)\|_{W^{-1}}^2 \right)} < \gamma,
\tag{4.29}
$$

where V and W are symmetric positive definite matrices chosen by the designer and γ is a small positive scalar, respectively.

The inequality (4.29) can be interpreted from the game theory point of view as follows [78, Section 11.3]. The $\mathcal{H}_\infty$ filter design is considered as a game between the nature and the designer. The nature's ultimate goal is to maximize the estimation error $(x(k) - \hat{x}(k))$ by clever choice of disturbances $v(k)$ and $w(k)$. The form of the cost function J prevents nature from using infinite magnitudes for $v(k)$ and $w(k)$. Instead, nature is assumed to actively seeks to degrade the state estimate as much as possible by choosing appropriate signals $v(k), w(k)$ maximizing the cost function J. Thus, the $\mathcal{H}_\infty$ filter is a worst case filter in the sense that $v(k), w(k)$ will be chosen by nature to maximize the cost function. Likewise, the designer must be clever in finding an estimation strategy to minimize $(x(k) - \hat{x}(k))$. This makes the $\mathcal{H}_\infty$ filter design a minimax problem.

The matrices V and W play an analogous role as the corresponding covariance matrices in Kalman filter design. If the process noise $v(k)$ and measurement noise $w(k)$ are zero-mean white noises with covariance matrices V and W, respectively, then these matrices may be used in the design of $\mathcal{H}_\infty$ filter. In the case of deterministic disturbances, the elements of V and W may be chosen according to the relative significance of the components of $v(k)$ and $w(k)$, respectively. If, for instance, the designer knows a priori that the second element of the disturbance $v(k)$ is large, then $V(2,2)$ should be chosen to be large relative to the other elements of V. The filter properties also depend on the value of the scalar parameter γ. Decreasing the value of γ increases the estimate accuracy but may lead to bad estimate response or even to filter instability.

The optimal $\mathcal{H}_\infty$ filter has the form

$$\hat{x}(k+1) = A\hat{x}(k) + Bu(k) + AL(y(k) - C\hat{x}(k)), \tag{4.30}$$

where the filter gain matrix L is determined as [78]

$$L = P[I - \gamma^{-1}P + C^T W^{-1} CP]^{-1} C^T W^{-1}$$

and P is the positive definite solution of the discrete-time matrix algebraic Riccati equation

$$APA^T - P + V - AP\left[(C^T W^{-1} C - \gamma^{-1} I)^{-1} + P\right]^{-1} PA^T = 0.$$

This equation may be solved numerically by the function `dare` from MATLAB. The matrix P will be a solution to the $\mathcal{H}_\infty$ filter problem if the following condition holds:

$$P^{-1} - \gamma^{-1} I + C^T W^{-1} C > 0.$$

This condition imposes a lower bound on γ which may be found numerically by an iterative process. The $\mathcal{H}_\infty$ filter found in this way is stable, i.e., the eigenvalues of the matrix $A(I - LC)$ are inside the unit circle.

It is easy to prove that if we set $\gamma \to \infty$, then the $\mathcal{H}_\infty$ filter reduces to the Kalman filter. That is why the Kalman filter may be considered as $\mathcal{H}_\infty$ filter when the performance index J in (4.29) is set equal to ∞. Hence, we come to the conclusion that although the Kalman filter minimizes the variance of the estimation error, it does not provide any guarantee as far as limiting the worst case estimation error.

Comparing the $\mathcal{H}_\infty$ filter with the Kalman filter, we can see that the $\mathcal{H}_\infty$ filter is a robust version of the Kalman filter. In this respect, the $\mathcal{H}_\infty$ filter theory shows the optimal way to robustify the Kalman filter [78].

Consider now the case of colored measurement noise $w(k)$. According to (4.24), the augmented plant equations in this case are represented as

$$\begin{aligned} x(k+1) &= Ax(k) + Bu(k) + Gv(k), \\ y(k) &= Cx(k) + Hv(k), \end{aligned} \tag{4.31}$$

where the elements of matrices G and H depend on the parameters of the shaping filters. Equation (4.31) has the form of (4.28) with process and measurement noises given by

$$\tilde{v}(k) = Gv(k), \quad \tilde{w}(k) = Hv(k).$$

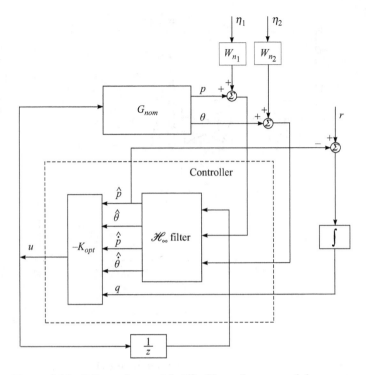

Figure 4.36 LQ regulator with $\mathcal{H}_\infty$ filter of cart–pendulum system

The covariance matrices of these noises are

$$\tilde{V} = E\left\{\tilde{v}(k)\tilde{v}(k)^T\right\} = GE\left\{v(k)v(k)^T\right\}G^T = GVG^T,$$
$$\tilde{W} = E\left\{\tilde{w}(k)\tilde{w}(k)^T\right\} = HE\left\{w(k)w(k)^T\right\}H^T = HVH^T,$$

respectively. If V is a positive definite matrix, but the matrices G and H are not of full row rank, then the matrices $\tilde{V}, \tilde{W}$ will be positive semidefinite and cannot be inverted as required by (4.29). This situation arises, for example, when the dimension of $v(k)$ is less than the dimension of $x(k)$ and $y(k)$. Provided the pair (A, G) is stabilizable and the pair (C, A) is detectable, in this case it is possible to find a stable $\mathcal{H}_\infty$ filter which corresponds to a modified cost function J.

Example 4.4. Design of LQ regulator with $\mathcal{H}_\infty$ filter of the cart–pendulum system

Consider again the cart–pendulum system whose controller design implementing Kalman filter was presented in Examples 4.2 and 4.3. In the given case, we shall make use of the LQ regulator with integral action designed in Example 4.2 but instead of the Kalman filter we shall implement the $\mathcal{H}_\infty$ filter.

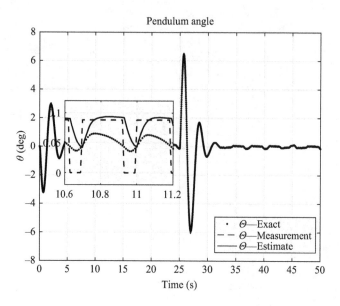

Figure 4.37   Pendulum angle for $\gamma = 0.6$

The block diagram of the closed-loop system with LQ regulator and $\mathcal{H}_\infty$ filter is given in Figure 4.36.

In the given case, the covariance matrix V which corresponds to the car position and pendulum angle encoder noises, is given by $V = I_2$, the matrix $\tilde{V} = GVG^T$ is positive semidefinite and the matrix $\tilde{W} = HVH^T$ is positive definite. The numerical experiments show that in the given case stable $\mathcal{H}_\infty$ filter exists for values of γ greater than $\gamma_{min} = 0.0270267230$.

In Figures 4.37–4.39, we show the transient responses of the pendulum angle θ for three $\mathcal{H}_\infty$ filters designed for $\gamma = 0.6, 0.035, 0.0276$. Due to the discontinuous character of the measurements, caused by the quantization, small transient responses appear in the computed estimate of θ. With the decreasing of γ, the magnitude of these estimate oscillations increases making the estimate unacceptable for $\gamma = 0.0276$. That is why as an appropriate value of γ, we take $\gamma = 0.035$. The $\mathcal{H}_\infty$ filter gain matrix for this value of γ is (to four digits)

$$L = \begin{bmatrix} 1.0296 \times 10^{-3} & 8.2516 \times 10^{-3} \\ 5.4689 \times 10^{-2} & 4.3830 \times 10^{-1} \\ 5.7713 \times 10^{-3} & 4.6253 \times 10^{-2} \\ 3.0655 \times 10^{-1} & 2.4568 \times 10^{0} \\ -3.5463 \times 10^{-1} & 3.9391 \times 10^{-3} \\ 9.4215 \times 10^{-3} & -5.0120 \times 10^{-2} \end{bmatrix}.$$

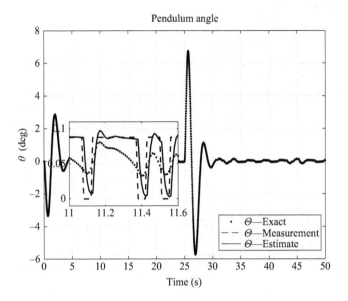

Figure 4.38   Pendulum angle for $\gamma = 0.035$

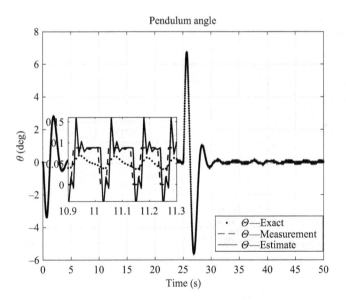

Figure 4.39   Pendulum angle for $\gamma = 0.0276$

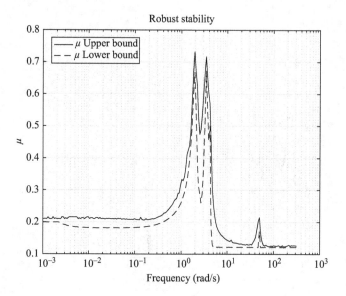

Figure 4.40   Robust stability for $\gamma = 0.035$

The filter poles [the eigenvalues of $A(I_6 - LC)$] are (to 15 digits)

0.642285107769573
0.270513164545753
0.286068842806314
0.999999996295013
0.994286462995327
0.945008902468039

Note that there is a pole very close to the unit circle which is usual for the $\mathscr{H}_\infty$ design.

The frequency plot of the structured singular value μ corresponding to the stability analysis of the uncertain closed-loop system is shown in Figure 4.40. Different to the case of Kalman filter implementation, the closed-loop system is robustly stable for the assumed plant uncertainty. The worst case stability margin is determined as

```
wcmarg =

    GainMargin: [0.7325 1.3652]
   PhaseMargin: [-17.5557 17.5557]
     Frequency: 1.8865
         WCUnc: [1x1 struct]
   Sensitivity: [1x1 struct]
```

4.3.2 $\mathcal{H}_\infty$ *Filter with bias compensation*

Likewise the Kalman filter, the $\mathcal{H}_\infty$ filter is designed for the nominal system model. In case of plant uncertainty, the estimate obtained by this filter may contain some bias which will lead to steady-state errors of the closed-loop system. Similarly to the technique used in Section 4.2, the bias may be estimated by augmenting the $\mathcal{H}_\infty$ filter equations with additional states in order to obtain accurate state estimate. For this aim, we may use a bias equation and augmented system equations of the form (4.26) and (4.27), respectively.

Example 4.5. Design of LQ regulator with $\mathcal{H}_\infty$ filter and bias compensation of the cart–pendulum system
The block diagram of LQ regulator with integral action and $\mathcal{H}_\infty$ filter of the cart–pendulum system is shown in Figure 4.41. The controller is designed using the M-file `LQR_Hinf_bias_design`. As in the case of Kalman filter implementation, the position estimate bias β is determined by using an additional state, which is added to the system state vector. The augmented state is estimated by a seventh order $\mathcal{H}_\infty$ filter. This allows to remove the position steady-state error that would appear when the plant model varies.

In Figures 4.42–4.44, we show the results of Monte-Carlo simulation of cart–pendulum closed-loop system with LQ regulator, $\mathcal{H}_\infty$ filter, and bias compensation for ten combinations of the uncertain parameters computed with the M-file

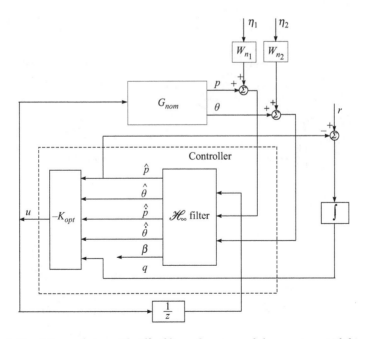

Figure 4.41   LQ regulator with $\mathcal{H}_\infty$ filter of cart–pendulum system with bias compensation

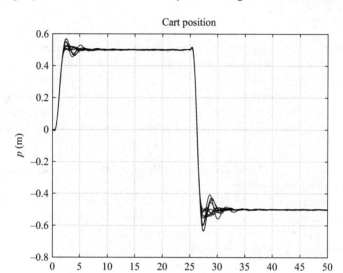

Figure 4.42   Monte-Carlo simulation in case of bias compensation. Cart position

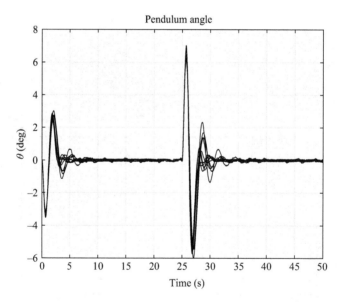

Figure 4.43   Monte-Carlo simulation in case of bias compensation. Pendulum angle

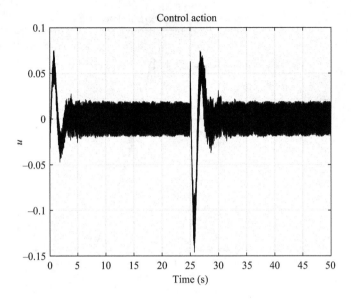

Figure 4.44 Monte-Carlo simulation in case of bias compensation. Control action

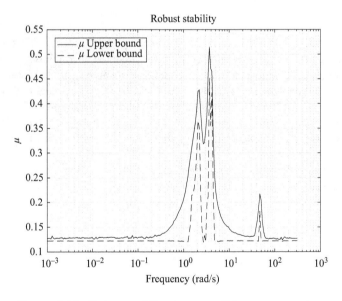

Figure 4.45 Robust stability in case of bias compensation

sim_MC_LQR_Hinf_bias. Obviously, the bias compensation allows to remove the cart position steady-state error.

The plot of the structured singular value μ corresponding to robust-stability analysis of the closed-loop system is shown in Figure 4.45. The closed-loop system

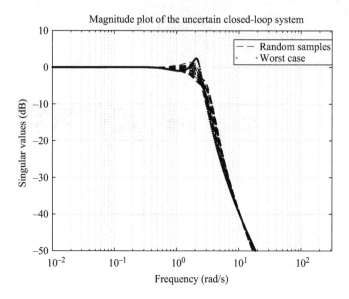

Figure 4.46   Worst case closed-loop system gain

achieves robust stability the peak value of μ being equal to 0.516. This means that the system can tolerate up to 194 percent of the modeled uncertainty.

The worst case frequency responses and time responses of the closed-loop system are computed by the M-file LQR_Hinf_bias_wcp. The magnitude plot of the cart position closed-loop transfer function for 30 random samples of the uncertain parameters and for the worst case gain is shown in Figure 4.46. The peak of the magnitude response is 2.5 dB which is acceptable. The worst case closed-loop bandwidth is 1.9 rad/s.

The worst case closed-loop transient responses of the cart position and pendulum angle are given in Figures 4.47 and 4.48, respectively.

The worst case gain and phase stability margins are computed as

```
wcmarg =

     GainMargin: [0.5670 1.7637]
    PhaseMargin: [-30.8951 30.8951]
      Frequency: 2.1331
          WCUnc: [1x1 struct]
    Sensitivity: [1x1 struct]
```

and are significantly larger than the corresponding margins of the system with Kalman filter.

The magnitude plot of control action sensitivity to the reference and encoder noises is shown in Figure 4.49. This plot is similar to the control sensitivity plot in case of Kalman filter implementation (Figure 4.35).   ❏

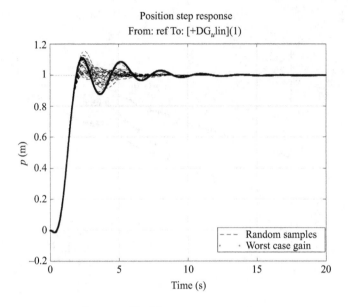

Figure 4.47 Worst case car position

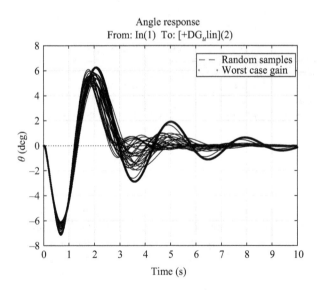

Figure 4.48 Worst case pendulum angle

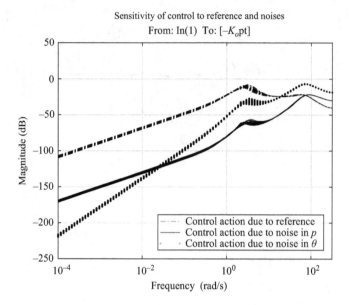

Figure 4.49   *Control sensitivity to reference and noises*

4.4 $\mathcal{H}_\infty$ Design

MATLAB files used in this section

Files for $\mathcal{H}_\infty$ design	Description
hinf_design	Design of $\mathcal{H}_\infty$ controller
hinf_robust_analysis	Robust-stability and robust-performance analysis
dfrs_hinf	Frequency responses
sim_hinf	Simulation
sim_MC_hinf	Monte-Carlo simulation
hinf_wcp	Worst case gain

In this section, we consider the application of $\mathcal{H}_\infty$ optimization to the design of stabilizing controllers which are capable to ensure efficient disturbance attenuation and noise suppression. For unity of presentation with the previous controllers, we give the formulas for the design of $\mathcal{H}_\infty$ discrete-time controllers.

4.4.1   *The $\mathcal{H}_\infty$ design problem*

To formulate the general $\mathcal{H}_\infty$ design problem, we shall use the block diagram represented in Figure 4.50. In this representation, the "exogenous input" w is an m_1 vector of all signals which are input to the system and the "error" z is a p_1 vector of all signals (errors) that are necessary to characterize the behavior of the closed-loop

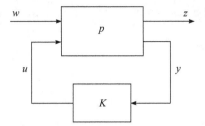

Figure 4.50 $\mathcal{H}_\infty$ Control problem

system. Both vectors may contain components which are abstract in the sense that they can be defined mathematically but do not represent signals that really exist at some system point. Here u is the m_2 vector of control signals and y is the p_2 vector of measurable outputs. P represents the generalized plant model of order n and K is the controller.

Let the state-space description of the generalized time-invariant plant model P is given by the linear difference equation

$$x(k+1) = Ax(k) + B_1w(k) + B_2u(k),$$
$$z(k) = C_1x(k) + D_{11}w(k) + D_{12}u(k), \qquad (4.32)$$
$$y(k) = C_2x(k) + D_{21}w(k) + D_{22}u(k).$$

The closed-loop system in Figure 4.50 is described by the following equations:

$$\begin{bmatrix} z \\ y \end{bmatrix} = P(z) \begin{bmatrix} w \\ u \end{bmatrix} \qquad (4.33)$$
$$u = K(z)y,$$

where the transfer function matrix $P(z)$ is partitioned as

$$P(z) = \begin{matrix} p_1\{ \\ p_2\{ \end{matrix} \begin{bmatrix} P_{11}(z) & P_{12}(z) \\ P_{21}(z) & P_{22}(z) \end{bmatrix}.$$
$$\underbrace{\phantom{P_{11}(z)}}_{m_1} \quad \underbrace{\phantom{P_{12}(z)}}_{m_2}$$

The plant transfer function matrix $P(z)$ is determined from the state-space description (4.32) by

$$P(z) = \begin{matrix} n\{ \\ p_1\{ \\ p_2\{ \end{matrix} \left[\begin{array}{c|cc} A & B_1 & B_2 \\ \hline C_1 & D_{11} & D_{12} \\ C_2 & D_{21} & D_{22} \end{array} \right].$$
$$\underbrace{}_{n} \quad \underbrace{}_{m_1} \underbrace{}_{m_2}$$

The transfer function $P(z)$ is derived from the nominal plant model but may contain some weighting functions depending on the problem under solution. The closed-loop system transfer function matrix from w to z is given by the lower linear fractional transformation (LFT)

$$z = F_\ell(P,K)w, \tag{4.34}$$

where

$$F_\ell(P,K) = P_{11} + P_{12}K(I - P_{22}K)^{-1}P_{21}.$$

The standard problem for $\mathcal{H}_\infty$ optimal control is to find all stabilizing controllers K, which minimize

$$\|F_\ell(P,K)\|_\infty = \max_{\theta \in (-\pi,\pi]} \bar{\sigma}(F_\ell(P,K)(e^{j\theta})),$$

i.e., to solve the optimization problem

$$\min_{K\text{stabilizing}} \|F_\ell(P,K)\|_\infty. \tag{4.35}$$

The $\mathcal{H}_\infty$ norm has several important interpretations in the terms of system performance. One of them is that the minimization of $\mathcal{H}_\infty$ norm minimizes the peak value of the maximum singular value of $F_\ell(P,K)(e^{j\theta})$. It also has an interpretation in the time domain as an induced 2-norm, characterizing the worst case. Let $z = F_\ell(P,K)w$. Then

$$\|F_\ell(P,K)\|_\infty = \max_{w\neq 0} \frac{\|z\|_2}{\|w\|_2}, \tag{4.36}$$

where

$$\|z\|_2 = \sqrt{\sum_{k=-\infty}^{\infty} z(k)^T z(k)}$$

is the 2-norm of the discrete-time vector signal z.

In practice, usually it is not necessary to obtain the strictly optimal controller for the $\mathcal{H}_\infty$ problem and it is theoretically and computationally simpler to design a *suboptimal controller*, i.e., a controller which is close to the optimal one in the sense of $\mathcal{H}_\infty$ norm. Let γ_{min} is the minimum value of $F_\ell(P,K)$ over all stabilizing controllers K. Then, the problem for $\mathcal{H}_\infty$ suboptimal control is for a given $\gamma > \gamma_{min}$ to find a stabilizing K, such that

$$F_\ell(P,K) < \gamma.$$

If we desire an optimal controller that achieves γ_{min} to within a specified tolerance, then it is possible to perform a bisection on γ until its value is sufficiently accurate. This leads to an iterative procedure which may require several steps.

Currently, there are two types of numerical methods to solve the $\mathcal{H}_\infty$ continuous-time or discrete-time suboptimal problem. The first type is an algorithm which is based on the solution of two nth order indefinite matrix algebraic Riccati equations. This approach has computational complexity of order n^3 per step and is the only practical choice for higher dimensional problems. It may be associated, however, with some numerical difficulties especially when γ approaches $\gamma_{\min}$. The second type of methods embeds the $\mathcal{H}_\infty$ problem into a linear matrix inequality (LMI) and then employs methods for semidefinite programing to find $\gamma_{\min}$ [112]. The LMI methods allow to remove some of the numerical difficulties associated with the Riccati-based methods, but their computational complexity is of order n^6, which makes them practical only for low-dimensional problems.

The algorithm for $\mathcal{H}_\infty$ suboptimal design [99, Section B.4], presented briefly below, is a discrete-time counterpart of the famous Glover–Doyle algorithm [113] used in the continuous-time $\mathcal{H}_\infty$ optimization. The solution of the $\mathcal{H}_\infty$ suboptimal control problem by this algorithm is found under the following assumptions:

A1 (A, B_2) is stabilizable and (C_2, A) is detectable;

A2 $\begin{bmatrix} A - e^{j\theta} I_n & B_2 \\ C_1 & D_{12} \end{bmatrix}$ has full column rank for all $\theta \in (-\pi, \pi]$;

A3 $\begin{bmatrix} A - e^{j\theta} I_n & B_1 \\ C_2 & D_{21} \end{bmatrix}$ has full row rank for all $\theta \in (-\pi, \pi]$;

A4 $D_{22} = 0$.

Assumption (A1) is required for the existence of stabilizing controllers K. Assumptions (A2) and (A3) ensure that the optimal controller does not cancel poles or zeros on the unit circle. Assumption (A4) is not required by the optimization, but it simplifies significantly the expressions used in the computation of the suboptimal controller. If the matrix D_{22} is nonzero, then it is possible to construct an equivalent $\mathcal{H}_\infty$ problem, in which it is zeroed.

To simplify the solution, it is also assumed that

A5 The matrix D_{12} is of full column rank, i.e., $D_{12}^T D_{12} > 0$,

A6 The matrix D_{21} is of full row rank, i.e., $D_{21} D_{21}^T > 0$.

Let

$$\bar{C} = \begin{bmatrix} C_1 \\ 0 \end{bmatrix}, \quad \bar{D} = \begin{bmatrix} D_{11} & D_{12} \\ I_{m_1} & 0 \end{bmatrix},$$

and define

$$J = \begin{bmatrix} I_{p_1} & 0 \\ 0 & -\gamma^2 I_{m_1} \end{bmatrix}, \quad \hat{J} = \begin{bmatrix} I_{m_1} & 0 \\ 0 & -\gamma^2 I_{m_2} \end{bmatrix}, \quad \tilde{J} = \begin{bmatrix} I_{m_1} & 0 \\ 0 & -\gamma^2 I_{p_1} \end{bmatrix}.$$

Let X_∞ be the solution to the discrete-time Riccati equation

$$X_\infty = \bar{C}^T J \bar{C} + A^T X_\infty A - L^T R^{-1} L, \tag{4.37}$$

where

$$R = \bar{D}^T J \bar{D} + B^T X_\infty B =: \begin{bmatrix} R_1 & R_2^T \\ R_2 & R_3 \end{bmatrix},$$

$$L = \bar{D}^T J \bar{C} + B^T X_\infty A =: \begin{bmatrix} L_1 \\ L_2 \end{bmatrix}.$$

Assume that there exists an $m_2 \times m_2$ matrix V_{12} such that

$$V_{12}^T V_{12} = R_3,$$

and an $m_1 \times m_1$ matrix V_{21} such that

$$V_{21}^T V_{21} = -\gamma^{-2}\nabla, \quad \nabla = R_1 - R_2^T R_3^{-1} R_2 < 0.$$

Define the matrices

$$
\begin{bmatrix} A_t & \tilde{B}_t \\ C_t & \tilde{D}_t \end{bmatrix} =:
\left[
\begin{array}{c|cc}
A_t & \tilde{B}_{t_1} & \tilde{B}_{t_2} \\
\hline
C_{t_1} & \tilde{D}_{t_{11}} & \tilde{D}_{t_{12}} \\
C_{t_2} & \tilde{D}_{t_{21}} & \tilde{D}_{t_{22}}
\end{array}
\right]
$$

$$
=
\left[
\begin{array}{c|cc}
A - B_1 \nabla^{-1} L_\nabla & B_1 V_{21}^{-1} & 0 \\
\hline
V_{12} R_3^{-1}(L_2 - R_2 \nabla^{-1} L_\nabla) & V_{12} R_3^{-1} R_2 V_{21}^{-1} & I \\
C_2 - D_{21} \nabla^{-1} L_\nabla & D_{21} V_{21}^{-1} & 0
\end{array}
\right],
$$

where

$$L_\nabla = L_1 - R_2^T R_3^{-1} L_2.$$

Let Z_∞ be the solution to the discrete-time Riccati equation

$$Z_\infty = \tilde{B}_t \hat{J} \tilde{B}_t^T + A_t Z_\infty A_t^T - M_t S_t^{-1} M_t^T \tag{4.38}$$

in which

$$S_t = \tilde{D}_t \hat{J} \tilde{D}_t^T + C_t Z_\infty C_t^T =: \begin{bmatrix} S_{t_1} & S_{t_2} \\ S_{t_2}^T & S_{t_3} \end{bmatrix},$$

$$M_t = \tilde{B}_t \hat{J} \tilde{D}_t^T + A_t Z_\infty C_t^T =: [M_{t_1} M_{t_2}].$$

Further on we shall refer to (4.37) and (4.38) as to *X-Riccati equation* and *Z-Riccati equation*, respectively.

As it is proved in [99], a stabilizing controller that satisfies

$$\|F_\ell(P, K)\|_\infty < \gamma$$

exists, if and only if

1. There exists a solution to the Riccati equation (4.37) satisfying

$$X_\infty \geq 0,$$

$$R_1 - R_2^T R_3^{-1} R_2 = \nabla < 0$$

    such that $A - BR^{-1}L$ is asymptotically stable.
2. There exists a solution to the Riccati equation (4.38) such that

$$Z_\infty \geq 0,$$

$$S_{t_1} - S_{t_2} S_{t_3}^{-1} S_{t_2}^T < 0$$

    with $A_t - M_t S_t^{-1} C_t$ asymptotically stable.

In this case, a controller that achieves the objective is

$$\hat{x}(k+1) = A_t\hat{x}(k) + B_2u(k) + M_{t_2}S_{t_3}^{-1}(y(k) - C_{t_2}\hat{x}(k))$$
$$V_{12}u(k) = -C_{t_1}\hat{x}(k) - S_{t_2}S_{t_3}^{-1}(y(k) - C_{t_2}\hat{x}(k)) \quad\quad (4.39)$$

which yields

$$K_0 = \left[\begin{array}{c|c} \dfrac{A_t - B_2V_{12}^{-1}(C_{t_1} - S_{t_2}S_{t_3}^{-1}C_{t_2}) - M_{t_2}S_{t_3}^{-1}C_{t_2}}{-V_{12}^{-1}(C_{t_1} - S_{t_2}S_{t_3}^{-1}C_{t_2})} & \dfrac{-B_2V_{12}^{-1}S_{t_2}S_{t_3}^{-1} + M_{t_2}S_{t_3}^{-1}}{-V_{12}^{-1}S_{t_2}S_{t_3}^{-1}} \end{array} \right].$$

This is the so-called central controller which has the same number of states as the generalized plant $P(z)$.

As it follows from (4.39), the $\mathcal{H}_\infty$ controller has an observer structure like in the continuous-time case [82, Section 16.8], [87, Section 9.3]. The vector $\hat{w}(k)^* = \nabla^{-1}L_\nabla\hat{x}(k)$ can be interpreted as an estimate of the worst case disturbance (exogenous input w) and the product

$$C_{t_2}\hat{x}(k) = C_2\hat{x}(k) - D_{21}\nabla^{-1}L_\nabla\hat{x}(k) = C_2\hat{x}(k) - D_{21}\hat{w}(k)^*$$

represents an estimate of the worst observer input.

In this way, the solution of the $\mathcal{H}_\infty$ suboptimal control problem requires solution of two matrix Riccati equations similarly to the solution of the LQG problem. If $\gamma \to \infty$, then the Riccati equations (4.37) and (4.38) tend to the corresponding Riccati equations in LQG problem.

If one wants a controller which achieves γ_{min} up to specified threshold, it is possible to use a bisection procedure on γ (the so-called γ-iterations), until obtaining a sufficiently accurate value. The algorithm presented gives a test for each value of γ to determine whether it is smaller or greater than γ_{min}.

It may be proved that if γ is increased to very high value, then the controller generated by the $\mathcal{H}_\infty$ optimization algorithm converges to an LQG controller.

The design of the $\mathcal{H}_\infty$ (sub)optimal controller in Robust Control Toolbox is done by the function `hinfsyn` which may be used in the synthesis of continuous-time as well as discrete-time systems. This function may determine the $\mathcal{H}_\infty$ controller implementing two Riccati equations (default algorithm) or LMI. It uses a bisection procedure to find γ_{min} and there is an option to display information about the current iteration step.

It should be noted that in principle, the discrete-time $\mathcal{H}_\infty$ suboptimal design problem may be solved by algorithms for continuous-time design converting the discrete-time generalized plant into equivalent continuous-time one by using the bilinear transformation

$$z = \frac{1 + 1/2sh}{1 - 1/2sh}.$$

This transformation maps the unit disk in the z-plane into the left half of the s-plane for any $h > 0$. (Usually, h is chosen equal to the sampling period of the discrete-time system). The continuous-time controller designed in this way is converted back to discrete-time by using the inverse transformation (Tustin approximation)

$$s = \frac{2}{h}\frac{z-1}{z+1}.$$

Such indirect method for solving the discrete-time $\mathcal{H}_\infty$ problem is not recommended from numerical point of view.

4.4.2   Mixed-sensitivity $\mathcal{H}_\infty$ control

Mixed-sensitivity is the name of a design procedure in which the sensitivity transfer function $S = (I + GK)^{-1}$ is shaped along with one or more other closed-loop transfer functions such as KS or the complementary sensitivity function $T = I - S$ [87]. This is an efficient method which allows to incorporate into controller design several frequency domain requirements in order to achieve desirable trade-off between performance and robustness.

Consider first the tracking problem shown in Figure 4.51. The exogenous input is the reference r and the error signals are $z_1 = -W_1 e = W_1(r - y)$ and $z_2 = W_2 u$. Taking into account that S is the transfer function matrix between r and $-e$, and KS is the transfer function between r and control u, we have that $z_1 = W_1 Sr$ and $z_2 = W_2 KSr$. The stable minimum-phase transfer functions W_1 and W_2 represent weighting filters used to shape S and KS. Shaping appropriately, the sensitivity function S allows to achieve small tracking error e in the low frequency range while shaping KS allows to limit the controller gain and bandwidth and hence the control energy. These aims can be achieved by choosing $W_1(z)$ as a low-pass filter and $W_2(z)$ as a high-pass filter. Note that in the general case, $W_1(z)$ and $W_2(z)$ may be set as transfer function matrices chosen in diagonal form for convenience. The controller design is done by solving the $\mathcal{H}_\infty$ optimization problem

$$\min_{K_{\text{stabilizing}}} \left\| \begin{bmatrix} W_1 S \\ W_2 KS \end{bmatrix} \right\|_\infty.$$

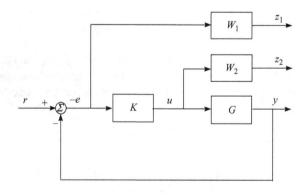

Figure 4.51   A mixed sensitivity setting

If one achieves

$$\left\| \begin{bmatrix} W_1 S \\ W_2 KS \end{bmatrix} \right\|_\infty < 1, \tag{4.40}$$

then it follows by necessity that

$$\bar{\sigma}(S(e^{j\theta})) \le \underline{\sigma}(W_1^{-1}(e^{j\theta})), \tag{4.41}$$

$$\bar{\sigma}(K(e^{j\theta})S(e^{j\theta})) \le \underline{\sigma}(W_2^{-1}(e^{j\theta})). \tag{4.42}$$

Thus, the transfer functions W_1^{-1} and W_2^{-1} determine the shape, i.e., the magnitude at different frequencies, of the sensitivity functions S and KS, respectively. Changing the weighting functions W_1 and W_2, it is possible to obtain the desired tracking accuracy using in the same time control actions with small magnitudes.

The mixed sensitivity problem can be formulated in general form as shown in Figure 4.52. The controller input is $v = r - y$ and the error signal is defined as $z = [z_1^T z_2^T]^T$, where $z_1 = W_1(r - y)$ and $z_2 = W_2 u$. It is possible to show that $z_1 = W_1 Sr$ and $z_2 = W_2 KSr$, as required. The elements of the generalized plant P are given by

$$P_{11} = \begin{bmatrix} W_1 \\ 0 \end{bmatrix}, \quad P_{12} = \begin{bmatrix} -W_1 G \\ W_2 \end{bmatrix},$$

$$P_{21} = I, \quad P_{22} = -G,$$

where the matrix P is partitioned such that

$$\begin{bmatrix} z_1 \\ z_2 \\ -- \\ v \end{bmatrix} = \begin{bmatrix} P_{11} & P_{12} \\ P_{21} & P_{22} \end{bmatrix} \begin{bmatrix} r \\ u \end{bmatrix},$$

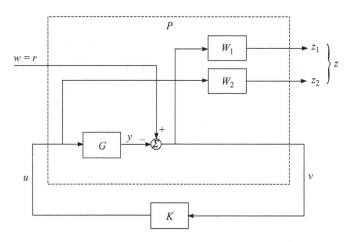

Figure 4.52   Minimization of S/KS sensitivity in standard form

The closed-loop system is described by $z = F_\ell(P, K)r$, where

$$F_\ell(P, K) = \begin{bmatrix} W_1 S \\ W_2 KS \end{bmatrix}.$$

Suppose now that it is desirable to achieve good tracking in the presence of unstructured multiplicative perturbations at the plant output. This means that we desire to make small the sensitivity S and the complementary sensitivity T in appropriate frequency ranges. This leads to the $\mathcal{H}_\infty$ optimization problem

$$\min_{K_{\text{stabilizing}}} \left\| \begin{bmatrix} W_1 S \\ W_2 T \end{bmatrix} \right\|_\infty. \tag{4.43}$$

for appropriately chosen weighting filters W_1 and W_2. The filter W_1 is chosen again as a low-pass filter. Since the uncertainty is increasing with the frequency, the filter W_2 is usually chosen as a high-pass filter.

The mixed sensitivity design problem (4.43) is represented as a standard $\mathcal{H}_\infty$ optimization problem in Figure 4.53. In the given case, we determine a controller $K(z)$ which stabilizes the closed-loop system and minimizes $\|F_\ell(P, K)\|_\infty$, where $F_\ell(P, K)$ is the closed-loop transfer function matrix from r to $[z_1^T z_2^T]^T$ given by

$$F_\ell(P, K) = \begin{bmatrix} W_1 S \\ W_2 (I - S) \end{bmatrix}.$$

The generalized plant P is determined as

$$P_{11} = \begin{bmatrix} W_1 \\ 0 \end{bmatrix}, \quad P_{12} = \begin{bmatrix} -W_1 G \\ W_2 G \end{bmatrix},$$

$$P_{21} = I, \quad P_{22} = -G.$$

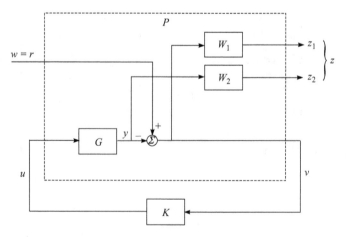

Figure 4.53   Minimization of S/T sensitivity in standard form

If we achieve $\|F_\ell(P,K)\|_\infty < 1$, then it is fulfilled that

$$\bar{\sigma}(S(e^{j\theta})) \le \underline{\sigma}(W_1^{-1}(e^{j\theta})),$$
$$\bar{\sigma}(T(e^{j\theta})S(e^{j\theta})) \le \underline{\sigma}(W_2^{-1}(e^{j\theta})). \tag{4.44}$$

In the more general case, it is possible to find a stabilizing controller K minimizing the $\mathscr{H}_\infty$ norm of the weighted mixed sensitivity

$$\left\|\begin{bmatrix} W_1 S \\ W_2 KS \\ W_3 T \end{bmatrix}\right\|_\infty$$

which allows to "penalize" simultaneously S, KS, and T. This design can be done by using the Robust Control Toolbox function `mixsyn`.

4.4.3 Two degrees-of-freedom controllers

The mixed-sensitivity approach, considered above, may be extended to the design of two degree-of-freedom controllers. One of the possible configuration with such controller is shown in Figure 4.54.

The system has a reference r, output disturbance d, and two output errors z_1 and z_2. The system M is the ideal model, to which the closed-loop system should match.

In the given case, the controller K consists of a feedback controller K_y for disturbance attenuation and a prefilter K_r to achieve the desired closed-loop performance and is represented as

$$K = [K_r \; K_y].$$

The transfer function matrices K_r and K_y may be obtained as follows.

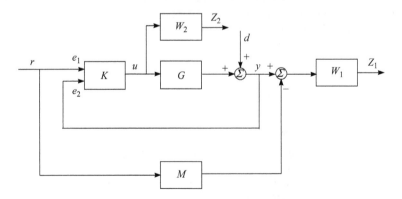

Figure 4.54 Two degrees-of-freedom controller

The closed-loop system may be represented in the form of the standard problem shown in Figure 4.50. The system is described by

$$
\begin{bmatrix} z_1 \\ z_2 \\ e_1 \\ e_2 \end{bmatrix} = \left[\begin{array}{cc|c} -W_1 M & W_1 & W_1 G \\ 0 & 0 & W_2 \\ \hline I & 0 & 0 \\ 0 & I & G \end{array} \right] \begin{bmatrix} r \\ d \\ u \end{bmatrix}.
$$

The closed-loop transfer function matrix is given by

$$
T_{zw} = \begin{bmatrix} W_1(S_o G K_r - M) & W_1 S_o \\ W_2 S_i K_r & W_2 K_y S_o \end{bmatrix},
$$

where the input and output sensitivities are equal to

$$
S_i = (I - K_y G)^{-1}, \quad S_o = (I - G K_y)^{-1},
$$

respectively. Again, the goal is to obtain stabilizing controller K that minimizes the $\mathcal{H}_\infty$ norm of T_{zw}. The four functions that should be minimized are described in Table 4.3.

In the table, W_1 and W_2 are frequency dependent weighting functions that are called weighting performance function and weighting control function, respectively. The function W_1 is chosen as a low-pass filter to ensure closeness between the system dynamics and the model in desired low frequency range and the function W_2 is chosen as a high-pass filter to limit the control actions in the high-frequency range.

The model M is usually set in the diagonal form

$$
M = \begin{bmatrix} M_1 & 0 & \cdots & 0 \\ 0 & M_2 & \cdots & 0 \\ \cdots & \cdots & \cdots & \cdots \\ 0 & 0 & \cdots & M_{p_2} \end{bmatrix},
$$

so that to achieve decoupling of the system outputs. The blocks $M_i, i = 1, \ldots, p_2$ are usually set as second-order lags with desired time constants and damping.

The design of two degree-of-freedom controller is done taking as input to the controller the vector $v = [r^T y^T]^T$.

Table 4.3 $\mathcal{H}_\infty$ function to be minimized

Function	Description
$W_1(S_o G K_r - M)$	Weighted difference between the real and ideal closed-loop system
$W_1 S_o$	Weighted output sensitivity
$W_2 S_i K_r$	Weighted control action due to reference
$W_2 K_y S_o$	Weighted control action due to disturbance

4.4.4 Numerical issues in $\mathcal{H}_\infty$ design

Numerical difficulties in $\mathcal{H}_\infty$ design may arise if some of the assumptions (A1)–(A6) are not satisfied. A frequent cause of troubles, especially in the mixed-sensitivity design, is the violation of assumptions (A2), (A3) and (A5), (A6). According to these assumptions we have that

- The matrix

$$
\begin{array}{c}
n\{ \\
p_1\{
\end{array}
\underbrace{\begin{bmatrix} A & B_2 \\ C_1 & D_{12} \end{bmatrix}}_{\displaystyle n \quad m_2}
$$

must have rank $n + m_2$ (full column rank),
- The matrix

$$
\begin{array}{c}
n\{ \\
p_2\{
\end{array}
\underbrace{\begin{bmatrix} A & B_1 \\ C_2 & D_{21} \end{bmatrix}}_{\displaystyle n \quad m_1}
$$

must have rank $n + p_2$ (full row rank).
- The $p_1 \times m_2$ matrix D_{12} must have rank m_2 (full column rank),
- The $p_2 \times m_1$ matrix D_{21} must have rank p_2 (full row rank).

It follows from the above conditions that the solution of the $\mathcal{H}_\infty$ suboptimal control problem requires

$$p_1 \geq m_2, \tag{4.45}$$

$$p_2 \leq m_1, \tag{4.46}$$

i.e., *the number of the error signals should be greater than the numbers of controls and the number of measurements should be smaller than the number of exogenous inputs.* Note that these are only necessary (but not sufficient) conditions to satisfy the assumptions. If conditions (4.45) and/or (4.46) are not satisfied, it is necessary to increase the number p_1 of error signals or to increase the number m_1 of exogenous signals, respectively.

$\mathcal{H}_\infty$ optimization problems, for which the matrix D_{12} is of full column rank and the matrix D_{21} is of full row rank, are called *regular*. If either of these matrices is rank deficient, then the corresponding $\mathcal{H}_\infty$ problem is called *singular* and cannot be solved by using the Riccati-based approach. In such case, it is necessary to use other methods, for instance the LMI approach. An alternative technique, frequently used in practice, is to regularize the problem by perturbing slightly the matrices D_{12} and/or D_{21} in order to increase their rank. Unfortunately, this may lead to ill-conditioning of some matrices used in the solution of the Riccati equations and may introduce large rounding errors in the result. Also, as noted in [114], such regularization technique may cause the value of γ to drop sharply below $\gamma_{\min}$ leading to $\mathcal{H}_\infty$ controllers yielding poor closed-loop stability.

Example 4.6. Design of $\mathscr{H}_\infty$ controller for cart–pendulum system

Consider the design of two degree-of-freedom $\mathscr{H}_\infty$ controller with integral action for the cart–pendulum system. Note that LQR and LQG controllers for this system were already designed in Examples 4.2–4.5.

The block diagram of the closed-loop system with performance and control weighting functions is shown in Figure 4.55. For clearer physical interpretation the design problem is formulated as a continuous-time problem but the controller is designed in discrete-time. Since the $\mathscr{H}_\infty$ design framework do not in general produce integral control, an integral of the cart position error is added to remove the position steady-state error. The design aim is to determine a two degree-of-freedom $\mathscr{H}_\infty$ controller so as to minimize the cart position error and pendulum angle oscillations. For this aim, we shall implement the mixed-sensitivity design using as feedback signals the cart position, the pendulum angle, and the integral of cart position error. The $\mathscr{H}_\infty$ design is done by the M-file `hinf_design` using the nominal plant model. The error signal z_1 is obtained by "penalizing" the cart position error, the pendulum angular velocity, and the integral of cart position error using the performance weighting function

$$W_p(s) = \begin{bmatrix} \dfrac{4s+1}{8s+0.05} & 0 & 0 \\[2ex] 0 & \dfrac{10s+2}{10s+0.01} & 0 \\[2ex] 0 & 0 & \dfrac{10s+2}{10s+0.01} \end{bmatrix}.$$

The error signal z_2 "penalizes" the control action using the weighting function

$$W_u(s) = \frac{0.08s+0.1}{0.01s+1}.$$

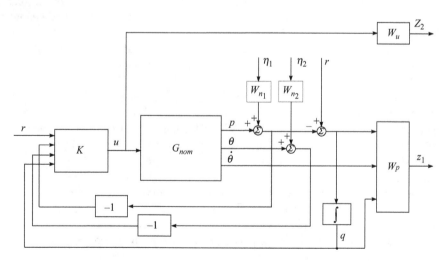

Figure 4.55   Closed-loop system with $\mathscr{H}_\infty$ controller and performance weighting functions

The magnitude plots of the inverse performance weighting functions are shown in Figure 4.56, and the inverse control weighting function is displayed in Figure 4.57. The generalized plant P is of order 11 with $m_1 = 3, m_2 = 1, p_1 = 4, p_2 = 4$. After discretization of the open-loop interconnection with sampling interval $T_s = 0.01$ s and implementing the function hinfsyn, one obtains the message

```
[a b1;c2 d21] does not have full row rank at s=0
```

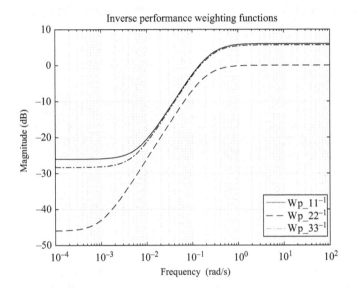

Figure 4.56   Inverse performance weighting functions

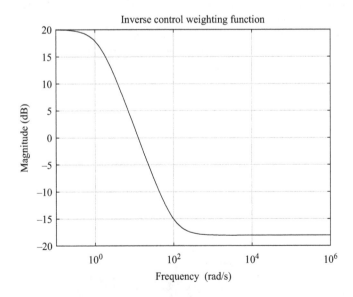

Figure 4.57   Inverse control weighting function

In the given case, the condition $p_2 \le m_1$ is violated so that assumption (A3) is not satisfied. To increase the row rank of the matrix

$$M_2 = \begin{bmatrix} A & B_1 \\ C_2 & D_{21} \end{bmatrix},$$

we will increase the number m_1 of the exogenous inputs adding to the control action an artificial small white noise n_3 with variance equal to 10^{-3}. The addition of this noise do not change significantly the $\mathcal{H}_\infty$ problem but leads to rank $M_2 = 15$ which allows to obtain a solution. Now we receive the following report from hinfsyn:

```
Resetting value of Gamma min based on D_11, D_12, D_21 terms

Test bounds:         0.4994 <  gamma  <=  67108864.0000


   gamma    hamx_eig xinf_eig hamy_eig   yinf_eig    nrho_xy   p/f
 6.711e+07  5.0e-06  0.0e+00  5.0e-06   -3.0e-05#   0.0000     f
Gamma max, 67108864.0000, is too small !!
Resetting value of Gamma min based on D_11, D_12, D_21 terms

Test bounds:         0.4994 <  gamma  <=     0.5158


   gamma    hamx_eig xinf_eig hamy_eig   yinf_eig    nrho_xy   p/f
   0.516    5.0e-06 -4.6e-08  5.0e-06   -9.5e-10    0.0112     p
   0.508    4.1e-18# ******** 5.0e-06   -3.5e-12   ********    f


 Gamma value achieved:     0.5158
```

The function hinfsyn determines a controller for which $\gamma_{min} = 0.5158$. However, for this controller, the closed-loop system has a pole at 1.0076, i.e., this system is unstable. In the given case, the $\mathcal{H}_\infty$ optimization problem is singular due to the rank deficiency of the matrix D_{21} (rank $D_{21} = 3 < p_2$). This causes γ_{min} to drop below the minimal value ensuring closed-loop stability.

The numerical difficulties in the given case can be avoided using the option to implement the LMI method for finding the $\mathcal{H}_\infty$ controller. With this option, the function hinfsyn produces the following report:

```
Minimization of gamma:

Solver for linear objective minimization under LMI constraints

Iterations   :   Best objective value so far

     1
     2
     .
     .
     .
    50                    0.974273
    51                    0.969438
    52                    0.969438
***                new lower bound:     0.964951
```

```
Result:    feasible solution of required accuracy
           best objective value:        0.969438
           guaranteed absolute accuracy:  4.49e-03
           f-radius saturation:  0.732% of R =  1.00e+08
```

In this case, the closed-loop system is stable and the true minimum value of γ is equal to 0.9694. Note that the same result is obtained if the LMI method is applied to the original problem, since this method does not impose the rank conditions given above. The controller found is of 11th order and has the disadvantage that it is unstable, one of his poles being at 1.0124.

Results from robust-stability and robust-performance analysis of the closed-loop system with $\mathcal{H}_\infty$ controller, obtained by the M-file `hinf_robust_analysis`, are shown in Figures 4.58 and 4.59, respectively. The closed-loop system is robustly stable for the plant uncertainties but does not achieve robust performance. This is a result from the fact that the $\mathcal{H}_\infty$ design does not take directly into account the plant uncertainty.

Results of Monte-Carlo simulation of the closed-loop system are represented in Figures 4.60–4.62. It is seen that the corresponding cart position and pendulum angle trajectories obtained for varying plant parameters are confined in relatively tight areas as a result of system robustness. The pendulum angle oscillations do not exceed 7.5 deg.

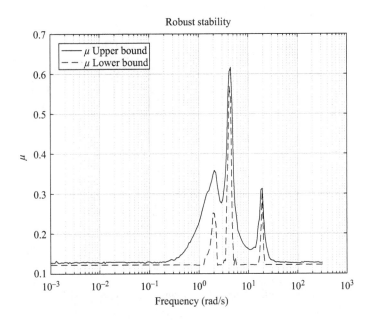

Figure 4.58   Robust stability

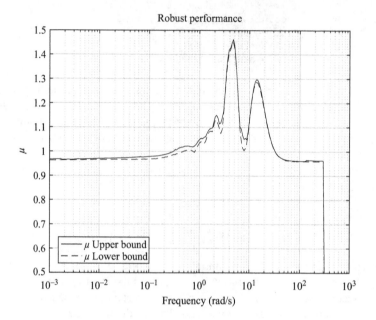

Figure 4.59   *Robust performance*

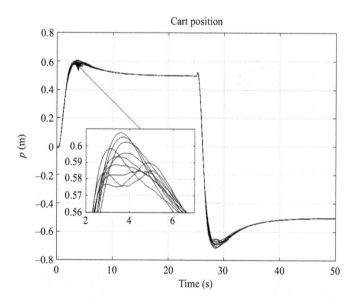

Figure 4.60   *Monte-Carlo simulation. Cart position*

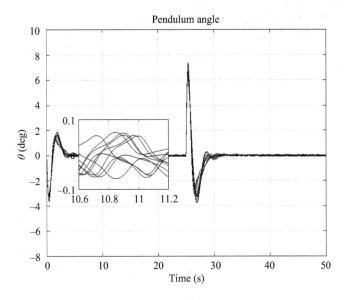

Figure 4.61   Monte-Carlo simulation. Pendulum angle

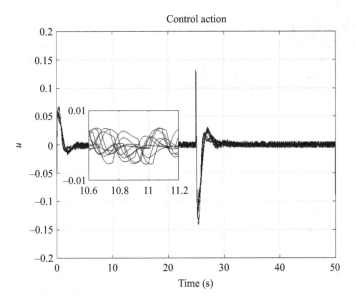

Figure 4.62   Monte-Carlo simulation. Control action

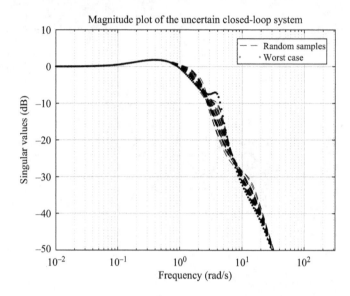

Figure 4.63   Worst case gain

The closed-loop system properties for the worst case uncertainty are analyzed by the M-file `hinf_wcp`. From the worst case magnitude plot shown in Figure 4.63, we see that the closed-loop bandwidth is 1.40 rad/s which is less than the bandwidths in cases of LQG and LQR controllers. The worst case cart position and pendulum angle are shown in Figures 4.64 and 4.65, respectively. The stability margins in the worst case are

```
wcmarg =

    GainMargin: [0.9194 1.0877]
   PhaseMargin: [-4.8091 4.8091]
     Frequency: 4.5597
         WCUnc: [1x1 struct]
   Sensitivity: [1x1 struct]
```

The control sensitivity to exogenous signals (reference and noises) is shown in Figure 4.66. As a result of the mixed-sensitivity design, this sensitivity is relatively low.   ❑

Position step response

From: ref To: [+$G_u$lin](1)

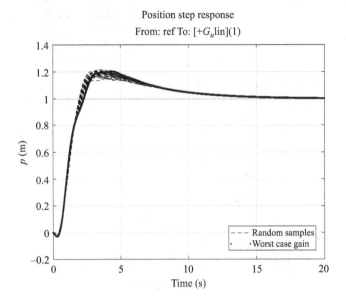

Figure 4.64 Worst case car position

Angle response

From: ref To: [+$G_u$lin](2)

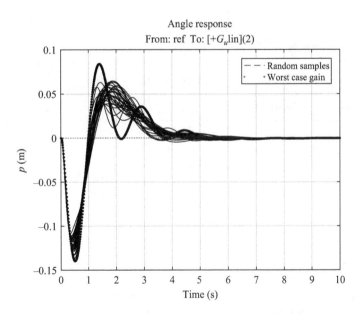

Figure 4.65 Worst case pendulum angle

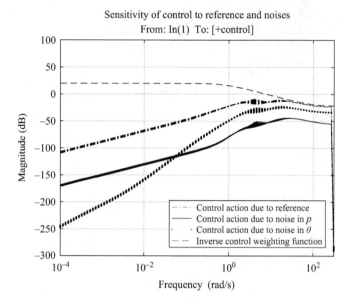

Figure 4.66   *Control sensitivity to reference and noises*

4.5   μ Synthesis

MATLAB files used in this section

Files for μ synthesis	Description
unc_approx_model	Uncertain model approximation
olp_mu_reg	Open-loop interconnection
mu_design	Controller μ synthesis
mu_robust_analysis	Robust-stability and robust-performance analysis
mu_reg_comparison	Comparison of different order μ controllers
dfrs_mu	Frequency responses
sim_mu	Simulation
sim_MC_mu	Monte-Carlo simulation
mu_wcp	Worst case gain

In this section, we consider the μ synthesis of stabilizing controllers of multivari-able plants. It is shown that this synthesis may be associated with serious problems related to the presence of large number of multiple parametric uncertainties. These problems may be removed by some simplification of uncertainty model which allow to obtain an acceptable solution. It is demonstrated on the example of cart–pendulum system that the closed-loop system with μ controller may achieve both robust stability and robust performance contrary to the other controller types, used in the previous sections of this chapter.

4.5.1 The μ synthesis problem

To apply the theory of structured singular value μ to the control system design, the control problem is represented in the form of LFT shown in Figure 3.27 and repeated for convenience in Figure 4.67. The system denoted by P is the open-loop interconnection which contains all known elements including the nominal plant model as well as the uncertainty weighting functions. The block Δ is the uncertain element of the set $\mathbf{\Delta}$, which parameterizes the whole model uncertainty. The controller is denoted by K. The input of P are three signal sets: inputs u_Δ from uncertainty, the reference, disturbance and noise signals collected in the n_w-vector w, and control actions u. Three output sets are generated: outputs y_Δ to uncertainty, n_z-vector of controlled outputs (errors) z, and measurements vector y (Figure 4.68).

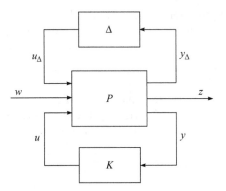

Figure 4.67   Standard representation of uncertain closed-loop system

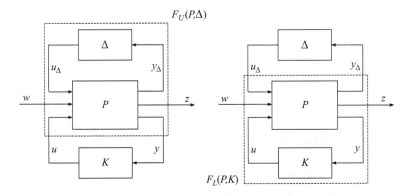

Figure 4.68   Two different transformations of the closed-loop system

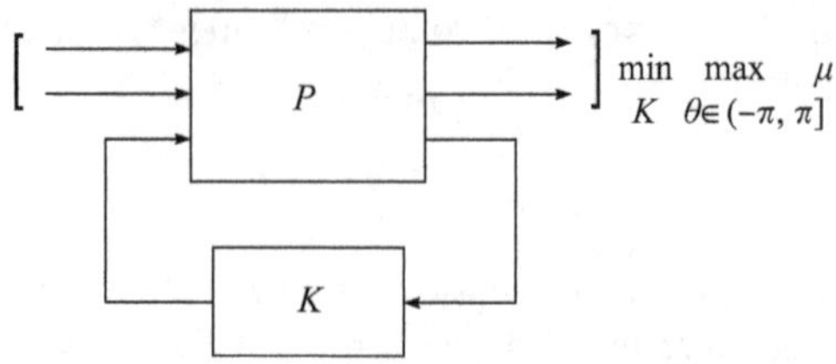

Figure 4.69   μ Synthesis

The set of systems, which have to be controlled, is described by the LFT (for definiteness, we assume the discrete-time case)

$$\left\{ F_U(P, \Delta) : \Delta \in \mathbf{\Delta}, \; \max_{\theta \in (-\pi,\pi]} \bar{\sigma}[\Delta(e^{j\theta})] \leq 1 \right\}.$$

The design aim is to determine a controller K, stabilizing the nominal system, such that for all $\Delta \in \mathbf{\Delta}$, $\max_{\theta \in (-\pi,\pi]} \bar{\sigma}[\Delta(e^{j\theta})] \leq 1$, the closed-loop system is stable and satisfies

$$\|F_U[F_L(P, K), \Delta]\|_\infty < 1.$$

For a given arbitrary K, this performance criterion can be checked by using robust-performance test on the lower LFT $F_L(P, K)$. The robust-performance test should be done in respect to the augmented uncertain structure

$$\Delta_P := \left\{ \begin{bmatrix} \Delta & 0 \\ 0 & \Delta_F \end{bmatrix} : \Delta \in \mathbf{\Delta}, \Delta_F \in \mathbb{C}^{n_w \times n_z} \right\}.$$

The system with controller K achieves robust performance if and only if

$$\mu_{\Delta_P}(F_L(P, K)(e^{j\theta})) < 1,$$

where $\mathbf{\Delta}_P$ is the set of all block-diagonal matrices Δ_P.

The aim of the *μ synthesis* is to minimize the peak value $\mu_{\Delta_P}(\cdot)$ of the closed-loop transfer function matrix $F_L(P, K)$ over the set of all stabilizing controllers K. This may be written as

$$\min_{\substack{K \\ \text{stabilizing}}} \; \max_{\theta \in (-\pi,\pi]} \mu_{\Delta_P}(F_L(P, K)(e^{j\theta})). \tag{4.47}$$

The optimization problem (4.47) is shown in Figure 4.69.

4.5.2   Replacing μ with its upper bound

One possible solution of the μ synthesis problem may be obtained replacing $\mu_{\Delta_P}(\cdot)$ with its upper bound. For a constant transfer function matrix $M = F_L(P, K)(e^{j\theta})$ and

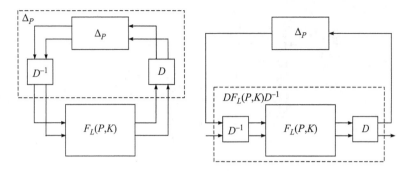

Figure 4.70 Introducing scaling matrices

uncertain structure $\mathbf{\Delta}_P$, an upper bound on $\mu_{\mathbf{\Delta}_P}(M)$ is the optimally scaled maximum singular value

$$\mu_{\mathbf{\Delta}_P}(M) \leq \inf_{D \in \mathbf{D}} \bar{\sigma}(DMD^{-1}).$$

In this inequality, $\mathbf{D}$ is the set of matrices with property $D\Delta_P = \Delta_P D$ for each $D \in \mathbf{D}$, $\Delta_P \in \mathbf{\Delta}_P$. Note that if the matrix Δ_P has a full $m_i \times m_i$ complex block Δ_i, then the corresponding block of the matrix D is a diagonal block of the form $d_i I_{m_i}$. If, however, Δ_P has a scalar block $\delta_j I_{m_j}$, then the corresponding block of D is a full $m_j \times m_j$ block D_j.

The usage of an upper bound on the structured singular value is equivalent to the introduction of scaling matrices D and D^{-1} in the system loop as shown in Figure 4.70. The introduction of these matrices gives additional freedom which may be used in controller design.

Using this upper bound, the optimization problem in (4.47) is reformulated as

$$\min_{\substack{K \\ \text{stabilizing}}} \max_{\theta \in (-\pi,\pi]} \min_{D_\omega \in \mathbf{D}} \bar{\sigma}[D_\omega F_L(P,K)(e^{j\theta})D_\omega^{-1}]. \tag{4.48}$$

In this equation, the minimization over D represents determination of approximation to the structured singular value $\mu[F_L(P,K)(e^{j\theta})]$. The scaling matrix D_ω is chosen from the set of scaling matrices $\mathbf{D}$, independently for each ω. Therefore, we have that

$$\min_{\substack{K \\ \text{stabilizing}}} \min_{D_\omega \in \mathbf{D}} \max_{\theta \in (-\pi,\pi]} \bar{\sigma}[D_\omega F_L(P,K)(e^{j\theta})D_\omega^{-1}]. \tag{4.49}$$

Using the notion for $\mathcal{H}_\infty$ norm, (4.49) is written as

$$\min_{\substack{K \\ \text{stabilizing}}} \min_{D_\omega \in \mathbf{D}} \left\| D_\omega F_L(P,K)D_\omega^{-1} \right\|_\infty \tag{4.50}$$

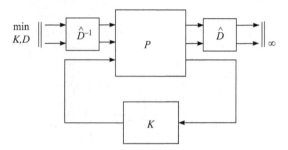

Figure 4.71   Replacing μ with its upper bound

Consider a single matrix $D \in \boldsymbol{D}$ and a complex matrix M. Assume that U is a complex matrix with the same structure as D, which satisfies $U^*U = UU^* = I$. Each block of U is unitary (orthogonal in the real case) matrix. The multiplication by orthogonal matrix does not affect the maximum singular value so that

$$\bar{\sigma}[(UD)M(UD)^{-1}] = \bar{\sigma}[(UD)MD^{-1}U^*] = \bar{\sigma}(DMD^{-1}).$$

In this way, the replacement of D by UD does not change the upper bound of μ. Using this freedom in the phase of each block D, we restrict the frequency dependent scaling matrix D_ω from (4.50) to be a real rational, stable, minimum phase transfer function $\hat{D}(z)$ without affecting the value of the minimum. Note that the computation of the upper bound on μ represents a convex optimization problem which can be efficiently solved.

The new optimization problem becomes

$$\min_{\substack{K \\ \text{stabilizing}}} \quad \min_{\substack{\hat{D}(z) \in \boldsymbol{D} \\ \text{stable,} \\ \text{minimum phase}}} \quad \left\| \hat{D} F_L(P,K) \hat{D}^{-1} \right\|_\infty. \tag{4.51}$$

This optimization problem is solved by an iterative approach which is called *DK iteration*. A block diagram, illustrating this optimization problem, is given in Figure 4.71.

4.5.3   DK iteration

4.5.3.1   First step of DK iteration: holding *D* fixed

The optimization problem (4.51) is difficult to solve since it has two independent matrix arguments K and $\hat{D}$. To find an approximate solution to this problem, first

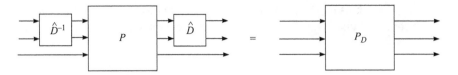

Figure 4.72 Absorbing the rational scaling matrices D

consider $D(z)$ as a fixed stable, minimum-phase real rational transfer function matrix $\hat{D}(z)$. This makes possible to solve the optimization problem

$$\min_{\substack{K \\ \text{stabilizing}}} \left\| \hat{D} F_L(P,K) \hat{D}^{-1} \right\|_\infty . \tag{4.52}$$

Define P_D to be the system, shown in Figure 4.72. The optimization problem (4.52) is equivalent to

$$\min_{\substack{K \\ \text{stabilizing}}} \left\| F_L(P_D,K) \right\|_\infty .$$

Since P_D is known at this step, this optimization problem is precisely an $\mathcal{H}_\infty$ optimization problem.

4.5.3.2 Second step of DK iteration: holding K fixed

With K held fixed, the optimization over D is done by the following two-step procedure:

1. Finding of optimal, frequently dependent scaling matrix D at a large, but finite set of frequencies (this is the computation of upper bound on μ).
2. Approximating of this optimal frequency-dependent scaling matrix by stable, minimum–phase, real, rational transfer function $\hat{D}$.

This two-step procedure is considered as a sufficiently reliable approach for finding D. The individual steps of this procedure are carried out efficiently with the existing approximation algorithms based on fast Fourier transform and least squares method.

The DK iterations are an efficient algorithm for μ synthesis which usually works successfully. The most serious limitation of these iterations is that they may converge to a local minimum, which is not necessarily the global one. This is because the μ synthesis is not a convex optimization problem and it is not guaranteed that the iterations will find the minimum μ. Despite these shortcomings, the DK iteration is the only available at the moment design method which allows to achieve robust performance of closed-loop systems with real and complex plant uncertainties.

The μ synthesis by DK iteration is implemented in Robust Control Toolbox by the function `dksyn` which can be used for continuous-time as well as discrete-time systems.

4.5.4 Numerical issues in μ synthesis

The DK iterations for μ synthesis works well when the number of uncertain blocks is sufficiently small. Unfortunately, the number of multiple real uncertainties in case of structured uncertainty is usually large, even for low-order systems with small number of parameters. This may require so large number of approximations that the corresponding computational problem becomes untraceable. In such case, it may become unavoidable to approximate some or all of the real uncertainties with a single additive or multiplicative complex uncertainty. Obviously, such approximation introduces some conservatism in the μ synthesis, but in most cases, it allows to obtain acceptable solution provided the performance weighting functions are carefully chosen. Another difficulty is related to the bad convergence of the DK iterations when the corresponding $\mathcal{H}_\infty$ design problem is singular or ill-conditioned. These issues are illustrated by the μ synthesis of the cart–pendulum system whose $\mathcal{H}_\infty$ design was presented in the previous section.

Example 4.7. μ Synthesis of the cart–pendulum system
Consider the μ synthesis of a two degree-of-freedom controller with integral action of the cart–pendulum system using mixed-sensitivity design.

The block diagram of the closed-loop system with the performance and control weighting functions is shown in Figure 4.73. This diagram is the same as the diagram used in $\mathcal{H}_\infty$ design except that instead of the nominal plant model G_{nom} we shall

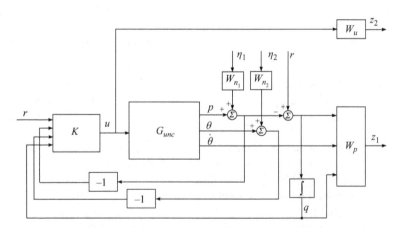

Figure 4.73   Closed-loop system with μ controller and weighting functions

make use of the uncertain model G_{unc}. In the design, we shall use the continuous-time weighting functions

$$W_p(s) = \begin{bmatrix} 0.95\dfrac{4s+1}{8s+0.05} & 0 & 0 \\ 0 & 1.6\dfrac{5s+1}{10s+0.01} & 0 \\ 0 & 0 & \dfrac{4s+1}{10s+0.05} \end{bmatrix},$$

$$W_u(s) = \dfrac{0.08s+0.1}{0.01s+1}$$

which are similar to the corresponding weighting functions used in the $\mathcal{H}_\infty$ design.

In the μ synthesis, we use the plant model of the cart–pendulum system which contains as uncertain parameters the equivalent cart mass M with multiplicity 10, the pendulum inertial moment I with multiplicity 10, the dynamic cart friction coefficient f_c with multiplicity 2, and the rotational friction coefficient f_p with multiplicity 2. This leads to an uncertainty matrix Δ_P with 24 real blocks and 1 complex block Δ_F. Hence at the second step of the DK iteration, it will be necessary to find approximations of the elements of two 10×10 repeated scalar blocks and two 2×2 of the scaling matrix D which requires to perform 208 approximations at each iteration. Since each approximation involves computations at several frequencies, the volume of the work makes the problem intractable. For the example under consideration, the program dksyn stops to work during the approximation of the first 10×10 block due to numerical difficulties. The attempts to approximate the individual blocks containing the uncertainties of M and I do not lead to result. That is why all scalar uncertainties of M, I, and f_c are approximated simultaneously by a single input multiplicative uncertainty. (Due to the small value of the parameter f_p it is substituted by its nominal value.)

The upper bound on the relative uncertainty of M, I, and f_c is shown in Figure 4.74. The uncertainties are computed by the M-file unc_approx_model for 4 values of each parameter which accounts for $4^3 = 64$ uncertainty values together. The upper bound on the relative uncertainty is approximated with sufficient accuracy by the magnitude response of a fourth order transfer function W_m which allows to obtain a plant uncertainty model in the form $G_{unc} = G_{nom}(1 + W_m\delta)$ where δ is a scalar complex parameter with $|\delta| \le 1$. In this way, the scalar repeated uncertainties of the plant model are replaced by a single complex input multiplicative uncertainty.

The Bode plots of the approximated transfer functions from control u to cart position p and from u to the pendulum angle θ are shown in Figures 4.75 and 4.76, respectively. It is seen that the plots corresponding to the approximated complex uncertainty are confined to slightly wider areas in comparison to the plots corresponding to the exact parametric uncertainty which reflects the approximation conservativeness.

The approximation of parametric uncertainties by an unstructured (complex) uncertainty may be done also for plants with several inputs and outputs. If the number of plant inputs n_u is larger than the number of plant outputs n_y, then it is advantageous

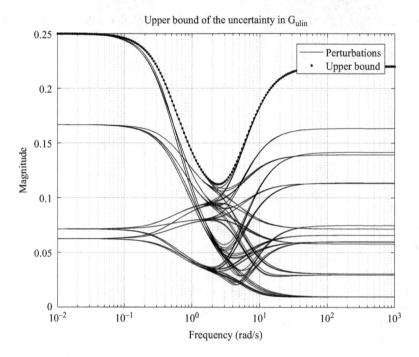

Figure 4.74   Upper bound on the real parametric uncertainties

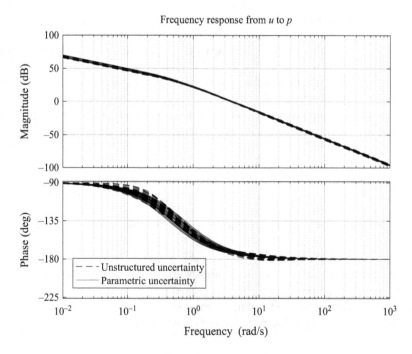

Figure 4.75   Bode plot of the transfer function from u to p

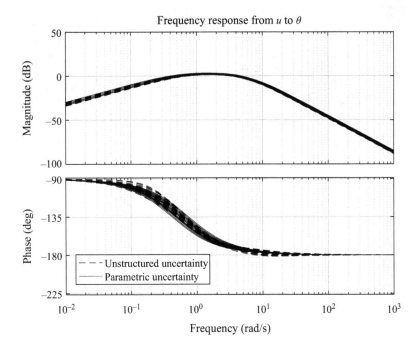

Figure 4.76   Bode plot of the transfer function from u to θ

to use as an approximation output multiplicative uncertainty; while in the case when n_u is smaller than n_y, it is better to use input multiplicative uncertainty.

The μ synthesis of the cart–pendulum system with the approximated uncertain plant is done by the M-files `olp_mu_reg` and `mu_design`. The first of these files determines the 11th order open-loop interconnection necessary for the design, while the second one implements the DK iterations using the function `dksyn`. To facilitate the iterations start, the first iteration is performed with a $\mathcal{H}_\infty$ controller designed by the function `hinfsyn` using the LMI option. The iteration progress is shown in the following report from `dksyn`:

```
Iteration Summary
------------------------------------------------------------------------
Iteration #                    2        3        4        5        6
Controller Order              25       25       21       21       25
Total D-Scale Order           10       10        6        6       10
Gamma Achieved             0.935    0.895    0.872    0.934    1.080
Peak mu-Value              0.933    0.894    0.871    0.920    1.073
```

Since the peak value of μ is minimal at the fourth step, it is justified to take the controller determined at this step. The stability test, however, reveals that the nominal closed-loop system with this controller is unstable. This surprising result is due to the fact that the matrix D_{21} of the state-space realization of the open-loop

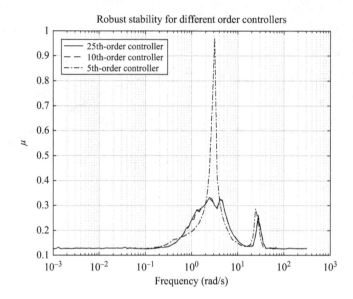

Figure 4.77   Robust stability for μ controllers of 25th, 10th, and 5th order

interconnection P is not of full row rank which violates assumption A(6) in the $\mathcal{H}_\infty$ design step. The floating point errors introduced in the computations make this matrix to appear as a matrix of full rank but it is ill-conditioned which causes the computed upper bound on μ to drop below the minimum possible value thus leading to non-stabilizing controller. That is why in the given case, we take the controller computed at the third step corresponding to minimum value of μ equal to 0.894. This controller stabilizes the closed-loop system and ensures robust stability and robust performance of the uncertain system. Note that while the open-loop interconnection is of 11th order, the controller designed is of 25 ft order. For this reason the controller order n_c is reduced obtaining controllers of tenth and fifth order, respectively.

In Figures 4.77 and 4.78, we give the plots of the structured singular values in case of robust-stability and robust-performance analysis, respectively, for the three controllers ($n_c = 25, 10, 5$). Note that the robustness analysis is done for the original uncertain model with real parametric uncertainty. Obviously, the controller of fifth order is not a suitable choice since it does not ensure robust performance of the closed-loop system. Further on, we shall use the controller of tenth order which leads to a closed-loop system with properties closed to the properties of the system with 25th-order controller. For the tenth-order controller, μ has value of 0.332 in case of stability robustness analysis and value of 0.963 in case of performance robustness analysis.

Results from Monte-Carlo analysis of the closed-loop system with μ controller are presented in Figures 4.79 and 4.80. The cart position and pendulum angle plots are confined to very narrow areas which is a consequence of the closed-loop system robust performance.

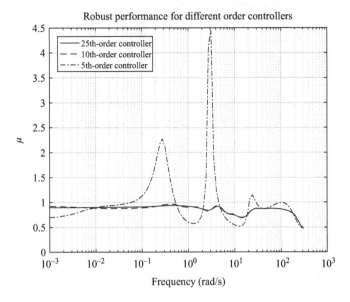

Figure 4.78   Robust performance for μ controllers of 25th, 10th, and 5th order

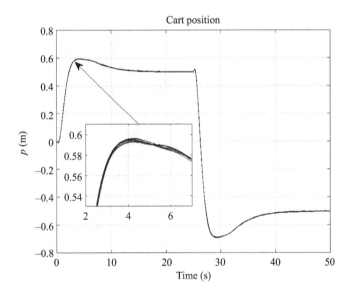

Figure 4.79   Monte-Carlo simulation. Cart position

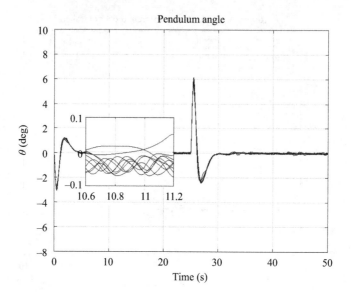

Figure 4.80   Monte-Carlo simulation. Pendulum angle

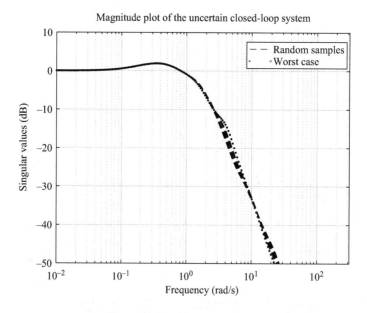

Figure 4.81   Worst case gain

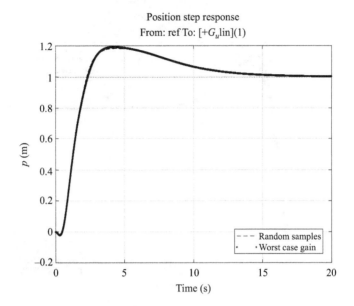

Figure 4.82   Worst case car position

The magnitude response of the closed-loop system for the worst case uncertainty is shown in Figure 4.81. The closed-loop bandwidth is equal to 1.40 rad/s, the same as in the case of $\mathcal{H}_\infty$ controller.

The worst case cart position step response and worst case pendulum angle responses along with the corresponding responses for 30 random values of the uncertain parameters are, respectively, shown in Figures 4.82 and 4.83. The worst case stability margins are

```
wcmarg =

    GainMargin: [0.6668 1.4997]
   PhaseMargin: [-22.6104 22.6104]
     Frequency: 4.3827
         WCUnc: [1x1 struct]
   Sensitivity: [1x1 struct]
```

Finally, in Figure 4.84, we show the control sensitivity to the cart position reference and measurement noises. It is seen from this figure that effect of noises at the plant input is sufficiently small.   ❐

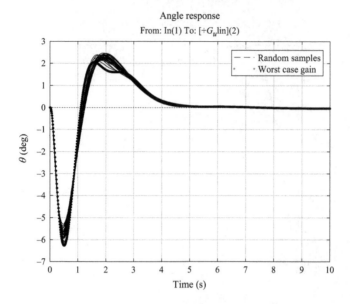

Figure 4.83   Worst case pendulum angle

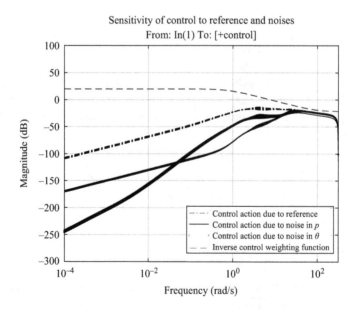

Figure 4.84   Control sensitivity to reference and noises

4.6 Controller comparison

A comparison of some closed-loop characteristics for different controllers in the case of cart–pendulum system is done in Table 4.4. (Note that the LQG controller and LQR controller plus $\mathcal{H}_\infty$ filter are taken in versions with bias compensation). The stability robustness margin shown in the third column of the table is computed by using μ-analysis.

An acceptable robustness in case of PID controller is obtained at the price of smallest closed-loop bandwidth leading to slowest transient response. In contrast to the PID controller, the LQG controller (LQ regulator+Kalman filter) leads to fastest transient response as a result of the largest bandwidth. The worst case margins, however, are relatively small for this controller. The LQ regulator plus $\mathcal{H}_\infty$ filter demonstrates better robustness properties than the LQG controller with comparable closed-loop performance for the plant under consideration. The worst case gain and phase margins for this controller are the largest margins for the five controllers under comparison. Surprisingly, the $\mathcal{H}_\infty$ controller has worse robustness properties in comparison to LQ regulator plus $\mathcal{H}_\infty$ filter and a smaller bandwidth. This leads to the conclusion that the controller combining LQR and $\mathcal{H}_\infty$ state estimation filter may be used as a successful alternative of the $\mathcal{H}_\infty$ controller. As expected, the best robustness properties of the closed-loop system are achieved by using μ controller. One should bear in mind, however, that such properties may be obtained at the price of large volume of computations necessary to find the appropriate weighting functions used in the μ synthesis. Note that both $\mathcal{H}_\infty$ and μ controllers have smaller bandwidths in comparison to linear quadratic controllers which reflects the trade-off between performance and robustness in controller design.

Table 4.4   Controller comparison for cart–pendulum system

Controller	Order	Stability robustness margin	Worst case gain margin	Worst case phase margin (deg)	Worst case bandwidth (rad/s)
PID	3	1.25	1.12	±6.24	0.42
LQG	7	1.12	1.28	±13.9	1.99
LQR + $\mathcal{H}_\infty$ filter	7	1.94	1.76	±30.9	1.90
$\mathcal{H}_\infty$	11	1.62	1.09	±4.8	1.40
μ	10	3.01	1.50	±22.6	1.40

4.7   HIL simulation

MATLAB files used in this section

Files for PID controller	Description
sim_HIL_SCI_PID.m	M-file for HIL simulation of PID controller
HIL_SCI_simulation_PID.slx	Host Simulink file for HIL simulation of PID controller
HIL_SCI_target_PID.slx	Target Simulink file for HIL simulation of PID controller

Files for LQG controller	Description
sim_HIL_SCI_LQG.m	M-file for HIL simulation of LQG controller
HIL_SCI_simulation_LQG.slx	Host Simulink file for HIL simulation of LQG controller
HIL_SCI_target_LQG.slx	Target Simulink file for HIL simulation of LQG controller

Files for LQ regulator and $\mathcal{H}_\infty$ filter	Description
sim_HIL_SCI_LQR_Hinf.m	M-file for HIL simulation of LQ regulator and $\mathcal{H}_\infty$ filter
HIL_SCI_simulation_LQR_Hinf.slx	Host Simulink file for HIL simulation of LQ regulator and $\mathcal{H}_\infty$ filter
HIL_SCI_target_LQR_Hinf.slx	Target Simulink file for HIL simulation of LQ regulator and $\mathcal{H}_\infty$ filter

Files for $\mathcal{H}_\infty$ controller	Description
sim_HIL_SCI_Hinf.m	M-file for HIL simulation of $\mathcal{H}_\infty$ controller
HIL_SCI_simulation_Hinf.slx	Host Simulink file for HIL simulation of $\mathcal{H}_\infty$ controller
HIL_SCI_target_Hinf.slx	Target Simulink file for HIL simulation of $\mathcal{H}_\infty$ controller

Files for μ controller	Description
`sim_HIL_SCI_mu.m`	M-file for HIL simulation of μ controller
`HIL_SCI_simulation_mu.slx`	Host Simulink file for HIL simulation of μ controller
`HIL_SCI_target_mu.slx`	Target Simulink file for HIL simulation of μ controller

The generalized block diagram of the closed-loop system HIL simulation is presented in Figure 4.85. The plant, actuator, and sensor dynamics with disturbances and noises is modeled in double precision on the host personal computer (PC). The corresponding controller is embedded in the target processor in single precision.

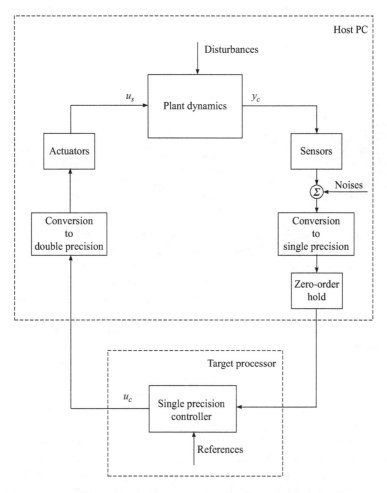

Figure 4.85   HIL simulation of the closed-loop system

290   *Design of embedded robust control systems using MATLAB®/Simulink®*

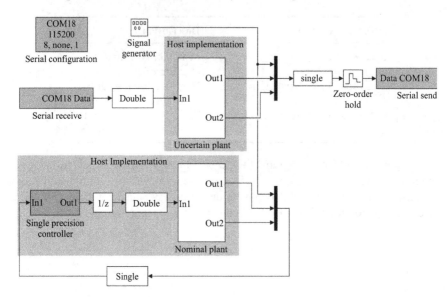

Figure 4.86   Host HIL Simulink model for cart–pendulum control

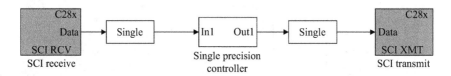

Figure 4.87   Target HIL Simulink model for cart–pendulum control

Consider the HIL simulation of the closed-loop cart–pendulum control system. The host Simulink model (Figure 4.86) contains two instances of the closed-loop system. At the bottom of the figure nominal plant model is controlled with host implemented single precision controller. Above is the worst case plant model controlled with target implemented controller with *TMS320F28335* Digital Signal Controller . In the given case, only plant model is host implemented, while its input and output signals are communicated over asynchronous serial connection with the target. So the target is represented as virtual or physical COM port accessible from Simulink model through driver blocks serial configuration, data sent and data receive.

While plant's model is implemented with double precision arithmetic its controller works with single precision or even fixed point arithmetic. A corresponding data type conversion blocks are the interface between different number representation formats. The user can select several rounding options and to observe its arithmetic effects. A zero-order hold block discretizes continuous-time plant output before sending it to the hardware target which works inherently in discrete time.

The controller implementation in a target hardware platform receives data for the plant output, calculates next control action and sends it back to the host (Figure 4.87).

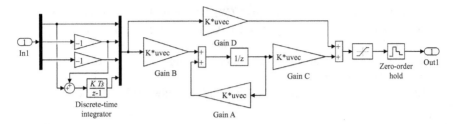

Figure 4.88 Data Flow model for code generation. Tenth-order μ controller for inverted pendulum and cart

The *TMS320F28335* microcontroller has a dedicated SCI module for asynchronous communication which is accessed from Simulink model with corresponding driver blocks. Host and target configurations for data rate, message length, stop bit, parity have to match for successful communication. Data rate for example is 115,200 bps. Therefore,

$$115{,}200 \text{ bps} = 14{,}400 \text{ bps} = 3{,}600 \text{ singles per second}$$
$$= 36 \text{ singles per tick} \qquad (4.53)$$

where $T_s = 0.01$ s. With this data rate one can transmit around 30 single precision numbers in each direction before sample interval to elapse. However, this is a theoretical maximum and practically achievable rate is around 15 signal values per second.

The control algorithms for inverted pendulum are represented in single precision as corresponding Simulink blocks. The model from Figure 4.88 represents the μ-controller with Simulink blocks appropriate for target code generation. Matrix gain, Unit delay and Sum blocks support control signal calculation. Saturation and zero-order hold blocks put amplitude and sample time constraints on the signal. Analogous to host implementation preprocessing applies to received data from serial port.

Further on, we represent the experimental data from HIL simulation of the cart–pendulum system for five different controllers—PID, LQG, LQR with $\mathcal{H}_\infty$ state estimator, $\mathcal{H}_\infty$ and μ-regulator. The LQG and LQR with $\mathcal{H}_\infty$ filter controllers are implemented with bias compensation. Consider for definiteness the HIL simulation of the cart–pendulum system in the case of LQG controller. The HIL experiment work flow in this case is the following one:

- Compile the target model HIL_SCI_target_LQG.slx to build executable .out file for deployment in embedded system. Intermediate .c and .h files are generated too, representing Simulink model functionality. All files are located in a separate folder named HIL_SCI_target_LQG_ticcs. The subfolder CustomMW contains the executable file HIL_SCI_target_LQG.out.

- Start the Code Composer Studio (CCS) Integrated Development Environment (IDE). Establish a connection between CCS and the target processor (first connect the platform to the USB port of the host PC and then logically access target processor memory from the CSS). Instruct the target processor to begin execution of the loaded program.
- Run the file `sim_HIL_SCI_LQG.m` which executes the host Simulink program `HIL_SCI_simulation_LQG.slx`.

The work flows for HIL simulation of the other controllers are similar to the described one.

As stated previously, the aim of control design for cart–pendulum system is pendulum stabilization in an upward position and minimization of tracking error between desired and measured cart position. On a signal level, the cart–pendulum system acts as a low frequency filter between control u and outputs cart position p and pendulum angle θ. So any sinusoidal oscillation in the closed loop will be more strongly evident in the signal u than in the signals θ and p. Also the dynamics of θ is faster than this of p because of lower inertial properties and unstable pole related to θ. Therefore, experimental results support that sensitivity of u to model perturbations is greater than sensitivity of θ which in turn is greater than sensitivity of p.

Consider first the results from simulation of PID controller. As noted in the previous section, the PID regulator is characterized by the slowest transient response. It is seen from Figure 4.89 that the settling time of the cart position is more than 10 s. This regulator has relatively good robustness properties. The deviation of the pendulum from vertical position is small and does not exceed 1.2 deg.

The LQG controller causes most oscillatory response for the perturbed system which is evident in cart position and pendulum angle measurement (Figure 4.90). At the same time, system reaction for the nominal parameters is not oscillatory at all. LQG control strategy assumes only zero mean Gaussian disturbances acting upon system ports and the underlying optimization problem doesn't account for deterministic disturbances which would be an equivalent signal representation of applied parametric perturbation.

The introduction of $\mathcal{H}_\infty$ state observer to LQR regulator lead to considerable attenuation of parametric perturbation (Figure 4.91). The state feedback matrix is derived from similar optimization problem as is the case with LQG; however, the $\mathcal{H}_\infty$ observer works to minimize maximal state estimation error rather than standard deviation of estimation error as in the case of Kalman filter and LQG control. As one can expect a more accurate state estimate can compensate to a certain level deviations between nominal and perturbed dynamics.

Design of $\mathcal{H}_\infty$ and μ controllers use different optimization cost functions contrasting to one used in linear quadratic control. The underlying structure of these controllers is again a state feedback with linear observer however tuned by different criteria. $\mathcal{H}_\infty$ controller minimizes the $\mathcal{H}_\infty$ norm for the transfer matrix from external disturbances to error signals describing deviation from nominal performance. Since parametric perturbation of the nominal plant can be represented as equivalent

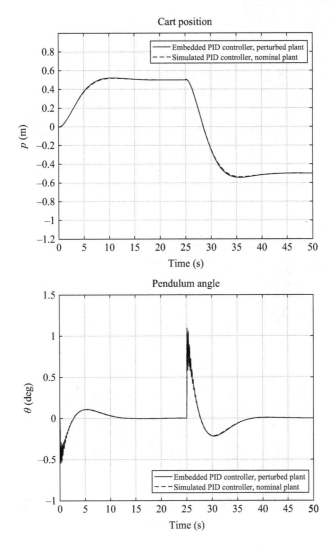

Figure 4.89   HIL-simulation of PID controller for cart–pendulum system

external disturbance, then the $\mathcal{H}_\infty$ controller will be effective to attenuate output error (Figure 4.92).

Since μ controller accounts directly for the uncertainty presented in the linear model of cart–pendulum system closed-loop responses are least oscillatory compared to other designed controllers (Figure 4.93). The controller minimize the structural singular value μ which describes the maximal parameter perturbation that makes the closed-loop system unstable.

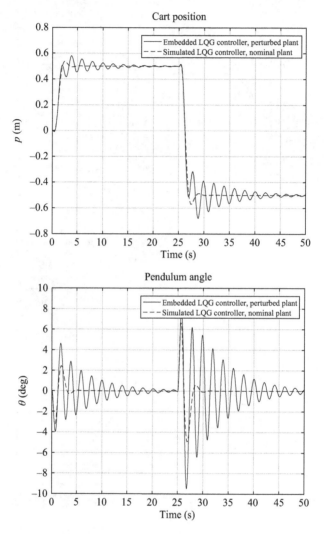

Figure 4.90   HIL-simulation of LQG controller for cart–pendulum system

4.8   Notes and references

Controller design is a fruitful area of control theory which is extremely rich of ideas and results. In this chapter, we present only a few results concerning the design of practically implemented controllers.

The PID controller is probably the most popular control device. A readable introduction to the theory of PID control is given in Åström and Murray [89, Ch. 10]. Detailed presentation of PID controller design and its practical implementation may

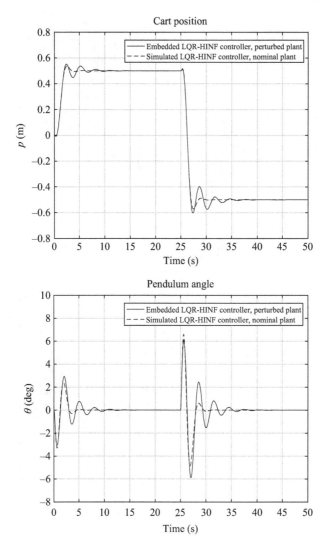

Figure 4.91   HIL-simulation of LQR controller with $\mathscr{H}_\infty$ observer for cart–pendulum system

be found in many books, see for instance, Åström and Hägglund [115], Johnson and Moradi [116], Visioli [117]. An immense collection of PI and PID tuning rules is presented by O'Dwyer [118] and an overview of the functionalities and tuning methods for PID controllers in patents, software packages and commercial hardware modules may be found in Ang, Chong, and Li [119]. Both decentralized and centralized forms of PID control for multivariable processes are described in Wang, Ye, Cai,

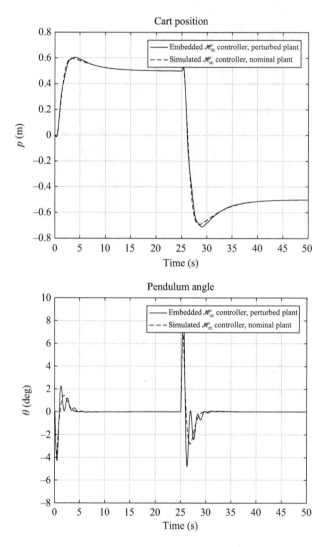

Figure 4.92 HIL-simulation of $\mathcal{H}_\infty$ controller for cart–pendulum system

and Hang [120]. The implementation of PID controllers by using FPGA is addressed in [121]. Finally, the perspectives of PID control in the third millennium are discussed in the book edited by Vilanova and Visioli [122].

As noted in [45, Section 5.5], the solution of LQR problem is probably one of the most important results in modern control theory and one that has far-reaching consequences for the design of optimal control systems. The LQR control theory begun with the seminal paper of Kalman [123] and its detailed presentation may be found in many excellent books such as Anderson and Moore [111], Bryson and

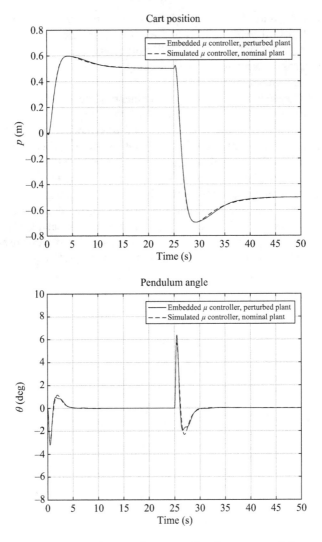

Figure 4.93   HIL-simulation of μ controller for cart–pendulum system

Ho [71], Kwakernaak and Sivan [124], Goodwin, Graebe, and Salgado [43], Green and Lamebeer [99] and Hendricks, Jannerup, and Sørensen [45].

The LQG theory for stochastic systems is based on linear quadratic optimization and Kalman filtering and is presented in lots of books, among them Anderson and Moore [75]; Åström [70]; Crassidis and Junkins [76]; Gibbs [77]; Lewis, Xie, and Popa [44]; Simon [78]; and Speyer and Chung [73].

Inclusion of Kalman filter in the LQG controller may lead to significant shrinking of the closed-loop stability margin in comparison to the full state LQR. In such case it is possible to use the so-called loop transfer recovery (LTR) procedure (see for instance [125,126]) which allows to "recover" to some extent the LQR stability margin. The practical implementation of this procedure is restricted to systems with transmission zeros in the left half part of the complex plane (minimum-phase systems) and square transfer functions.

Systematic presentation of $\mathcal{H}_\infty$ filtering theory is done in Green and Limebeer [99]; Lewis, Xie, and Popa [44]; and Simon [78].

Detailed exposition of the $\mathcal{H}_\infty$ design and μ synthesis may be found in the books of Zhou, Doyle, and Glover [82,102]; Green and Limebeer [99]; Skogestad and Postlethwaite [87]; and Sánchez-Peña and Sznaier [103]. A loop shaping design procedure using $\mathcal{H}_\infty$ synthesis is presented in [127]. The theory and numerical algorithms behind the LMI approach to the $\mathcal{H}_\infty$ design are presented in Boyd, El Ghaoui, Feron, and Balakrishnan [128]; El Ghaoui and Niculescu [129]; Nesterov and Nemirovskii [130]; Gahinet, Nemirovski, Laub, and Chilali [112]; and Gahinet and Apkarian [114]. The numerical problems in $\mathcal{H}_\infty$ design are addressed in Gahinet [131], Gahinet and Laub [132], and Stoorvogel [133].

Chapter 5

Case study 1: embedded control of tank physical model

MATLAB® files used in this section

File	Description
ident_tank.m	Tank plant identification
experiment_ident_tank.mat	Measured input–output data for tank identification
tank_static_characteristic.slx	Code generation for measurement of static characteristic
tank_LQR_LQG_design.m	Design of water level LQG controller
tank_Hinf_design.m	Design of water level $\mathcal{H}_\infty$ controller
frs_and_time_response_LQG.m	Plots of frequency response and time response of closed-loop system with LQG controller
frs_and_time_response_Hinf.m	Plots of frequency response and time response of closed-loop system with $\mathcal{H}_\infty$ controller
LQR_sim.slx	Simulation of closed-loop system with LQR controller
LQG_sim.slx	Simulation of closed-loop system with LQG controller
Hinf_sim.slx	Simulation of closed-loop system with $\mathcal{H}_\infty$ controller
Kalman_Hinf.slx	Off-line filtration of output signal of closed-loop system with $\mathcal{H}_\infty$ controller
LQG_cl_data.mat	Measured data from experiment with LQG controller
Hinf_cl_data.mat	Measured data from experiment with $\mathcal{H}_\infty$ controller

This chapter presents development and experimental evaluation of low-cost embedded system for control of liquid level in a model of tank. The plant is a physical laboratory model of water tank produced by Lucas Nülle Company [134]. The liquid level is controlled in wide range by designed and $\mathcal{H}_\infty$ controllers. The control algorithms are implemented in low-cost control kit Arduino Mega 2560 [135]. Software in MATLAB/Simulink® environment is developed for generation of control code. Some additional simple hardware devices are developed too. These devices provide appropriate voltage level of analogue signals which are exchanging between physical model of tank and control kit. Controllers are designed on the basis of the linear discrete-time black-box model derived from experimental data by one of the identification techniques described in Appendix D. The main advantage of this technique is that we obtain low-order models of plant and noise. The noise model is used to design appropriate Kalman filter that reduces significantly the sensitivity of control signal to the noise, which is very important for correct exploitation of the actuator. Results from simulation of the closed-loop system as well as experimental results obtained from real-time implementation of designed controllers are given. They confirm embedded control system performance in the whole working range.

The chapter is organized as follows. In Section 5.1, we give information about hardware configuration of embedded system for control of water level in tank. The derivation and validation of water tank model are considered in Section 5.2. The model obtained is used to design the linear-quadratic-Gaussian (LQG) and $\mathcal{H}_\infty$ controllers in Sections 5.3 and 5.4. The experimental results and comparison with simulation results are given in Section 5.5. The process of controller design, code generation, and controller evaluation is done by using MATLAB and Simulink.

5.1   Hardware configuration of embedded control system

The scheme of embedded system for control of water level in tank is shown in Figure 5.1. The system is comprised of water tank, voltage divider, DIP reed relay, and ARDUINO Mega 2560 kit. The aim of control is to set the water level to desired one regardless of amount of water which flows out through the outlet valve. The desired

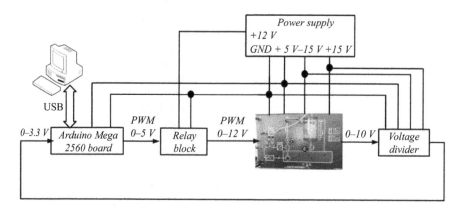

Figure 5.1   System for control of water level in tank

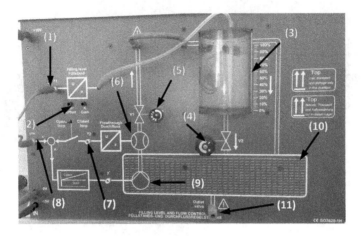

Figure 5.2 Water tank model

level is set by reference signal, and it is achieved via manipulating inlet water flow by water pump. The water level is measured by filling level sensor.

5.1.1 Water tank

The water tank produced by Lucas Nülle Company is shown in Figure 5.2, where (1) is the output of liquid level sensor, (2) knobs for tuning the offset and gain of liquid level sensor, (3) tank with liquid level scale, (4) manual outlet valve which can be used to produce the load disturbance, (5) manual inlet valve, (6) flow rate sensor, (7) output of flow rate sensor, (8) input for manipulated variable of water pump, (9) water pump which pumps the water in tank from the water reservoir located at a lower altitude, (10) water reservoir, and (11) is the outlet valve to empty model. To control the water level, we change the inlet water flow by manipulating water pump voltage via input (8) by pulse width modulation (PWM) signal. This signal should have high level of 10 V and low level of 0 V. The water level in tank is measured by liquid level sensor (1), which generates output signal in range 0–10 V.

5.1.2 ARDUINO MEGA 2560

The Arduino Mega 2560 board is presented in Figure 5.3. It is one of the most popular low-cost microcontroller kit [135], which can be used for automatic control of various plants. The Arduino Mega 2560 is a kit based on the microcontroller ATmega2560. It has 54 digital input/output pins, 16 analog inputs, 4 UARTs (hardware serial ports), a 16-MHz crystal oscillator, a USB connection, a power jack, an ICSP header, and a reset button. A total of 15 digital outputs can be used as PWM ones. This kit contains everything needed to support the microcontroller: simply connect it to a computer with a USB cable; power it through HOST computer USB port or with an AC-to-DC adapter or battery; provides various communication interfaces. The Mega 2560 board

Figure 5.3   Arduino Mega 2560 kit

Table 5.1   Technical specification of Arduino Mega 2560 kit

Microcontroller	ATmega2560
Operating voltage	5 V
Input voltage (recommended)	7–12 V
Input voltage limits	6–20 V
Digital I/O pins	54 (15 PWM outputs)
Analog inputs	16
DC current per I/O Pin	20 mA
DC current for 3.3-V Pin	50 mA
Flash memory	256 kb
SRAM	8 kb
EEPROM	4 kb
Clock speed	16 MHz

is compatible with most shields designed for the Uno and the former boards. A brief technical specification of controller kit is given in Table 5.1.

The ATmega2560 microcontroller has 256 kb of flash memory for storing code, 8 kb of SRAM, and 4 kb of EEPROM. Each of the 54 digital pins on the Mega can be used as an input or output, using appropriate configuration functions. They operate at 5 V. Each pin can provide or receive current of 20 mA as recommended operating condition and has an internal pull-up resistor (disconnected by default) of 20–50 kΩ. The maximum current of 40 mA must not be exceeded to avoid permanent damage of the microcontroller. Some of digital pins have specialized functions provided serial communication, SPI communication, external interrupts, and PWM output. The Mega 2560 has 16 analog inputs, each of which provide 10 b of resolution (i.e., 1,024 different values). By default they measure from ground to 5 V.

To control the water level in tank by manipulating the water pump voltage via PWM signal, we should have signal with high level of 10 V and low level of 0 V.

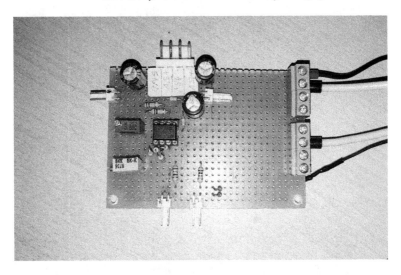

Figure 5.4   Voltage divider

The 8-bit PWM signal generated by digital pin of Mega 2560 has frequency of 490 Hz, high level of 5 V and low level of 0 V. Thus, it should be amplified to high level of 10 V. This is done by DIP reed relay. The control signal of embedded system is the integer number which is written to the PWM output. It can take values in range 0–255.

The output signal of water level sensor takes values in the range 0–10 V, but the range of ADC converter of Arduino Mega 2560 is 0–5 V. Thus, the sensor voltage signal should be divided. This is done by the voltage divider, which linearly transform the sensor voltage from the range of 0–10 V to the range of 0–3.3 V. The 10-bit ADC generates integer numbers in the range 0–1,023, but the value of voltage divider's output signal which corresponds to the water level of 100 percent is approximately 3.3 V. This means that the maximum integer number obtained from ADC is 676. Thus, the control variable can take integer values in the range 0–676.

The Arduino Mega 2560 board can be programed with the free Arduino Software (IDE) or the source code can be automatically generated and embedded in board from MATLAB/Simulink environment with the aid of Simulink Coder™ [9] and Simulink Support package for Arduino Hardware [136].

5.1.3   Voltage divider

The photo of voltage divider is shown in Figure 5.4. It linearly divides the output signal of water level sensor from the range of 0–10 V to the range of 0–3.3 V. The circuit diagram is presented in Figure 5.5. Voltage divider is designed on the basis of TL-082 JFET-Input Operational Amplifiers and passive electronic elements which provide the desired characteristics. The output of voltage divider is connected to ADC input of Arduino Mega 2560 board.

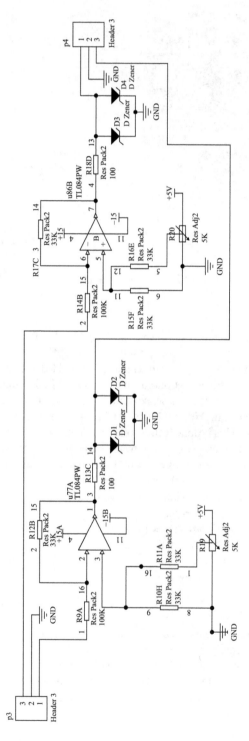

Figure 5.5   Circuit diagram of voltage divider

Figure 5.6   Relay block

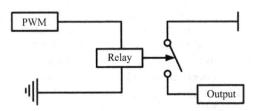

Figure 5.7   Diagram of relay block

5.1.4   Relay block

The photo of designed relay block is shown in Figure 5.6. It is based on DIP reed relay 1A72-12L. Relay switches the supply voltage of 12 V according to PWM signal of 5 V generated by Arduino Mega 2560 board as it is shown in Figure 5.7. The output of Relay block is connected to the manipulating input of water pump.

5.2   Plant identification

To determine the mathematical model of tank, one may apply physical modeling or identification. Physical modeling requires profound knowledge about physics of the plant and a lot of a priori information such as pump characteristics and hydraulic resistance of outlet valve. Due to the lack of a priori information, in this study, we prefer to use a numerical model obtained by identification procedure. Another reason to use an identification model is that such model in addition to the tank dynamics takes into account the dynamics of water level sensor and water pump which facilitates the

Measurement of tank static characteristic

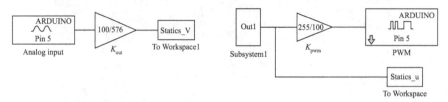

Figure 5.8   Simulink diagram for measuring of tank static characteristic

plant description. Moreover, by identification, we may obtain model of the noise, which can be used to design an appropriate optimal filter such as Kalman filter.

Before performing identification procedure, we should measure the plant static characteristics. It will show the range in which we can control the water level by manipulating water pump voltage via the 8-b PWM signal. To measure static characteristic, a specialized simulation diagram in MATLAB/Simulink environment is developed. This diagram is shown in Figure 5.8. It uses Analog Input block and PWM block from the library "Simulink support package for Arduino Hardware." With the aid of Simulink Coder [9] and Simulink support package for Arduino Hardware [136], the code is generated from this software which is embedded in Arduino Mega 2560 board. Configuration of properties for code generation and configuration of Arduino blocks will be discussed in more detail in Section 5.4. The input signal is the duty cycle of PWM signal, which can take values in the range of 0–100 percent. Thus, to transform the input signal to 8 b integer number, we should multiply it by scaling coefficient $K_{pwm} = 255/100$. The plant output signal is the water level in tank, which can take values in the range of 0–100 percent too. Due to the characteristic of voltage divider, the maximum integer value obtained from 10 b ADC is equal to 676. This means that we should multiply it by scaling coefficient $K_{out} = 100/676$. The static characteristic of plant is shown in Figure 5.9, and the steady-state values of input and output signals are given in Table 5.2.

As can be seen from Figure 5.9, the plant has linear characteristics for values of input signal in the range 58–100 percent. The water pump has wide dead zone of 58 percent. To use linear plant model and to control water level by linear controller, we should avoid the actuator dead zone. This may be done by adding the constant value of 58 percent to input signal. Static characteristic shows that we can control water level in the whole working range with duty coefficients between 59 percent and 100 percent.

Plant identification is done by some of the methods described in Appendix D. The goal is to obtain a linear black-box model of tank which sufficiently well describes plant dynamics and noises in wide working range. To obtain such model first, we design the open-loop identification experiments. In case of step input signal water level achieves its steady-state value for time of approximately 300 s. Thus, we choose the sample time $T_s = 0.5$ s which is sufficiently small. We apply the constant input signal of 75 percent for first 630 s of the identification experiment. This value is

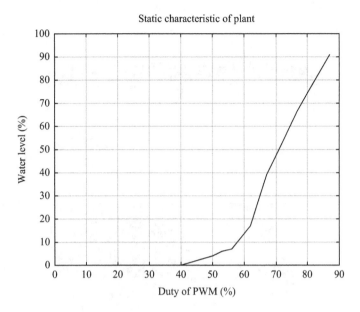

Figure 5.9   *Tank static characteristic*

Table 5.2   Steady-state values of plant input and output

Input signal (%)	0	20	40	50	53	56	59	62	67	77	87
Output signal (%)	0	0	0	4	6	7	12	17	39	67	91

approximately equal to the middle of actuator linear range. As a result, the water level achieves steady-state value of approximately 62 percent. After first 630 s, the random binary signal (RBS) is added to constant input signal which provides persistent excitation of identification signal. RBS is obtained from filtered through relay white Gaussian noise. The amplitude of RBS is chosen to be ±15 percent, so the input signal takes two values of 60 percent and 90 percent. In this manner, the whole linear range of input signal is used. The Simulink scheme for identification experiment code generation is `Ident_tank.slx`. The measured input–output data are shown in Figure 5.10. Nevertheless, we apply only constant input signal for the first 630 s, it is seen that plant output is corrupted by significant noise. This noise comprises power supply noise, measurement noise and noise generated from A/D conversion of output signal.

Before using measured input–output data for identification, we should cut the part of data which corresponds to the constant value of input signal, and the remaining one should be centered. Then, we can form two data sets first for model estimation and second for model validation. Next steps are to check persistence excitation level

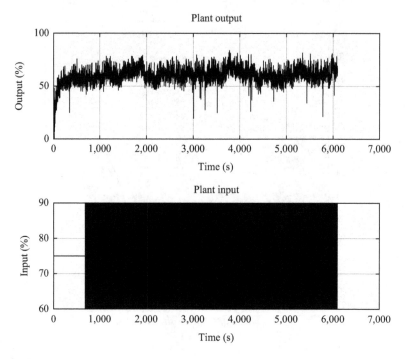

Figure 5.10   Measured input–output data

of input signal and to plot data sets (Figures 5.11 and 5.12). These are done by the command lines

```
load('experiment_ident_tank.mat')
y = y(3000:end);
u = u(3000:end);
y = y-mean(y);
u = u-mean(u);
yestimate = (y(1000:2000));
uestimate = (u(1000:2000));
yval = (y(2001:3000));
uval = (u(2001:3000));
IdentData = iddata(yestimate,uestimate,0.5);
ValidateData = iddata(yval,uval,0.5);
figure()
idplot(IdentData), grid
title('Estimation Data');
figure()
idplot(ValidateData), grid
title('Validation Data');
pexcit(IdentData,1000)
```

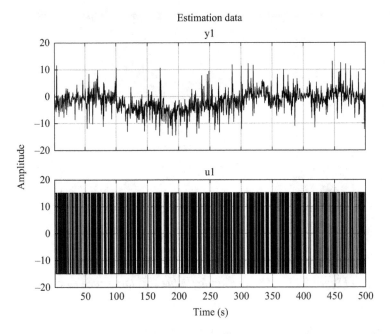

Figure 5.11 Estimation input–output data

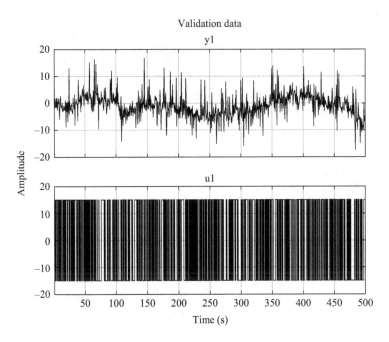

Figure 5.12 Validation input–output data

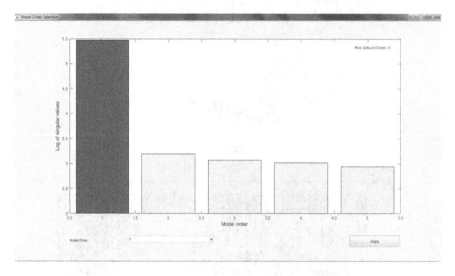

Figure 5.13   Hankel singular values of state-space models

The excitation level of input signal is 500, which means that we can estimate up to 500 parameters from estimation data set.

The procedure may start with estimation of simple black-box model with few design parameters. Such model is state-space model (D.65) in Appendix D with free parametrization. We choose to estimate a state-space model (D.65) because it has a form which can be directly used for Kalman filter design with MATLAB. Assuming that the possible model order is between 1 and 5, we form the model set of five state-space models. The order of the best model from this set is determined by MATLAB function n4sid with the command line

```
n4sid(IdentData,1:5)
```

The resulting plot of Hankel singular values is depicted in Figure 5.13. As it is seen, the best model order is 1. Then, the first-order state-space model (D.65) from Appendix D is estimated by the command lines

```
opt = ssestOptions('SearchMethod','lm');
sys_ss = ssest(IdentData,1,'Ts',0.5,opt)
```

The obtained model is in innovation form and has matrices

$$A = 0.9442, \quad B = 1.887 \times 10^{-4}, \quad D = 0, \quad K = 2.89 \times 10^{-4}. \tag{5.1}$$

The next step is to perform validation test of estimated state-space model. The test of residuals is done by the command lines

```
figure()
resid(sys_ss,ValidateData), grid
title('Residuals correlation of state space model')
figure()
```

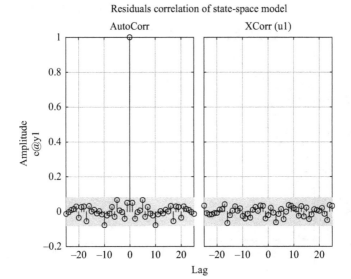

Figure 5.14   Residual test of state-space models

```
resid(sys_ss,ValidateData,'fr'), grid
title('Residuals frequency response of state space model')
```

The results from whitening and independence tests are shown in Figure 5.14. The frequency response of the estimated high-order finite impulse response model between control signal and residuals along with 99 percent confidence region is depicted in Figure 5.15. As can be seen from Figure 5.14, the obtained model passes both the tests, which means that the model captures sufficiently well tank dynamics and the noise model is adequate. Figure 5.15 shows that there is not significant dynamics between input signal and residuals in the whole interested frequency range. The comparison between identified model output and measured water level is shown in Figure 5.16. It is done by command line

```
figure()
compare(sys_ss,ValidateData),grid
title('Comparison between model and measured output signals')
```

The value of metric (D.71) for water level is $FIT = 15.33$ percent, which at a first glance is not so good result, but it is seen from Figure 5.16 that the estimated model captures sufficiently well slow plant dynamics. This is confirmed again from the comparison between model step response and measured step response, which is shown in Figure 5.17. The measured step response is obtained for input signal of 75 percent and the comparison is done by the command lines

```
load('experiment_ident_tank.mat')
t1 = 0:0.5:0.5*999;
y_sim = sim(sys_ss,u(1:1000)-58.5);
```

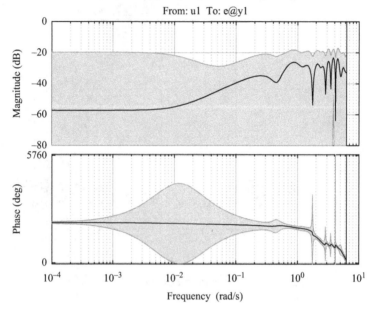

Figure 5.15   Residual to input signal frequency response

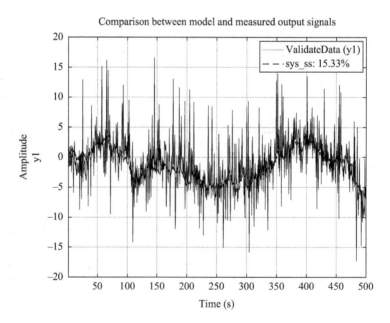

Figure 5.16   Model water level and measured water level

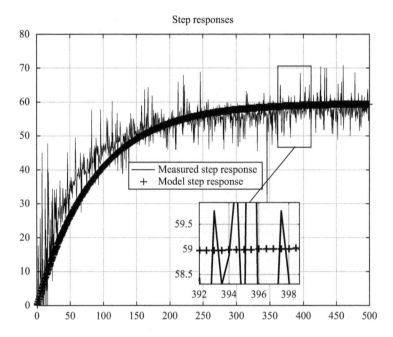

Figure 5.17   Model step response and measured step response

```
t_sim = 0:0.5:0.5*(length(y_sim)-1)
figure(23)
plot(t1,y(1:1000),'k',t_sim,y_sim,'k+'),grid
legend('Measured step response','Model step response')
```

Figure 5.17 shows again that we have a significant noise in measured data. This is the reason for low value of metric (D.71). The Bode plot from noise (model residual) to model output is shown in Figure 5.18. It is done by the command lines

```
sys_aug = ss(sys_ss,'augmented')
figure()
bodemag(sys_aug(1,2)),grid
title('Noise to model output frequency response')
```

It is seen that the noise is amplified in the whole interesting frequency range.

As a result from identification, we obtain a first-order state-space model which describes sufficiently well both the plant and the noise dynamics. As can be seen, the plant is corrupted by significant noise, but by using the noise model, we can reduce the influence of noise on the plant behavior. The plant dynamic model will be used for controller design, whereas the noise model will be used for Kalman filter design in the next section.

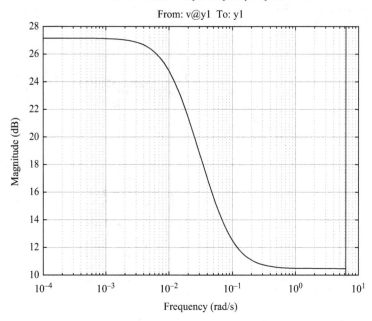

Figure 5.18   *Noise to model output frequency response*

5.3   LQR and LQG controllers design

The structure diagram of water level control system with LQR controller is shown in Figure 5.19. The design of LQR controller with integral action is done on the basis of deterministic part of estimated first-order tank model:

$$
\begin{aligned}
x(k+1) &= Ax(k) + Bu(k) + K_n e(k), \\
y(k) &= Cx(k) + e(k),
\end{aligned}
\tag{5.2}
$$

where matrices A, B, and C have values from (5.1). To include integral action in LQR controller, the approximation of discrete-time integral of system error is regarded:

$$
x_i(k+1) = x_i(k) + T_s err(k) = x_i(k) + T_s(r(k) - y(k)),
\tag{5.3}
$$

where $x_i(k)$ is the integral of system error $err(k) = r(k) - y(k)$, $T_s = 0.5$ s is the sample time, and $r(k)$ is the reference. Combining (5.2) and (5.3), one obtains the augmented system

$$
\begin{aligned}
\bar{x}(k+1) &= \bar{A}\bar{x}(k) + \bar{B}u(k) + \bar{G}r(k), \\
y(k) &= \bar{C}\bar{x}(k),
\end{aligned}
\tag{5.4}
$$

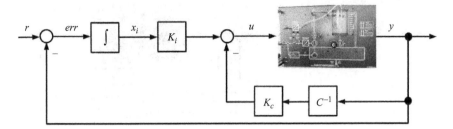

Figure 5.19   Water level control system with LQR controller

where

$$\bar{x}(k) = \begin{bmatrix} x(k) \\ x_i(k) \end{bmatrix}, \quad \bar{A} = \begin{bmatrix} A & 0 \\ -T_sC & 1 \end{bmatrix}, \quad \bar{B} = \begin{bmatrix} B \\ 0 \end{bmatrix}, \quad \bar{C} = [C0], \quad \bar{G} = \begin{bmatrix} 0 \\ T_s \end{bmatrix}.$$

The optimal control law is obtained in the form

$$u(k) = -K_cx(k) + K_ix_i(k), \tag{5.5}$$

where K_c and K_i are the controller coefficients. Note that there is no need to design a state observer because the model (5.2) is of first order, and the state can be obtained as

$$x(k) = C^{-1}y(k). \tag{5.6}$$

The design of LQR controller is done for

$$Q = \begin{bmatrix} 1.5 & 0 \\ 0 & C^TC \end{bmatrix}, \quad R = 1$$

by MATLAB file `Tank_lqr_lqg_design.m`.

The presence of significant noise in measured output determines the need to reduce influence of noise to water level and especially to actuator. This may be done by Kalman filter. The structure diagram of water level control system with LQG controller, which comprises LQ regulator and Kalman filter, is shown in Figure 5.20. The designed Kalman filter is described by the following equations:

$$\begin{aligned} \hat{x}(k) &= (A - K_fCA)\hat{x}(k-1) + (B - K_fCB)u(k-1) + K_fy(k), \\ \hat{y}(k) &= C\hat{x}(k), \end{aligned} \tag{5.7}$$

where $\hat{x}(k)$ is the state estimate, $\hat{y}(k)$ is the estimate of water level, and K_f is the Kalman filter gain. The LQG controller forms control signal according to

$$\begin{aligned} u(k) &= -K_c\hat{x}(k) + K_i\hat{x}_i(k), \\ \hat{x}_i(k+1) &= \hat{x}_i(k) + T_s(r(k) - \hat{y}(k)). \end{aligned} \tag{5.8}$$

Note that the design of Kalman filter (5.7) is performed for fictive output noise with small variance of 0.001 because there is no additional output noise in the

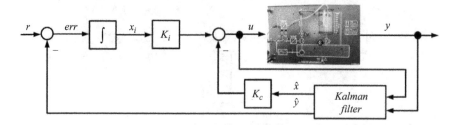

Figure 5.20   *Water level control system with LQG controller*

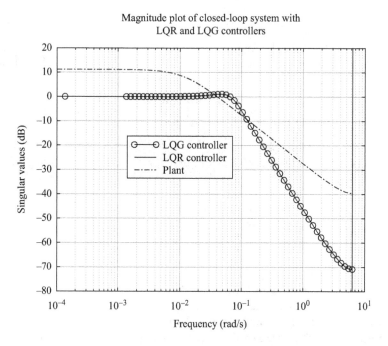

Figure 5.21   *Magnitude plot of closed-loop system*

model (5.2). The Kalman filter design is done using MATLAB function `kalman` by MATLAB file `tank_lqr_lqg_design.m`.

Now we compare the properties of designed LQR and LQG controllers in both frequency and time domains. The magnitude plot of the closed-loop system is shown in Figure 5.21 and the sensitivity of control signal to model noise is depicted in Figure 5.22. As should have been expected, the closed-loop systems with both controllers have the same characteristics. This is due to the fact that the frequency responses are calculated for zero initial conditions. It is seen that the closed-loop

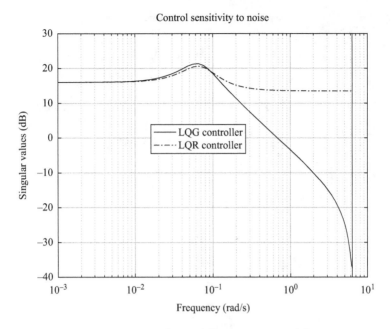

Figure 5.22   Control signal sensitivity to model noise

bandwidth is approximately 0.085 rad/s, which is sufficiently larger than plant band-width. The closed-loop system will track without steady-state error reference signal with frequency up to 0.085 rad/s. The control signal of closed-loop system with LQR controller is very sensitive to noise in high frequencies. Thus, for this system, the noise will be amplified by controller. As a result, undesirable oscillations with signif-icant amplitude will be occurred in control signal. In real application, this may cause damage of actuator. On the other hand, the control signal for closed-loop system with LQG controller is not sensitive to model noise in high frequencies due to the filtration properties of Kalman filter.

The reference signal, exact (without noise) output signal, and estimation by Kalman filter output signal of closed-loop system with LQG controller are presented in Figure 5.23. In Figure 5.24, the exact plant state and estimated one are presented. In Figure 5.25, control signals of closed-loop systems with LQR and LQG controllers are shown. It can be seen that the system with LQG controller tracks the reference signal in the whole working range. The settling time of approximately 100 s is achieved which is three times less than the one for plant step response. The overshooting is negligible. The Kalman filter estimates correctly plant output and state. Figure 5.25 shows the advantages of system with Kalman filter. The control signal of LQG controller does not exceed 42 percent which is the maximum allowed value due to the water pump dead zone of 58 percent. In real experiment, we will add 58 percent to the control signal produced by controller. In this way, we will work only in linear zone of actuator. In accordance with frequency response shown in Figure 5.22, the

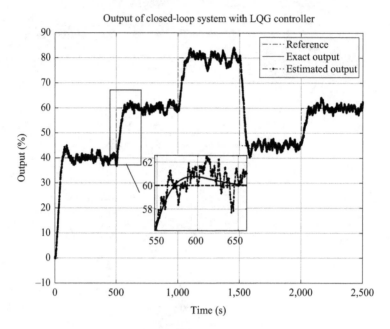

*Figure 5.23   Exact output and estimated output signal of system with LQG
controller*

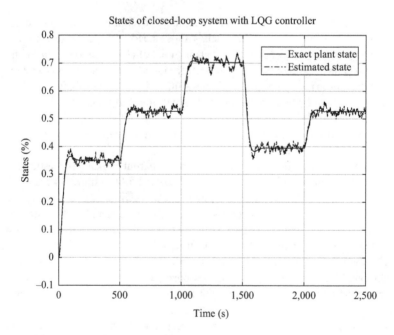

Figure 5.24   Exact plant state and state estimate of system with LQG controller

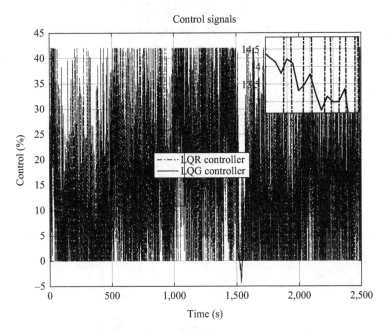

Figure 5.25   Control signals of system with LQG controller and LQR controller

control signal of LQR controller has many high frequency oscillations and exceeds the maximum allowed value. As a result in real application, the damage of actuator can occur.

5.4 $\mathcal{H}_\infty$ Controller design

To obtain good performance of closed-loop system, we shall design a two-degree of freedom $\mathcal{H}_\infty$ controller. The block diagram of control system is presented in Figure 5.26, where e_p and e_u are the weighted closed-loop system outputs and d is the output disturbance. The aim of controller determined by $\mathcal{H}_\infty$ synthesis is to ensure stability and performance of closed-loop system in the presence of disturbance and to provide good tracking of reference in whole working range. For simplification of design procedure, we do not take into account the model noise $e(k)$, but this noise will be used in performance analysis of the closed-loop system. In order to obtain better tracking of reference, the integral of system error is introduced in controller. To embedded the designed controller in Arduino Mega 2560 board the discrete-time $\mathcal{H}_\infty$ synthesis is done with sample time 0.5 s.

Let

$$y_c = [y_m \ \ err_{\text{int}}]^T \tag{5.9}$$

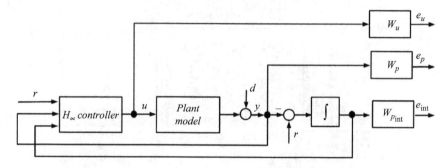

Figure 5.26   Block diagram of closed-loop system in case of $\mathcal{H}_\infty$ controller design

is the output feedback vector, where $y_m = y + d$ is the measured output and the quantity

$$err_{int} = W_{int}(r - y_m),$$
$$W_{int} = \frac{T_s}{z - 1}$$

(5.10)

is the discrete-time integral of system error. Assume that the controller transfer matrix K is represented as

$$K = \begin{bmatrix} K_r & K_{y_c} \end{bmatrix},$$

(5.11)

where K_r is the controller prefilter transfer matrix and

$$K_{y_c} = \begin{bmatrix} K_y & K_{int} \end{bmatrix},$$

(5.12)

is the output feedback transfer matrix. The matrix K_c is partitioned with respect to the dimensions of y_m and err_{int}. Then, the control action and measured output are calculated from the following expressions:

$$u = K \begin{bmatrix} r \\ y_c \end{bmatrix} = K_r r + K_y y_m + K_{int} err_{int},$$
$$y_m = Gu + d.$$

(5.13)

It is easy to show that the weighted closed-loop system outputs satisfies the equation

$$\begin{bmatrix} e_p \\ e_{int} \\ e_u \end{bmatrix} = T_{cl} \begin{bmatrix} r \\ d \end{bmatrix},$$

$$T_{cl} = \begin{bmatrix} W_p S_1 & W_p(1 + S_2) \\ W_{p_{int}} W_{int}(1 - S_1) & -W_{p_{int}} W_{int}(1 + S_2) \\ W_u(K_r + K_{int} W_{int}(1 + S_1)) & W_u(K_y + K_{int} W_{int})(1 + S_2) \end{bmatrix},$$

(5.14)

where

$$S_1 = \frac{G(K_r + K_{int} W_{int})}{1 + G(K_{int} W_{int} - K_y)}, \quad S_2 = \frac{G(K_y - K_{int} W_{int})}{1 + G(K_{int} W_{int} - K_y)}.$$

(5.15)

As usual, the $\mathcal{H}_\infty$ controller design problem is to find stabilizing suboptimal controller K, such that

$$\|T_{cl}\|_\infty < \gamma \tag{5.16}$$

where T_{cl} is the transfer matrix from exogenous input signals r and d to the output signals e_p, e_{int} and e_u, and $\gamma > \gamma_{min}$. The positive scalar γ_{min} is the minimum value of $\|T_{cl}\|_\infty$ over all stabilizing controllers K. The weighting transfer function matrices W_p, $W_{p_{int}}$, and W_u reflect the relative importance of the different frequency ranges for which the performance requirements should be fulfilled. Determination of stabilizing controller K for $\gamma \leq 1$ means that the performance requirements imposed by transfer matrices W_p, $W_{p_{int}}$, and W_u are satisfied. The $\mathcal{H}_\infty$ controller design is done with MATLAB file `Tank_Hinf_design.m` using the MATLAB function `hinfsyn`. The procedure is performed for various choices of performance weighting functions. After several trials, the performance weighting functions (in the continuous-time case) are chosen as

$$W_p = \frac{0.5s + 0.8}{5s + 0.24}, \quad W_{p_{int}} = 0.03, \quad W_u = \frac{2(s + 100)}{s + 1,000}. \tag{5.17}$$

As can be seen from expression (5.17), the transfer function W_p is chosen as low-pass filter. The transfer function $W_{p_{int}}$ is chosen as constant, which does not increase the controller order. The transfer function W_u is chosen as high-pass filter with appropriate bandwidth in order to impose constraints on the spectrum of control action. After minimization of norm (5.16), we obtain a fourth-order stabilizing $\mathcal{H}_\infty$ controller for $\gamma = 0.9997$. This means that the performance requirements are satisfied.

The comparison of designed $\mathcal{H}_\infty$ and LQG controllers in both frequency domain and time domain is performed. The magnitude plot of the closed-loop system for both controllers is shown in Figure 5.27. The sensitivity of control signals to model noise for both controllers is depicted in Figure 5.28. It is seen that the bandwidth of system with $\mathcal{H}_\infty$ controller is approximately 0.11 rad/s, which is higher than one of the system with LQG controller. This means that the step responses of system with $\mathcal{H}_\infty$ controller will be faster than one of the system with LQG controller. The control signal of closed-loop system with $\mathcal{H}_\infty$ controller is more sensitive to noise than one of the system with LQG controller. This may cause oscillations of control signal.

The reference and the exact (without noise) output signals of closed-loop system with $\mathcal{H}_\infty$ and LQG controllers are presented in Figure 5.29. In Figure 5.30, control signals of the same closed-loop systems in the case of model noise $e(k)$ according to (5.2) are shown.

It can be seen that both systems track very well reference signal in whole working range. The step response of $\mathcal{H}_\infty$ system is faster than one of the LQG system which is in accordance with system bandwidth. The $\mathcal{H}_\infty$ system achieves settling time of approximately 70 s without overshooting, which is important according to plant physics. The control actions of both systems are similar. They do not exceed the maximum allowed value of 42 percent. The control action of $\mathcal{H}_\infty$ system has larger oscillations than one of the LQG system which is due to sensitivity demonstrated in Figure 5.28.

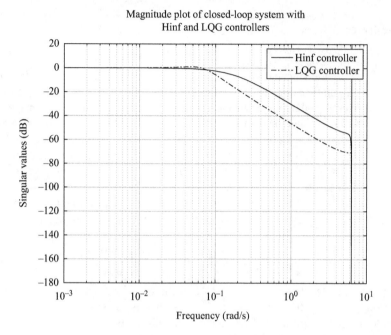

Figure 5.27   Magnitude plot of closed-loop system

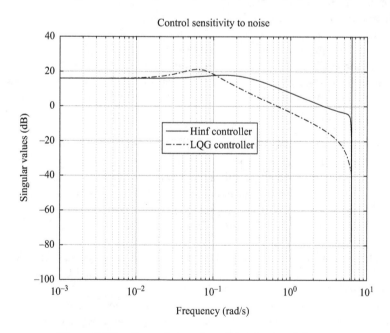

Figure 5.28   Control signal sensitivity to model noise

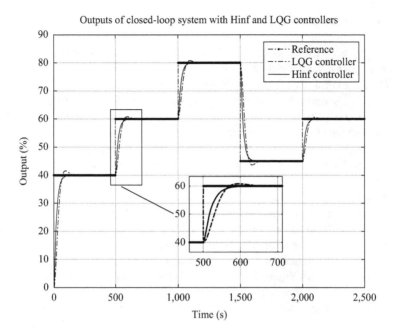

Figure 5.29 Exact output signals of system with $\mathcal{H}_\infty$ and LQG controllers

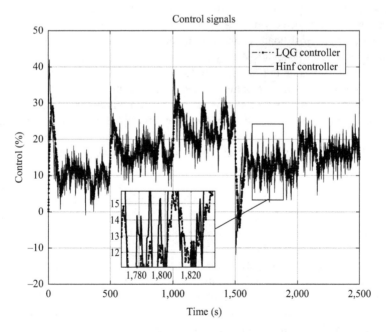

Figure 5.30 Noised control signals of system with $\mathcal{H}_\infty$ and LQG controllers

5.5 Experimental evaluation

The specialized software in MATLAB/Simulink environment is developed to implement the water level control code for the designed LQG and $\mathscr{H}_\infty$ controllers. With the aid of Simulink Coder [9], Embedded Coder [34], Arduino IDE 1.6.12 [137], and Simulink Support Package for Arduino Hardware v.16.1.2 [136] a code is generated from this software which is embedded into the microcontroller ATmega2560. The main advantages of code generation technology are the very short time for coding of control algorithm, reducing the overall time for testing and verification of the developed algorithm, and the relatively easy implementation of complex control algorithms. The block diagram of the software to generate code for water level LQG control is shown in Figure 5.31.

The diagram in Figure 5.31 includes the blocks "Analog input" and "PWM" which are taken from the specialized library "Simulink Support Package for Arduino Hardware." To generate code for Arduino hardware some additional configurations of Simulink diagram should be done. They are performed via "Configuration parameters" menu shown in Figure 5.32, which is achieved through "Simulation menu." First the user should choose a hardware board, for example, "Arduino Mega 2560" as it is shown in Figure 5.32. Next step is to configure some specific properties such as build options, host-board connection, external mode, and others. To generate and embed code in microcontroller and to run this code a build options should be chosen as "Build, load, and run." The host-board connection should be configured too. In our case connection is via USB interface and Windows discovers a hardware board as device which is connected via some virtual COM port. Then, the number of virtual COM port from "Host board connection menu" (Figure 5.33) should be the same as one from Windows Device Manager. To exchange the data between host PC and Arduino board, the Simulink model should be run in "External mode." This mode allows to change in real-time values of variables in control program and to obtain real-time data from microcontroller. When the hardware is connected to host PC via USB interface, the communication mode of property "External mode" should be set to "Serial" as it is shown in Figure 5.34. Simulink Coder uses information from "Configuration parameters" menu to properly simulate model and to generate executable code for the particular work environment.

The analog input block in the diagram depicted in Figure 5.31, whose interface is shown in Figure 5.35, is used to set the ADC output channel number (pin number) and sample time. The PWM block, whose interface is shown in Figure 5.36, is used to set the PWM output number. The generated signal has the frequency of approximately 490 Hz. The analog input block gives the number of type `uint10`, i.e., a number between 0 and 1,023. This number presented information of actual water level. To use it in control algorithm, it should be converted to double type and appropriate scaling should be performed. Due to static characteristic of voltage divider, the obtained `uint10` number, which corresponds to water level of 100 percent, is 676. Thus, the scaling coefficient which transform obtained integer number to water level is $K_{wl} = 100/676$. The PWM block accepts as input the integer numbers between 0 and

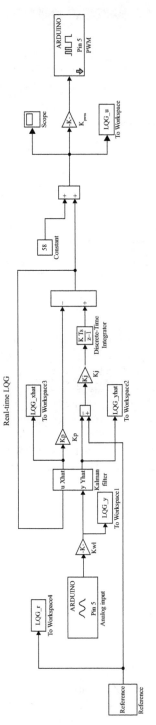

Figure 5.31   Simulink block diagram to generate code for LQG controller

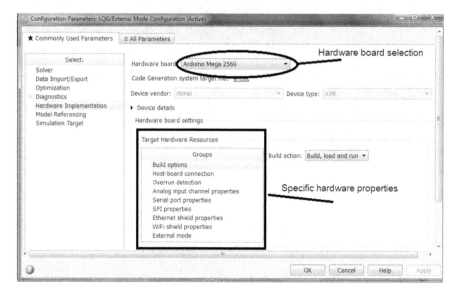

Figure 5.32   Hardware configurations in Simulink environment

255. An input value 0 produces 0 percent duty cycle, and an input value 255 produces 100 percent duty cycle. The control signal is number of type double between 0 percent and 100 percent. This number should be transformed to `uint8` type and should be scaled. Thus, the scaling coefficient is $K_{pwm} = 255/100$.

Note that the maximum number obtained from ADC is 676, which means that in feedback we use only approximately 9 b instead of the possible 10 b. As mentioned previously, the static characteristic of water pump has dead zone of approximately 58 percent. To work in linear range of the actuator, a constant value of 58 percent is added to control signal produced by the controller. Then, the controller produces signal in the range 0–42 percent which corresponds to 107 different values of duty cycle. Thus, we control a water level by approximately 7 b PWM instead of 8 b one.

Several experiments with controller designed are performed. Duration of every experiment is 2,500 s. The reference signal is varying in wide working range according to expression

$$r = \begin{cases} 40\%, 0 \leq t < 500, \\ 60\%, 500 \leq t < 1,000, \\ 80\%, 1,000 \leq t < 1,500, \\ 45\%, 1,500 \leq t < 2,000, \\ 60\%, 2,000 \leq t < 2,500. \end{cases} \tag{5.18}$$

First consider experiment with LQG controller. The comparison between results from experiment and simulation is done. The reference signal, the estimated by

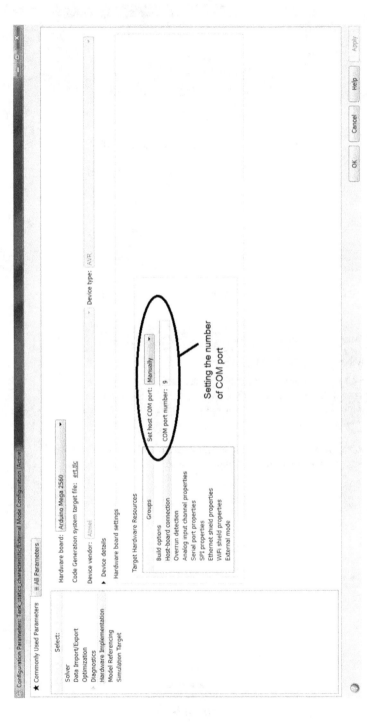

Figure 5.33   Configuration of host-board connection

Figure 5.34   Configuration of external mode communication interface

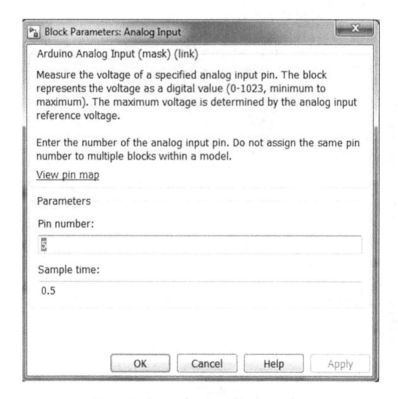

Figure 5.35   Analog input block interface

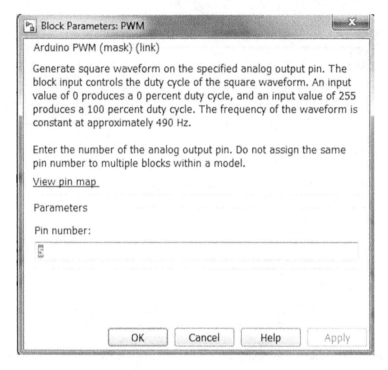

Figure 5.36   PWM block interface

Kalman filter water level from experiment and exact (without noise) water level from simulation, are shown in Figure 5.37. The exact control signal from simulation and the control signal from experiment are depicted in Figure 5.38. In Figure 5.39, the measured output and estimated output from experiment are presented. It is seen that there are very significant noise in measured output signal. This noise comprises of measurement noise, noise from power supply, and noise from ADC conversion. Nevertheless, the Kalman filter smooths out very well system output (see Figure 5.39), which means that the obtained by identification model is adequate. This statement is confirmed by excellent closeness between exact output from simulation and estimated output from experiment (see Figure 5.37). The control system with LQG controller has very good performance for the whole working range. The settling times of all transient responses depicted in Figure 5.37 are approximately 70 s, which is four times shorter than duration of plant step response. The overshoot is negligible. This is very important for tank, because overshoot means that extra quantity of water will flow through tank. As a result from filtration of output signal and plant state, the control signal has sufficiently small oscillations. Thus, the actuator works properly and safely. The control signal does not achieve maximum value of 100 percent.

Let us consider the results from experiment with $\mathcal{H}_\infty$ controller. In Figure 5.40, the reference, the output signal from experiment and exact (without noise) output

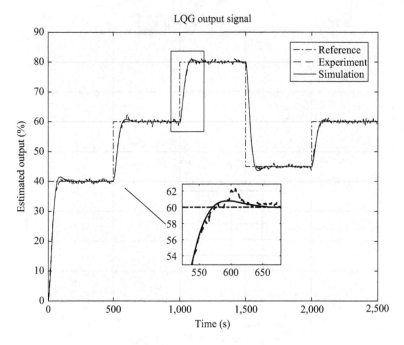

Figure 5.37   Output signals from experiment and simulation for LQG system

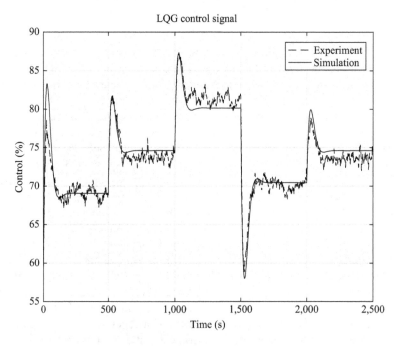

Figure 5.38   Control signals from experiment and simulation for LQG system

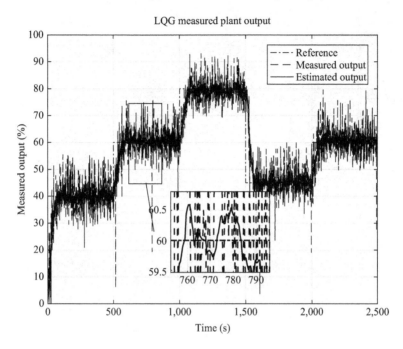

Figure 5.39   Measured output and estimated output for LQG system

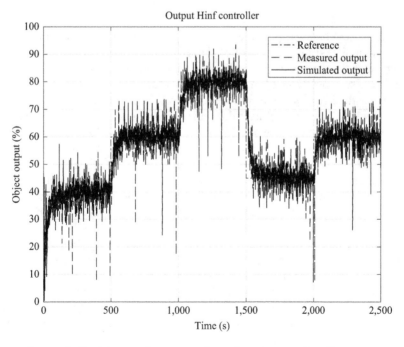

Figure 5.40   Measured output and simulated output for $\mathcal{H}_\infty$ system

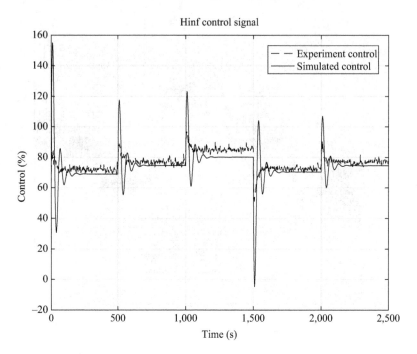

Figure 5.41   Measured control action and simulated control action for $\mathcal{H}_\infty$ system

signal from simulation are presented. In Figure 5.41, the control signals from experiment and simulation are depicted. As can be seen again, the measurement noise is very strong. Nevertheless, the output tracks very well reference signal and again the settling time is sufficiently small, and there is no overshoot of transient response. Again the experimental results are very close to the ones from simulation without noise. Though the system has no Kalman filter, control signal has no large oscillations. This is due to the fact that we have observer which is part of $\mathcal{H}_\infty$ controller. It estimates plant states and uses these estimates in feedback. This observer has similar equation as one of the Kalman filters [see (4.39)]. For comparison purpose, the measured output signal of system with $\mathcal{H}_\infty$ controller is filtered off-line (after experiment) with the Kalman filter used for LQG control. The comparison between estimated output signal for system with LQG controller and filtered output signal of system with $\mathcal{H}_\infty$ controller is presented in Figure 5.42.

As can be seen from Figure 5.42, the output signals are very similar. For most of the transient responses, the settling time of $\mathcal{H}_\infty$ system is slightly less than the one of LQG system. For example, the transient responses in the time range 500–1,000 s have settling times of approximately 60 s for $\mathcal{H}_\infty$ system and of approximately 70 s for LQG system. Moreover, all the transient responses of $\mathcal{H}_\infty$ system have not overshoot. Comparing the orders of LQG and $\mathcal{H}_\infty$ controllers, the LQG controller has advantage. The order of LQG controller is 2, while the order of $\mathcal{H}_\infty$ controller is equal to 4.

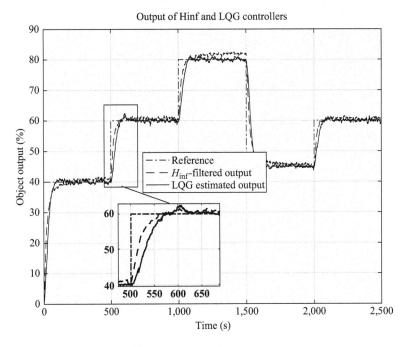

Figure 5.42 $\mathcal{H}_\infty$ system-filtered output and LQG system estimated

5.6   Notes and references

The control of liquid level in tank is a problem that commonly occurs in lot of industrial applications. These applications are related to the food processing, chemical industrial processes, agriculture, and nutrition. Due to that, the system for automatic liquid level control is one of the popular experimental setups in industrial control and embedded control laboratories, see for instance Ogata [68] and Slavov and Puleva [138]. The popularity and practical significance of liquid level control problem motivated the authors to design a low-cost embedded system for control of liquid level in a physical model of tank.

Chapter 6

Case study 2: robust control
of miniature helicopter

The aim of this case study is to present in detail the μ synthesis of a high-order integral attitude controller of a miniature helicopter and to demonstrate results from the hardware-in-the-loop simulation of the helicopter control system. The μ controller designed for hovering allows to suppress efficiently strong wind disturbances in the presence of 15 percent input multiplicative uncertainty. A simple position controller is added to ensure tracking of a desired trajectory in the 3D space. The results from hardware-in-the-loop simulation are close to the results from the double-precision simulation of helicopter control system in Simulink®. It is shown that even for large deviations of the helicopter variables from their trim values in hovering the control system has acceptable performance. The software platform developed allows to implement easily different sensors, servoactuators, and control laws and to investigate the closed-loop system behavior in the presence of different disturbances, noises, and parameter variations.

The helicopter controller is designed under the following assumptions:

- The helicopter model is linearized around the trim conditions for hovering so that the resulting linear system is time invariant, i.e., the plant parameters are considered as constant.
- There is an uncertainty in the mathematical description of the helicopter due to approximation errors, neglected dynamics, and nonlinearities which is taken into account by adding multiplicative uncertainty at the plant input.
- It is assumed that the angles and rate measurements are obtained from an inertial navigation system (INS) and smoothed by Kalman filter so that the effect of measurement noises on the control system is small in comparison with the wind disturbances effect. That is why these noises are neglected in the design but are taken into account in the simulation of the closed-loop nonlinear system.

The controller design is done on the basis of the sophisticated nonlinear model of the miniature helicopter X-Cell 60 SE presented in [139]. An analytically linearized model of the miniature helicopter is derived which is verified by comparison of its frequency responses with the responses of the numerically linearized model. The μ controller is implemented by Digital Signal Processor (DSP) whose performance is tested on a Simulink model of the nonlinear helicopter plant under the action of wind disturbances and sensor noises. The results from hardware-in-the-loop simulation of the helicopter control system confirms the controller ability to ensure good transient responses in the presence of parameter changes and sufficiently strong disturbances.

The chapter consists of three sections. In Section 6.1, we present the helicopter model that is used in the controller design and in the simulation of the closed-loop system. First, a nonlinear 13th-order model is described after that its analytical and numerical linearization are discussed. An uncertain linear helicopter model is considered with input multiplicative uncertainty that may reach 15 percent. This model is used in Section 6.2 to design a μ controller that stabilizes the angular helicopter motion. The μ controller ensures robust stability and robust performance of the uncertain closed-loop system. A simple proportional-derivative regulator is designed to implement the position control of helicopter center of mass. Results from hardware-in-the-loop simulation of the integral helicopter controller under the action of moderate wind disturbances are presented in Section 6.3 for a circular motion in the 3D space. The HIL simulation is done by using a digital signal controller and allows to reveal the accuracy of the actual control algorithms as well as to check the nonlinear system performance for different disturbances, noises, and parameter values.

The presentation in this chapter is based on the paper [140].

6.1   Helicopter model

MATLAB®   files used in this section

File	Description
heli_model.m	S-function to model the nonlinear helicopter dynamics
trim_val_heli.m	Finds the trim values of helicopter variables
lin_heli_model.m	Computes analytically linearized helicopter model
num_lin_heli.m.m	Numerical linearization using Simulink helicopter model
comp_freq_heli.m.m	Comparison of frequency responses of linearized helicopter models
mod_heli.m	Uncertain linear helicopter model

6.1.1   Nonlinear helicopter model

In this chapter, we use the analytic nonlinear model of highly maneuverable miniature helicopter with a main rotor and a tail rotor, presented in [139]. It contains the rigid-body helicopter dynamics, dynamics of the longitudinal, and lateral main rotor flapping, the rotor speed, and an integral of the rotor-speed tracking error used in the engine governor. Simplified expressions are used to determine the component forces and moments acting on the helicopter. A momentum theory-based iterative scheme adopted from [141] is used to compute the thrust coefficient and inflow ratio as a function of the airspeed, rotor speed, and collective pitch angle. The resulting helicopter model is of 13th order with notion given in Table 6.1. In this section, we present

Table 6.1 Notation

Symbol	Quantity	Units
ϕ, θ, ψ	Roll, pitch, and yaw angles	rad
p, q, r	Roll, pitch, and yaw rates	rad/s
u, v, w	Longitudinal, lateral, and normal velocities	m/s
V_x, V_y, V_z	Velocities in Earth fixed north-east-down (NED) frame	m/s
$V = \sqrt{V_x^2 + V_y^2 + V_z^2}$	Total velocity	m/s
x, y, z	Position coordinates in NED frame	m
C_b^e	Direction cosine matrix from body frame to NED frame	
h	Flight altitude, $h = |z|$	m
a_1, b_1	Longitudinal and lateral flapping angles	rad
$\Omega_{mr}, \Omega_{nom}, \Omega_c$	Main rotor speed, nominal main rotor speed, and main rotor-speed reference	rad/s
ω_i	Integral of the main rotor-speed tracking error	rad
T_{mr}	Main rotor thrust	N
X, Y, Z	Forces applied along the $X, Y,$ and Z body frame axes	N
L, M, N	Moments about the $X, Y,$ and Z body frame axes	N m
Q_e, Q_{mr}, Q_{tr}	Engine torque, main rotor torque, and tail rotor torque	N m
P_e^{max}	Engine max power	W
δ_{col}	Collective pitch angle of the main rotor blades	rad
$\delta_{lon}, \delta_{lat}$	Cyclic pitch angles of the main rotor blades	rad
δ_{tr}	Pitch angle of the tail rotor blades	rad
δ_t	Engine throttle setting	rad
$u_{wind}, v_{wind}, w_{wind}$	Wind velocities along $X, Y,$ and Z body frame axes	m/s
m	Helicopter mass	kg
g	Acceleration of gravity	m/s^2
ρ	Air density	kg/m^3
I_{xx}, I_{yy}, I_{zz}	Rolling, pitching, and yawing moments of inertia	kg m^2
γ_{fb}	Stabilizer bar Lock number	
R_{mr}	Main rotor radius	m
I_{rot}	Total main rotor moment of inertia	kg m^2
R_{tr}	Tail rotor radius	m
n_{tr}	Gear ratio of tail rotor to main rotor	
μ_a	Advance ratio	
μ_z	Normal airflow component	
$A_{\delta_{lon}}^{nom}, B_{\delta_{lat}}^{nom}$	Steady-state longitudinal and lateral gains from the cyclic inputs to the main rotor flapping angles at nominal main rotor speed	rad/rad

Table 6.1   (Continued)

Symbol	Quantity	Units
τ_e	Rotor time constant for a rotor with stabilizer bar	
K_p	Proportional engine governor gain	s/rad
K_i	Integral engine governor gain	1/rad
$w_{servo}^{col}, w_{servo}^{lon}, w_{servo}^{lat}$	Transfer functions of collective and cyclic servos	
w_{servo}^{tr}	Transfer function of the tail collective servo	
x_h	State vector of the linearized model	
y_h	Output vector of the linearized model	
u_c	Controller output vector	
u_s	Servoactuator output vector	
K_{p1}, K_{p2}, K_{p3}	Proportional coefficients of the position PD regulators	
K_{d1}, K_{d2}, K_{d3}	Derivative coefficients of the position PD regulators	
u_z, u_x, u_y	Position controller outputs	

only the main equations describing the nonlinear helicopter dynamics referring the reader to [139,142] where detailed description of the miniature helicopter model and its parameters identification are given.

The main variables which characterize the helicopter motion in body frame coordinate system are shown in Figure 6.1. The Newton–Euler equations that describe the rigid-body translational and angular helicopter motions are taken neglecting the cross products of inertia.

Equations for translational velocities.

$$\dot{u} = vr - wq - g\sin(\theta) + X/m$$
$$\dot{v} = wp - ur + g\sin(\phi)\cos(\theta) + Y/m \tag{6.1}$$
$$\dot{w} = uq - vp + g\cos(\phi)\cos(\theta) + Z/m.$$

Equations for angular rates.

$$\dot{p} = qr(I_{yy} - I_{zz})/I_{xx} + L/I_{xx}$$
$$\dot{q} = pr(I_{zz} - I_{xx})/I_{yy} + M/I_{yy} \tag{6.2}$$
$$\dot{r} = pq(I_{xx} - I_{yy})/I_{zz} + N/I_{zz}.$$

Equations for Euler angles.

$$\dot{\phi} = p + \sin(\phi)\tan(\theta)q + \cos(\phi)\tan(\theta)r$$
$$\dot{\theta} = \cos(\phi)q - \sin(\phi)r \tag{6.3}$$
$$\dot{\psi} = (\sin(\phi)/\cos(\theta))q + (\cos(\phi)/\cos(\theta))r.$$

The forces X, Y, Z are expressed as

$$X = X_{mr} + X_{fus}, \quad Y = Y_{mr} + Y_{fus} + Y_{tr} + Y_{vf}, \quad Z = Z_{mr} + Z_{fus} + Z_{ht},$$

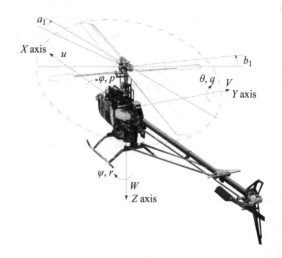

Figure 6.1   Helicopter variables in body frame coordinate system

where the subscripts *mr, fus, tr, vf,* and *ht* show that the corresponding force component pertains to the *main rotor, fuselage, tail rotor, vertical fin,* and *horizontal stabilizer,* respectively. The force components pertaining to the main rotor are functions of the main rotor thrust T_{mr} and the flapping angles a_1, b_1,

$$X_{mr} = -T_{mr} \sin(a_1), \quad Y_{mr} = T_{mr} \sin(b_1), \quad Z_{mr} = -T_{mr} \cos(a_1) \cos(b_1).$$

The main rotor thrust is determined from

$$T_{mr} = C_T \rho (\Omega_{mr} R_{mr})^2 \pi R_{mr}^2,$$

where the thrust coefficient C_T is a function of the collective pitch angle δ_{col}, the advance ratio

$$\mu_a = \sqrt{(u - u_{wind})^2 + (v - v_{wind})^2}/(\Omega_{mr} R_{mr})$$

and the normal airflow component

$$\mu_z = (v - v_{wind})/(\Omega_{mr} R_{mr}).$$

The above expressions show that the main rotor thrust and hence the associated force components depend strongly on the wind velocities $u_{wind}, v_{wind}, w_{wind}$. The fuselage forces $X_{fus}, Y_{fus}, Z_{fus}$ as well as the sideforce Y_{vf} produced by the vertical fin and the force Z_{ht} generated by the horizontal stabilizer depend on the fuselage center of pressure velocities $u - u_{wind}, v - v_{wind}, w - w_{wind}$. The tail rotor thrust is determined similarly to the case of main rotor thrust and depends on the pitch angle δ_{tr} of the tail rotor blades as well as on the wind velocities.

The moments along the X, Y, Z axes are represented as

$$L = L_{mr} + L_{vf} + L_{tr}, \quad M = M_{mr} + M_{ht}, \quad N = -Q_e + N_{vf} + N_{tr},$$

where the main rotor rolling moment L_{mr} and the pitching moment M_{mr} are functions of the main rotor thrust and the flapping angles, Q_e is the engine torque (positive clockwise), the rolling moment L_{vf} and the yawing moment N_{vf} are proportional to the sideforce Y_{vf} produced by the vertical fin, the rolling moment L_{tr} and the yawing moment N_{tr} are proportional to the tail rotor thrust Y_{tr} and the pitching moment M_{ht} is proportional to the force Z_{ht} generated by the horizontal stabilizer.

Detailed description of the component forces and moments that act on the helicopter body may be found in [139,141,143].

The main rotor dynamics is modeled by the following equation:

$$\dot{\Omega}_{mr} = \dot{r} + (Q_e - Q_{mr} - n_{tr}Q_{tr})/I_{rot}. \tag{6.4}$$

In this equation, the engine torque

$$Q_e = \left(\frac{P_e^{max}}{\Omega_{mr}}\right)\delta_t$$

can be changed by varying the engine throttle setting δ_t, the yawing moment produced by the main rotor being given by

$$Q_{mr} = C_Q\rho(\Omega_{mr}R_{mr})^2\pi R_{mr}^3,$$

where C_Q is the torque coefficient and the tail rotor torque is determined as

$$Q_{tr} = C_{Qtr}\rho(\Omega_{tr}R_{tr})^2\pi R_{tr}^3,$$

where C_{Qtr} is the tail rotor torque coefficient.

The rotor speed is stabilized by using an engine governor that implements proportional-integral feedback controller described by

$$\begin{aligned} \delta_t &= K_p.(\Omega_c - \Omega_{mr}) + K_i.\omega_i, \\ \dot{\omega}_i &= \Omega_c - \Omega_{mr}. \end{aligned} \tag{6.5}$$

The longitudinal and lateral flapping dynamics of the main rotor, taking into account the effect of the wind, is represented by the two first-order differential equations

$$\begin{aligned} \dot{a}_1 &= -q + (-a_1 + \frac{\partial a_1}{\partial\mu_a}\frac{u - u_{wind}}{\Omega_{mr}R_{mr}} + \frac{\partial a_1}{\partial\mu_z}\frac{w - w_{wind}}{\Omega_{mr}R_{mr}} + A_{\delta_{lon}}\delta_{lon})/\tau_e \\ \dot{b}_1 &= -p + (-b_1 - \frac{\partial b_1}{\partial\mu_v}\frac{v - v_{wind}}{\Omega_{mr}R_{mr}} + B_{\delta_{lat}}\delta_{lat})/\tau_e, \end{aligned} \tag{6.6}$$

where the steady-state longitudinal and lateral gains from the cyclic inputs to the main rotor flap angles are determined by

$$A_{\delta_{lon}} = A_{\delta_{lon}}^{nom}(\Omega_{mr}/\Omega_{nom})^2,$$

$$B_{\delta_{lat}} = B_{\delta_{lat}}^{nom}(\Omega_{mr}/\Omega_{nom})^2,$$

The flapping derivatives $(\partial a_1 / \partial \mu_a), (\partial b_1 / \partial \mu_v) = (-\partial a_1 / \partial \mu_a), (\partial a_1 \partial \mu_z)$ in respect to the advance ratio μ_a and normal airflow component μ_z, respectively, are found by the simplified expressions given in [139] and

$$\tau_e = \frac{16.0}{(\gamma_{fb} \Omega_{\mathrm{mr}})}.$$

To control the helicopter velocity and position, we shall need also the relationship between the velocities in body fixed reference frame and Earth fixed (i.e., inertial) reference frame. Define the direction cosine matrix, used to transform a vector expressed in the body frame to a vector expressed in the inertial frame, as

$$C_b^e = \begin{bmatrix} c\theta c\psi & s\phi s\theta c\psi - c\phi s\psi & c\phi s\theta c\psi + s\phi s\psi \\ c\theta s\psi & s\phi s\theta s\psi + c\phi c\psi & c\phi s\theta s\psi - s\phi c\psi \\ -s\theta & s\phi c\theta & c\phi c\theta \end{bmatrix},$$

where s and c denote $\sin(\cdot)$ and $\cos(\cdot)$, respectively. Then, the necessary relationship is written as

$$\begin{bmatrix} V_x \\ V_y \\ V_z \end{bmatrix} = C_b^e \begin{bmatrix} u \\ v \\ w \end{bmatrix}. \tag{6.7}$$

The position coordinates in north-east-down (NED) frame are determined from

$$\dot{x} = V_x, \quad \dot{y} = V_y, \quad \dot{z} = V_z.$$

The nonlinear helicopter model represented by (6.2)–(6.7) is implemented as the MATLAB S-function `heli_model` which is used in trimming and numerical plant linearization.

The helicopter control is done by signals to servoactuators which change in appropriate way the control actions $\delta_{\mathrm{col}}, \delta_{\mathrm{lon}}, \delta_{\mathrm{lat}}$, and δ_{tr}. The servos for collective and cyclic angles are assumed to have the same transfer functions

$$w_{\mathrm{servo}}^{\mathrm{col}}(s) = w_{\mathrm{servo}}^{\mathrm{lon}}(s) = w_{\mathrm{servo}}^{\mathrm{lat}}(s) = \frac{s/T_z + 1}{s/T_p + 1} \frac{\omega_n^2}{s^2 + 2\zeta \omega_n s + \omega_n^2}, \tag{6.8}$$

while the tail servo transfer function $w_{\mathrm{servo}}^{\mathrm{tr}}$ is taken as

$$w_{\mathrm{servo}}^{\mathrm{tr}}(s) = \frac{\omega_{\mathrm{tr}}^2}{s^2 + 2\zeta_{\mathrm{tr}} \omega_{\mathrm{tr}} s + \omega_{\mathrm{tr}}^2}. \tag{6.9}$$

The parameters of the servoactuators are taken from [139]:

$$T_z = 104, \quad T_p = 33, \quad \omega_n = 36, \quad \omega_{\mathrm{tr}} = 14\pi, \quad \zeta = 0.5, \quad \zeta_{\mathrm{tr}} = 0.6.$$

The helicopter dynamics, described by (6.2)–(6.7), is combined with the dynamics of the servoactuators to obtain the extended plant dynamics shown in Figure 6.2.

6.1.2 Linearized model

Equations (6.2)–(6.6) are linearized analytically to produce a time-invariant state-space model, needed in the attitude controller design. The linearization is done by

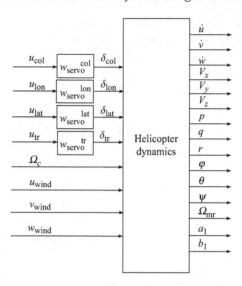

Figure 6.2   Extended helicopter dynamics

taking the linear terms of the Taylor series expansion in the arguments of the deviation $\Delta v = v - v^{trim}$ of each variable v from its trim value v^{trim}. This requires computation of the partial derivatives that determine the linear terms of the expansion in respect to the corresponding arguments. The linearization is done for hovering after finding the trim values of the state and input vector components from the nonlinear helicopter model using the MATLAB function `trim_val_heli` utilizing the nonlinear model `heli_model`. Since there is some freedom in determining the trim values, we set

$$u^{trim} = 0.001 \text{ m/s}, \quad v^{trim} = 0.001 \text{ m/s}, \quad w^{trim} = 0 \text{ m/s}, \quad \Omega_{mr}^{trim} = 167 \text{ rad/s}.$$

Note that the values of u^{trim} and v^{trim} are set slightly greater than zero to avoid numerical difficulties during trimming. As a result, we obtain the following trim values corresponding to hovering:

$$p^{trim} = 0 \text{ rad/s}, \quad q^{trim} = 0 \text{ rad/s}, \quad r^{trim} = 0 \text{ rad/s}$$

$$\phi^{trim} = 0.0873 \text{ rad}, \quad \theta^{trim} = -3.3 \times 10^{-6} \text{ rad}, \quad \psi^{trim} = 0 \text{ rad}$$

$$\omega_i^{trim} = 29.17 \text{ rad}, \quad a_1^{trim} = 0 \text{ rad}, \quad b_1^{trim} = 8.48 \times 10^{-3} \text{ rad}$$

and

$$\delta_{col}^{trim} = 0.0967 \text{ rad}, \quad \delta_{lon}^{trim} = 0.0 \text{ rad}, \quad \delta_{lat}^{trim} = 0.020 \text{ rad}, \quad \delta_{tr}^{trim} = -0.2488 \text{ rad}.$$

We note that the large value of the trim value of the tail rotor blades pitch angle is due to the necessity to counteract the main rotor torque in order to stabilize the helicopter body along Z axis.

Simplifying the presentation, the linearization of (6.2)–(6.6) is done as follows. Taking into account that in steady state it is fulfilled that $p = q = r = 0$, (6.2) is linearized as

$$\Delta\dot{u} = v\Delta r - w\Delta q - g\cos(\theta)\Delta\theta + \Delta X/m$$
$$\Delta\dot{v} = w\Delta p - u\Delta r + g\cos(\phi)\cos(\theta)\Delta\phi - g\sin(\phi)\sin(\theta)\Delta\theta + \Delta Y/m \qquad (6.10)$$
$$\Delta\dot{w} = u\Delta q - v\Delta p - g\sin(\phi)\cos(\theta)\Delta\phi - g\cos(\phi)\sin(\theta)\Delta\theta + \Delta Z/m$$

where the linear terms in the forces expansion are given by

$$\begin{aligned}
\Delta X ={}& X^u\Delta u + X^{u\text{wind}}\Delta u_{\text{wind}} + X^v\Delta v + X^{v\text{wind}}\Delta v_{\text{wind}} + X^w\Delta w + X^{w\text{wind}}\Delta w_{\text{wind}} \\
&+ X^{\Omega\text{mr}}\Delta\Omega_{\text{mr}} + X^{\delta\text{col}}\Delta\delta_{\text{col}} + X^{a_1}\Delta a_1 \\
\Delta Y ={}& Y^u\Delta u + Y^{u\text{wind}}\Delta u_{\text{wind}} + Y^v\Delta v + Y^{v\text{wind}}\Delta v_{\text{wind}} + Y^w\Delta w + Y^{w\text{wind}}\Delta w_{\text{wind}} \\
&+ Y^p\Delta p + Y^q\delta q + Y^r\Delta r + Y^{\Omega\text{mr}}\Delta\Omega_{\text{mr}} + Y^{\delta\text{col}}\Delta\delta_{\text{col}} + Y^{\delta\text{tr}}\Delta\delta_{\text{tr}} + Y^{b_1}\Delta b_1 \\
\Delta Z ={}& Z^u\Delta u + Z^{u\text{wind}}\Delta u_{\text{wind}} + Z^v\Delta v + Z^{v\text{wind}}\Delta v_{\text{wind}} + Z^w\Delta w + Z^{w\text{wind}}\Delta w_{\text{wind}} \\
&+ Z^q\delta q + Z^{\Omega\text{mr}}\Delta\Omega_{\text{mr}} + Z^{\delta\text{col}}\Delta\delta_{\text{col}} + Z^{a_1}\Delta a_1 + Z^{b_1}\Delta b_1
\end{aligned}$$

and the partial derivatives are computed for the trim values of the variables shown in superscripts. The expressions for these derivatives are involved and may be found in the file `lin_heli_model`.

After linearization, (6.3) takes the simple form

$$\begin{aligned}
\Delta\dot{p} &= \frac{\Delta L}{I_{xx}} \\
\Delta\dot{q} &= \frac{\Delta M}{I_{yy}} \qquad\qquad (6.11) \\
\Delta\dot{r} &= \frac{\Delta N}{I_{zz}}
\end{aligned}$$

where

$$\begin{aligned}
\Delta L ={}& L^u\Delta u + L^{u\text{wind}}\Delta u_{\text{wind}} + L^v\Delta v + L^{v\text{wind}}\Delta v_{\text{wind}} + L^w\Delta w + L^{w\text{wind}}\Delta w_{\text{wind}} \\
&+ L^p\Delta p + L^q\delta q + L^r\Delta r + L^{\Omega\text{mr}}\Delta\Omega_{\text{mr}} + L^{\delta\text{col}}\Delta\delta_{\text{col}} + L^{\delta\text{tr}}\Delta\delta_{\text{tr}} + L^{b_1}\Delta b_1, \\
\Delta M ={}& M^u\Delta u + M^{u\text{wind}}\Delta u_{\text{wind}} + M^v\Delta v + M^{v\text{wind}}\Delta v_{\text{wind}} + M^w\Delta w + M^{w\text{wind}}\Delta w_{\text{wind}} \\
&+ M^q\delta q + M^{\Omega\text{mr}}\Delta\Omega_{\text{mr}} + M^{\delta\text{col}}\Delta\delta_{\text{col}} + M^{a_1}\Delta a_1, \\
\Delta N ={}& N^u\Delta u + N^{u\text{wind}}\Delta u_{\text{wind}} + N^v\Delta v + N^{v\text{wind}}\Delta v_{\text{wind}} + N^w\Delta w + N^{w\text{wind}}\Delta w_{\text{wind}} \\
&+ N^p\Delta p + N^q\delta q + N^r\Delta r + N^{\Omega_c}\Delta\Omega_c + N^{\Omega\text{mr}}\Delta\Omega_{\text{mr}} + N^{\delta\text{col}}\Delta\delta_{\text{col}} + N^{\delta\text{tr}}\Delta\delta_{\text{tr}}.
\end{aligned}$$

Equation (6.4) for the Euler angles is linearized as

$$\Delta\dot{\phi} = \Delta p + \sin(\phi)\tan(\theta)\Delta q + \cos(\phi)\tan(\theta)\Delta r$$

$$\Delta\dot{\theta} = \cos(\phi)\Delta q - \sin(\phi)\Delta r \qquad (6.12)$$

$$\Delta\dot{\psi} = \left(\frac{\sin(\phi)}{\cos(\theta)}\right)\Delta q + \left(\frac{\cos(\phi)}{\cos(\theta)}\right)\Delta r$$

After linearization, (6.4) takes the form

$$\Delta\dot{\Omega}_{mr} = \Omega^u\Delta u + \Omega^{u_{wind}}\Delta u_{wind} + \Omega^v\Delta v + \Omega^{v_{wind}}\Delta v_{wind}$$

$$+ \Omega^w\Delta w + \Omega^{w_{wind}}\Delta w_{wind} + \Omega^p\Delta p + \Omega^q\Delta q + \Omega^r_{mr}\Delta r \qquad (6.13)$$

$$+ \Omega^{\Omega_{mr}}\Delta\Omega_{mr} + \Omega^{\omega_i}\Delta\omega_i + \Omega^{\Omega_c}\Delta\Omega_c + \Omega^{\delta_{col}}\Delta\delta_{col} + \Omega^{\delta_{tr}}\Delta\delta_{tr}.$$

Finally, the linearized equations (6.6) for the flapping angles are obtained as

$$\Delta\dot{a}_1 = -\Delta q - \Delta a_1/\tau_e + \left(\frac{\partial a_1}{\partial\mu_a}\frac{\Delta u - \Delta u_{wind}}{\Omega_{mr}R_{mr}} + \frac{\partial a_1}{\partial\mu_z}\frac{\Delta w - \Delta w_{wind}}{\Omega_{mr}R_{mr}}\right)/\tau_e$$

$$- \left(\frac{\partial a_1}{\partial\mu_a}\frac{u - u_{wind}}{\Omega_{mr}^2 R_{mr}} + \frac{\partial a_1}{\partial mu_z}\frac{w - w_{wind}}{\Omega_{mr}^2 R_{mr}}\right)\Delta\Omega_{mr}/\tau_e + A^{nom}_{\delta_{lon}}\Delta\delta_{lon}/\tau_e$$

$$+ 2A^{nom}_{\delta_{lon}}\Omega_{mr}\delta_{lon}/(\tau_e\Omega^2_{nom})\Delta\Omega, \qquad (6.14)$$

$$\Delta\dot{b}_1 = -\Delta p - \Delta b_1/\tau_e - \left(\frac{\partial b_1}{\partial\mu_v}\frac{\Delta v - \Delta v_{wind}}{\Omega_{mr}R_{mr}} - \frac{\partial b_1}{\partial\mu_v}\frac{v - v_{wind}}{\Omega_{mr}^2 R_{mr}}\right)\Delta\Omega_{mr}/\tau_e$$

$$+ B^{nom}_{\delta_{lat}}\Delta\delta_{lat}/\tau_e + 2B^{nom}_{\delta_{lat}}\Omega_{mr}\delta_{lat}/(\tau_e\Omega^2_{nom})\Delta\Omega.$$

Define the state vector as

$$\Delta x_h = [\Delta u, \Delta v, \Delta w, \Delta p, \Delta q, \Delta r, \Delta\phi, \Delta\theta, \Delta\psi, \Delta\Omega_{mr}, \Delta\omega_i, \Delta a_1, \Delta b_1]^T,$$

the control vector as

$$\Delta u_s = [\Delta\delta_{col}, \Delta\delta_{lon}, \Delta\delta_{lat}, \Delta\delta_{tr}]^T,$$

the disturbance vector as

$$\Delta dist = [\Delta u_{wind}, \Delta v_{wind}, \Delta w_{wind}]^T,$$

and the output vector as

$$\Delta y_h = [\Delta\dot{u}, \Delta\dot{v}, \Delta\dot{w}, \Delta u, \Delta v, \Delta w, \Delta p, \Delta q, \Delta r, \Delta\phi, \Delta\theta, \Delta\psi, a_1, b_1]^T.$$

Then, (6.10)–(6.14) are represented as the linearized helicopter model

$$\Delta\dot{x}_h = A\Delta x_h + B\begin{bmatrix}\Delta u_s \\ \Omega_c \\ \Delta dist\end{bmatrix},$$

$$\qquad (6.15)$$

$$\Delta y_h = C\Delta x_h + D\begin{bmatrix}\Delta u_s \\ \Delta\Omega_c \\ \Delta dist\end{bmatrix},$$

Table 6.2   Basic helicopter parameters

Parameter	Value	Parameter	Value
m	8.2 kg	R_{mr}	0.775 m
I_{xx}	0.18 kg m^2	R_{tr}	0.13 m
I_{yy}	0.34 kg m^2	n_{tr}	4.66
I_{zz}	0.28 kg m^2	$A_{\delta_{lon}}^{nom}$	4.2 rad/rad
I_{rot}	0.095 kg m^2	$B_{\delta_{lat}}^{nom}$	4.2 rad/rad
γ_{fb}	0.8	K_p	0.01 s/rad
Ω_{nom}	167 rad/s	K_i	0.02 1/rad

where A, B, C, and D are matrices with dimensions 13×13, 13×8, 14×13, and 14×8, respectively. In this way, for small deviations from trim conditions, the helicopter dynamics is described by a linear model of 13th order with 8 inputs (5 control actions and 3 disturbances) and 14 outputs. For simplicity, further on the symbol Δ will be dropped out remembering that the elements of the vectors x_h, u_s, y_h represent the deviations of the corresponding variables from their trim values and that the elements of the matrices A, B, C, D are computed for the trim values of the variables involved. The linearized helicopter model is obtained by the M-function `lin_heli_model`. The computations are done for the parameters of the X-Cell 60 SE Helicopter given in [139]. The basic helicopter parameters are presented in Table 6.2.

Equation (6.15) may be represented in the frequency domain as

$$y_h(s) = G_{heli} \begin{bmatrix} u_s(s) \\ \Omega_c(s) \\ dist(s) \end{bmatrix}, \tag{6.16}$$

where the helicopter transfer function matrix G_{heli} is determined by the matrices A, B, C, D as

$$G_{heli}(s) = C(sI - A)^{-1}B + D.$$

The linearization of the helicopter model requires a large volume of computations by hand which may be accompanied with errors. That is why the model obtained is checked by a numerical linearization using the Simulink model `linmod_heli.slx` build on the basis of the S-function `heli_model`. The numerical linearization is performed using the M-function `num_lin_heli` which implements the MATLAB function `linmod`.

The advantage of the analytical linear model over the numerical one is that it allows to compute easily the model matrices for a given set of the helicopter parameters. Also, this description may be used to obtain the so-called parameter-dependent model [112, Chapter 2] which depends on parameters like velocity and altitude that may undergo large variations along the time.

The analytically linearized helicopter model (6.15) is verified by comparing the frequency responses of the individual input/output channels with the corresponding frequency responses of the numerically linearized model.

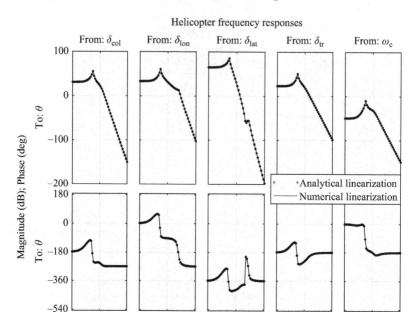

Figure 6.3   Frequency responses of analytical and numerical linearized models—control actions to θ

The frequency responses corresponding to the transfer functions from control actions and disturbances to the Euler angles for the analytical and numerical linearized models are obtained with the M-file comp_freq_heli. In Figures 6.3 and 6.4, we show a comparison of the frequency responses in respect to angle θ for both models. It is seen that the frequency responses of the models practically coincide.

It is necessary to point out that the analytically linearized helicopter model may be used not only for hovering but also for different translational velocities provided that the corresponding trim conditions are preliminary found.

6.1.3   Uncertain model

Equation (6.16) is considered as the nominal model of the helicopter dynamics. To account for the neglected helicopter dynamics, approximation errors, and changes of the trim conditions, we add multiplicative uncertainty to each component of the control vector u_s as shown in Figure 6.5. Each quantity δ_i, $i = 1,\ldots,4$ is supposed to be a gain bounded uncertain linear time-invariant object that satisfy

$$|\delta_i| \leq 0.15, \quad i = 1,\ldots,4.$$

In this way, the uncertainty in each control component may reach 15 percent. More uncertainty allows to take into account larger model errors, parameter changes,

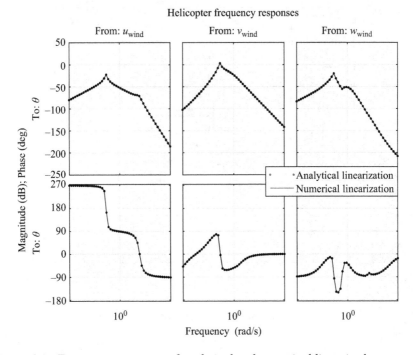

Figure 6.4 *Frequency responses of analytical and numerical linearized models—disturbances to θ*

and nonlinearities, but it becomes more difficult to confront it with a single time-invariant controller. Note that the uncertainties do not vary over the frequency which is somewhat a conservative assumption. The reason for this assumption is the lack of knowledge about the model accuracy over the frequency range. The resulting uncertain plant is described by the transfer function matrix

$$G(s) = G_{\text{heli}}(s) \begin{bmatrix} 1+\delta_1 & 0 & 0 & 0 & 0\,0\,0\,0 \\ 0 & 1+\delta_2 & 0 & 0 & 0\,0\,0\,0 \\ 0 & 0 & 1+\delta_3 & 0 & 0\,0\,0\,0 \\ 0 & 0 & 0 & 1+\delta_4 & 0\,0\,0\,0 \\ 0 & 0 & 0 & 0 & 1\,0\,0\,0 \\ 0 & 0 & 0 & 0 & 0\,1\,0\,0 \\ 0 & 0 & 0 & 0 & 0\,0\,1\,0 \\ 0 & 0 & 0 & 0 & 0\,0\,0\,1 \end{bmatrix}. \tag{6.17}$$

In this way, the uncertain helicopter model is described as a continuous-time-invariant control plant by the following equation:

$$y_h = G(s) \begin{bmatrix} u_s \\ \Omega_c \\ dist \end{bmatrix}. \tag{6.18}$$

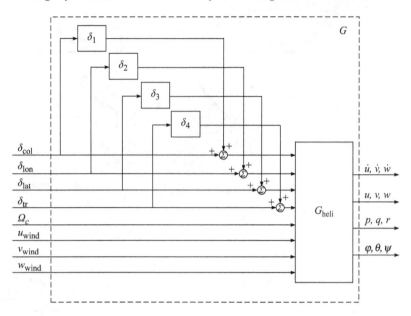

Figure 6.5   Structure of the uncertain plant

Similar to the nominal model, the uncertain helicopter model has 8 inputs, 14 outputs, and 13 states. This model is found by using the M-file `mod_heli`.

The frequency response plot of the uncertain plant singular values, obtained by using the transfer function between control actions u_s and Euler angles for 30 random values of the input uncertainties δ_i, $i = 1, \ldots, 4$, is presented in Figure 6.6. The frequency responses of the singular values show the presence of an integrator in the helicopter transfer function matrix.

6.2 μ Synthesis of attitude controller

MATLAB   files used in this section

File	Description
dlp_heli.m	Generates the open-loop connection for helicopter control system
design_heli	μ Synthesis of attitude helicopter controller
dmu_heli	Robust stability and robust performance analysis
dfrs_heli	Frequency responses of the closed-loop system
dmcs_heli	Time responses of the uncertain closed-loop system

The helicopter control system is designed at two levels taking into account that the dynamics of the helicopter translational motion is slower than the dynamics of its

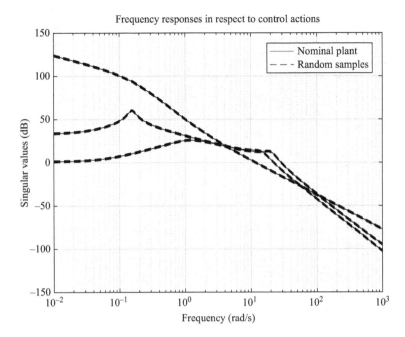

Figure 6.6 Plant singular values

angular motion, see for instance [144]. First, an integral three channel μ controller is designed to ensure robust stability and robust performance of the angular motion. This controller ensures small deviations of the Euler angles from their trim values in the presence of wind disturbances and plant uncertainty. Then, a simplified positional controller is designed which consists of three proportional-derivative (PD) regulators in the position and altitude channels. Thus, the helicopter control is decomposed in two loops: a fast loop for attitude control and a relatively slow loop for translational motion control.

6.2.1 Performance requirements

The block diagram of the closed-loop system for attitude helicopter control that includes the uncertain helicopter model, the feedback, and the controller, as well as the elements which reflect the performance requirements, is shown in Figure 6.7. The aim of this control is to keep the attitude angles ϕ, θ, ψ close to their reference values ϕ_{ref}, θ_{ref}, ψ_{ref} in the presence of input disturbances *dist* and model uncertainties. The design diagram includes a reference model M, which is used to set the desired helicopter response. The helicopter control is done by using a feedback from the Euler angles ϕ, θ, ψ and from the angular rates p, q, r. It is supposed that the information about the measurement variables is obtained from an onboard INS equipped with a Kalman filter to improve the accuracy. As mentioned in the beginning of this chapter,

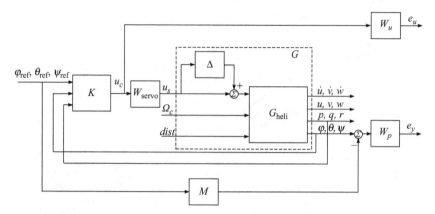

Figure 6.7   Block diagram of the closed-loop system with performance requirements

it is assumed that the effect of measurement noises on the control system is negligible in comparison to the effect of wind disturbances which is confirmed by the simulations done latter on. That is why the noise suppression is not considered as a performance requirement in the given case.

The transfer function matrix G_{heli} represents the nominal helicopter model. The matrix

$$\Delta = \begin{bmatrix} \delta_1 & 0 & 0 & 0 \\ 0 & \delta_2 & 0 & 0 \\ 0 & 0 & \delta_3 & 0 \\ 0 & 0 & 0 & \delta_4 \end{bmatrix}$$

contains the gain bounded uncertainties that represent the input multiplicative model uncertainty.

The controller output

$$u_c = \begin{bmatrix} u_c^{\text{col}}, u_c^{\text{lon}}, u_c^{\text{lat}}, u_c^{\text{tr}} \end{bmatrix}^T$$

is the input to the servoactuators. The transfer function matrix

$$W_{\text{servo}} = \begin{bmatrix} w_{\text{servo}}^{\text{col}}(s) & 0 & 0 & 0 \\ 0 & w_{\text{servo}}^{\text{lon}}(s) & 0 & 0 \\ 0 & 0 & w_{\text{servo}}^{\text{lat}}(s) & 0 \\ 0 & 0 & 0 & w_{\text{servo}}^{\text{tr}}(s) \end{bmatrix}$$

contains the transfer functions of the corresponding servoactuators.

To obtain good performance of the closed-loop system, we shall implement a two-degree-of-freedom controller. The control actions are generated according to the expression

$$u_c = [K_r \ K_y] \begin{bmatrix} r_c \\ y_c \end{bmatrix} = K_r r_c + K_y y_c, \tag{6.19}$$

where

$$r_c = [\phi_{\text{ref}}, \theta_{\text{ref}}, \psi_{\text{ref}}]^T$$

is the reference vector,

$$y_c = [\phi, \theta, \psi, p, q, r]^T$$

is the output feedback vector, K_r is the 4×3 prefilter transfer function matrix, and K_y is the 4×6 output feedback transfer function matrix. The control actions should be such that the servoactuator outputs do not exceed the values

$$\delta_{\text{col}}^{\max} = 0.183 \text{ rad}, \quad \delta_{\text{lon}}^{\max} = 0.096 \text{ rad}, \quad \delta_{\text{lat}}^{\max} = 0.096 \text{ rad}, \quad \delta_{\text{tr}}^{\max} = 0.38 \text{ rad}$$

which correspond to the servo saturations.

The system has two output vector signals (e_y and e_u). The block M is the 3×3 transfer function matrix of the ideal dynamics model that the designed closed-loop system should match to.

The equation for the feedback variables may be represented as

$$y_c = G_u u_s + G_\Omega \Omega_c + G_d d, \tag{6.20}$$

where G_u is the transfer function matrix with respect to the control signals from servoactuators, G_Ω is the transfer function matrix with respect to the engine speed reference, and G_d is the plant transfer function matrix with respect to disturbances. (Here and further on, the disturbance vector is denoted for brevity by d.) The matrices G_u, G_Ω, G_d are obtained from the transfer function matrix G of the uncertain helicopter model (6.18).

Taking into account the structure of the feedback vector y_c, one obtains that

$$e_y = W_p^c(y_c - M^c r_c),$$

where

$$W_p^c = \begin{bmatrix} W_p & 0_{3 \times 3} \end{bmatrix}, \quad M^c = \begin{bmatrix} M \\ 0_{3 \times 3} \end{bmatrix}.$$

A simplified block diagram of the closed-loop system with performance requirements, utilizing (6.19) and (6.20), is shown in Figure 6.8. The weighted closed-loop system outputs e_y and e_u satisfy the equation

$$\begin{bmatrix} e_y \\ e_u \end{bmatrix} = \begin{bmatrix} W_p^c(S_o G_u W_{\text{servo}} K_r - M^c) & W_p^c S_o G_d & W_p^c S_o G_\Omega \\ W_u S_i K_r & W_u S_i K_y G_d & W_u S_i K_y G_\Omega \end{bmatrix} \begin{bmatrix} r_c \\ d \\ \Omega_c \end{bmatrix}, \tag{6.21}$$

where the matrix $S_i = (I - K_y G_u W_{\text{servo}})^{-1}$ is the input sensitivity transfer function matrix, and $S_o = (I - G_u W_{\text{servo}} K_y)^{-1}$ is the output sensitivity transfer function matrix.

Further on, the rotor-speed reference Ω_c is considered as constant and its value is taken as $\Omega_c = 167$ rad/s for the helicopter under consideration.

The performance criterion requires the transfer function matrix from the exogenous input signals r_c, d, and Ω_c to the output signals e_y and e_u to be small in the sense of $\|\cdot\|_\infty$, for all possible uncertain plant models G. The transfer function matrices W_p and W_u are used to reflect the relative importance of the different frequency ranges

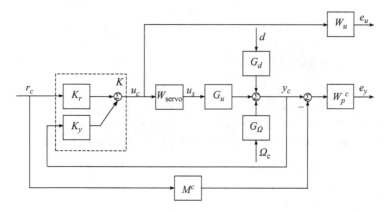

Figure 6.8   Simplified block diagram of the closed-loop system

Table 6.3 $\mathcal{H}_\infty$ *functions to be minimized*

Function	Description
$W_p^c(S_o G_u W_{\text{servo}} K_r - M^c)$	Weighted difference between the real and the ideal closed-loop system
$W_p^c S_o G_d$	Weighted sensitivity to disturbance
$W_p^c S_o G_\Omega$	Weighted sensitivity to rotor-speed reference
$W_u S_i K_r$	Weighted control action due to angle references
$W_u S_i K_y G_d$	Weighted control action due to disturbance
$W_u S_i K_y G_\Omega$	Weighted control action due to rotor-speed reference

for which the performance requirements should be fulfilled. The six transfer function matrices which constitute the transfer function matrix between the inputs and outputs of the extended system are described in Table 6.3.

The design problem for the attitude helicopter control is to find a linear controller $K(s)$ in the reference and the measurable output

$$K = [K_r \ K_y]$$

that has to ensure the following properties of the closed-loop system:

Robust stability: The closed-loop system achieves robust stability if this system is internally stable for all possible plant models G.

Robust performance: The closed-loop system should remain internally stable for all G and in addition the performance criterion

$$\left\| \begin{bmatrix} W_p^c(S_o G_u W_{\text{servo}} K_r - M^c) & W_p^c S_o G_d & W_p^c S_o G_\Omega \\ W_u S_i K_r & W_u S_i K_y G_d & W_u S_i K_y G_\Omega \end{bmatrix} \right\|_\infty < 1 \qquad (6.22)$$

should be satisfied for each G.

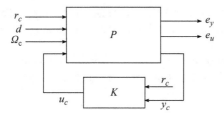

Figure 6.9 Block diagram of μ synthesis

6.2.2 Controller design

The closed-loop system block diagram associated with the μ synthesis problem is shown in Figure 6.9. The matrix P is the transfer function matrix of the extended open-loop system that consists of the uncertain plant model plus the servoactuators, the matching model, and the performance weighting functions. It has 11 inputs (r_c, d, Ω_c, u_c) and 13 outputs (e_y, e_u, ϕ, θ, ψ, p, q, r).

The control actions to the plant are realized by a Digital Signal Controller in real time with sampling frequency $f_s = 100$ Hz so that the μ synthesis is implemented to design a discrete-time controller at this sampling frequency. For this aim, the transfer function matrix of the nominal open-loop system is discretized for the same frequency assuming that there are zero-order holds at the plant inputs.

Let $N_d(z)$ be the transfer function matrix of the discretized nominal open-loop system, and denote by $P_d(z) = F_U(N_d, \Delta)$ the transfer function matrix of the uncertain open-loop system. Let the block structure Δ_{P_d} is defined as

$$\Delta_{P_d} := \left\{ \begin{bmatrix} \Delta & 0 \\ 0 & \Delta_F \end{bmatrix} : \Delta \in \mathbb{C}^{4\times4},\ \Delta_F \in \mathbb{C}^{7\times7} \right\}.$$

The first block of the matrix Δ_{P_d} corresponds to the 4×4 input multiplicative uncertainty, included in the helicopter model. The second block Δ_F is a 7×7 fictitious uncertainty block, used to include the performance requirements into the framework of the μ approach. The inputs of this block are the weighted error signals e_y and e_u, and the outputs are the exogenous signals r_c, d, and Ω_c.

The aim of the μ synthesis is to find a discrete-time stabilizing controller K_d, such that for each frequency $\omega \in [0, \pi/T_s]$, where $T_s = 2\pi/f_s$, the structured singular value μ satisfies the condition

$$\mu_{\Delta_{P_d}}[F_L(N_d, K_d)(j\omega)] < 1,$$

where $F_L(N_d, K_d)$ is the transfer function matrix of the discrete-time closed-loop system. The fulfillment of this condition guarantees robust performance of the closed-loop system, i.e.,

$$\left\| F_U \left[F_L(N_d, K_d), \Delta_{P_d} \right] \right\|_\infty < 1$$

for all uncertainties Δ_{P_d} with $\|\Delta_{P_d}\|_\infty < 1$. Due to the usage of a discrete-time model of the open-loop system, the stability of the closed-loop system with the controller K_d is guaranteed.

The transfer function matrix M of the ideal matching model is chosen as diagonal in order to suppress the interaction between the three channels and is taken as

$$M(s) = \begin{bmatrix} w_{m1} & 0 & 0 \\ 0 & w_{m2} & 0 \\ 0 & 0 & w_{m3} \end{bmatrix},$$

where

$$w_{m1} = \frac{1}{0.40^2 s^2 + 2 \times 0.40 \times 0.7s + 1},$$

$$w_{m2} = \frac{1}{0.35^2 s^2 + 2 \times 0.35 \times 0.7s + 1},$$

$$w_{m3} = \frac{1}{0.25^2 s^2 + 2 \times 0.25 \times 0.7s + 1}.$$

The model transfer functions corresponds to desired closed-loop bandwidths of the roll, pitch and yaw channels equal to 2.5, 2.86, and 4.0 rad/s, respectively.

The μ synthesis is done for several performance weighting functions that ensure a good balance between system performance and robustness. On the basis of the experimental results, we choose the performance weighting function

$$W_p(s) = \begin{bmatrix} 4.5\dfrac{10^{-3}s + 1}{10^{-2}s + 1} & 0 & 0 \\ 0 & 5.0\dfrac{10^{-3}s + 1}{10^{-2}s + 1} & 0 \\ 0 & 0 & 10.0\dfrac{10^{-3}s + 1}{10^{-2}s + 1} \end{bmatrix}$$

and the control weighting function

$$W_u(s) = \begin{bmatrix} \dfrac{0.02s + 1}{10^{-4}s + 1} & 0 & 0 & 0 \\ 0 & \dfrac{0.02s + 1}{10^{-4}s + 1} & 0 & 0 \\ 0 & 0 & \dfrac{0.02s + 1}{10^{-4}s + 1} & 0 \\ 0 & 0 & 0 & \dfrac{0.02s + 1}{10^{-4}s + 1} \end{bmatrix}.$$

The performance weighting functions are chosen as low-pass filters to suppress the difference between the system and model for frequencies up to 20 rad/s (Figure 6.10). The control weighting functions are chosen as high-pass filters to impose constraints on the control action components with frequencies above 10 rad/s (Figure 6.11).

The transfer function matrices $G(s)$, $M(s)$, $W_p(s)$, and $W_u(s)$ are converted to discrete-time form when finding the discretized model $N_d(z)$.

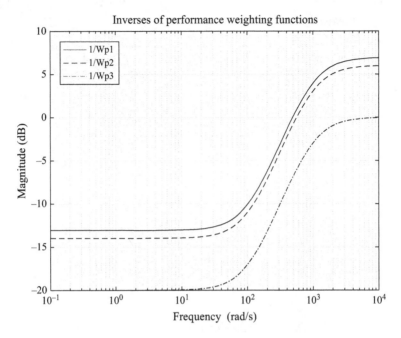

Figure 6.10 Magnitude responses of performance weighting functions

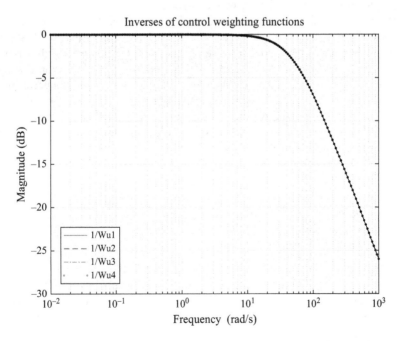

Figure 6.11 Magnitude responses of control weighting functions

Table 6.4   Results of the μ synthesis

Iteration	Controller order	Maximum value of μ
1	37	1.089
2	47	0.993
3	41	0.993
4	41	0.991

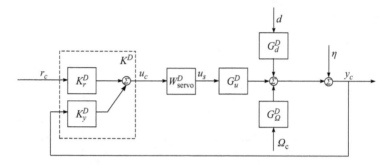

Figure 6.12   Block diagram of the discrete-time closed-loop system

The open-loop interconnection used in the design is done by the file `dlp_heli`. The μ synthesis is performed by using the M-file `design_heli` which implements the MATLAB function `dksyn`. The results from the D–K iterations are shown in Table 6.4. The best controller is obtained after the fourth iteration and is of 41st order. (In the given case, the function `dksyn` stops automatically after the fourth iteration.) It should be pointed out that controllers of such order are typical for the μ synthesis method and result from the necessity to satisfy simultaneously the performance and robustness requirements. The usual practice is to decrease the controller order implementing some method for order reduction. In the given case, it is possible to reduce the controller order to 30 without sacrificing the closed-loop performance and robustness. This is done by the command line

```
Kd_30 = reduce(Kd,30)
```

Further on, we use the 30th-order controller in the investigation of the closed-loop system properties.

6.2.3   Frequency responses

The block diagram of the discrete-time system, used in the closed-loop analysis, is shown in Figure 6.12. Note that the transfer function matrices with superscript D are discrete-time counterparts of the transfer function matrices of the continuous-time system model. In the analysis, we take into account the effect of measurement noise

$$\eta = \left[\eta_\phi, \ \eta_\theta, \ \eta_\psi, \ \eta_p, \ \eta_q, \ \eta_r \right]^T$$

in the angles and angular rates which was neglected in the design.

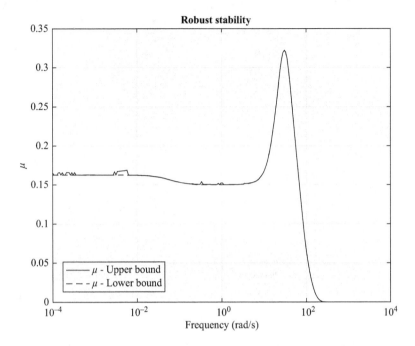

Figure 6.13   Robust stability of the closed-loop system

Consider first the robustness properties of the closed-loop discrete-time system determined by the M-file dmu_heli. In Figure 6.13, we show the frequency response of the structured singular value $\mu_\Delta F_L(N_d, K_d)$ corresponding to the robust stability analysis of the closed-loop system. Since the maximum value of μ over the frequency is 0.3223, the uncertain system can tolerate up to 310 percent of the modeled uncertainty, i.e., the robust stability margin of the closed-loop system is 310 percent.

The frequency response of the structured singular value corresponding to the robust performance analysis is shown in Figure 6.14. The closed-loop system achieves robust performance for all uncertain plant models and disturbances.

The frequency responses of the discrete-time closed-loop system are obtained using the M-file dfrs_heli.

The singular value frequency response plots of the 30th-order μ controller are shown in Figure 6.15. Since the controller has nine inputs and four outputs, its behavior in the frequency domain is represented by the responses of four singular values.

The frequency responses of the singular values of the closed-loop transfer function matrix for 30 random values of the plant uncertainty are shown in Figure 6.16. It is seen that the closed-loop system frequency responses are close to these of the model M for frequencies up to 10 rad/s which is a result of achieving robust performance. The closed-loop bandwidths of the roll, pitch, and yaw channels are equal to 2.3, 2.8, and 3.9 rad/s, respectively, which are close to the corresponding values prescribed

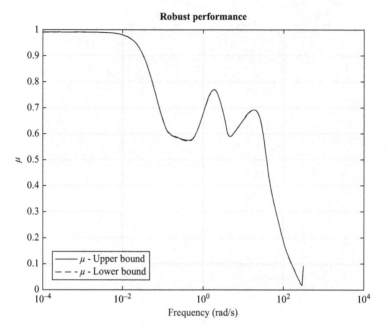

Figure 6.14   *Robust performance of the closed-loop system*

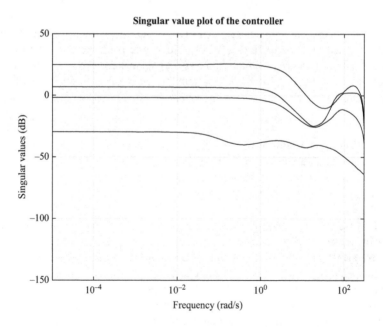

Figure 6.15   *Frequency responses of the μ controller*

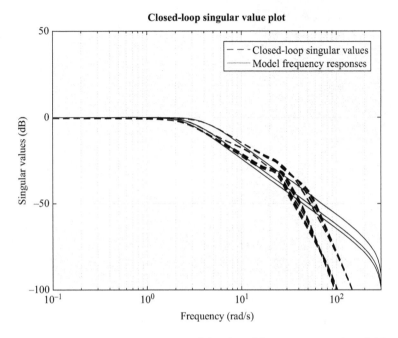

Figure 6.16   Frequency responses of the closed-loop system and model M

by the model. The frequency response plot of the singular values of disturbance-to-output transfer function matrix is shown in Figure 6.17. It is seen from the figure that the disturbance attenuation at low frequencies is more than 300 times (50 dB). This guarantees good response of the closed-loop system in the presence of significant wind disturbances.

As it is seen from Figures 6.18 and 6.19, the measurement noises have weak effect on the output and control of closed-loop system. For this reason, the noise suppression was not included in the performance requirements during the controller design.

The worst case loop-at-a-time gain and phase margins of the uncertain closed-loop system are determined by using the Robust Control Toolbox™ function wcmargin eliminating the reference channel of the controller. This function computes the disk margins at the input and output such that for all gain and phase variations inside the disk the nominal closed-loop system remains stable. From the worst case bounds at the output, we obtain for roll, pitch, and yaw channels the following results:

```
phi_loop_margin =

    GainMargin: [0.1210 8.2662]
    PhaseMargin: [-76.2044 76.2044]
```

```
Frequency: 5.6285
      WCUnc: [1x1 struct]
Sensitivity: [1x1 struct]


theta_loop_margin =

  GainMargin: [0.1129 8.8548]
 PhaseMargin: [-77.1134 77.1134]
   Frequency: 8.7824
       WCUnc: [1x1 struct]
 Sensitivity: [1x1 struct]


psi_loop_margin =

  GainMargin: [0.0777 12.8780]
 PhaseMargin: [-81.1196 81.1196]
   Frequency: 6.2048
       WCUnc: [1x1 struct]
 Sensitivity: [1x1 struct]
```

Clearly, the largest gain and phase margins have the yaw channel.

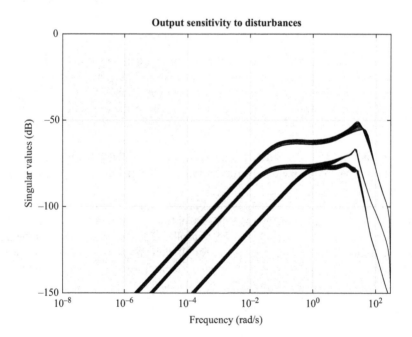

Figure 6.17   Output sensitivity to disturbances

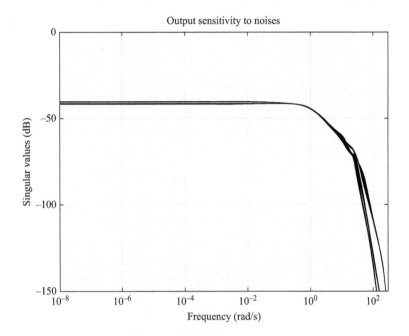

Figure 6.18   *Output sensitivity to measurement noises*

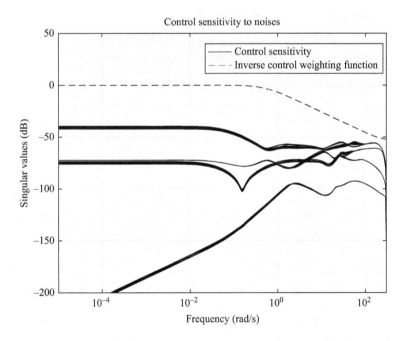

Figure 6.19   *Sensitivity of control action to measurement noises*

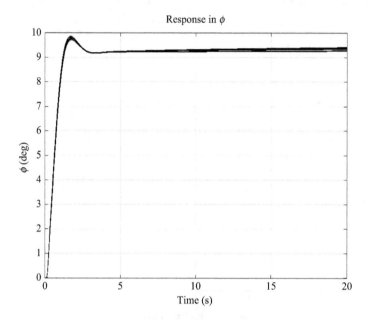

Figure 6.20   Response in roll angle ϕ

6.2.4   Transient responses of the linear system

Consider now the transient responses of the uncertain closed-loop system due to step reference inputs. The following reference amplitudes are chosen:

$$\phi_{\text{ref}} = 10 \ \text{deg}, \quad \theta_{\text{ref}} = -10 \ \text{deg}, \quad \psi_{\text{ref}} = 15 \ \text{deg}$$

The step responses of the sampled-data closed-loop system, obtained by the M-file dmcs_heli for 30 random values of the uncertainty, are shown in Figures 6.20–6.22. Clearly, the closed-loop system is not too sensitive to the uncertainty. The responses have small overshoots and the steady-state errors are less than 0.9 deg which is acceptable in practice. Smaller steady-state errors may be obtained if the uncertainty level assumed in the design is smaller. Note that the angles ϕ, θ, ψ, shown in the figures, are not the corresponding full angles characterizing the nonlinear plant but represent the deviations of these angles from their respective trim values.

6.2.5   Position controller design

Once the attitude motion of the helicopter is stabilized, it is possible to translate the helicopter to the desired point of space by using three simple PD regulators which produce the necessary control actions. To design such regulators, we shall implement a technique similar to the one, presented in [145] and based on approximate description of the helicopter translational motion.

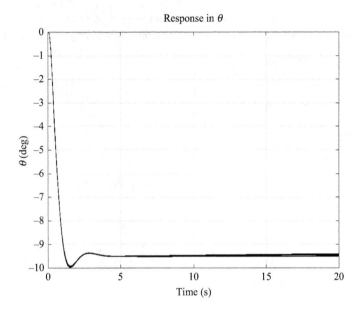

Figure 6.21   Response in pitch angle θ

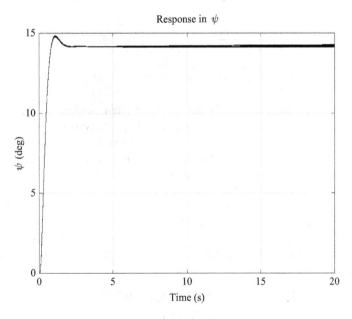

Figure 6.22   Response in yaw angle ψ

Assuming that the angles ϕ and θ are small (i.e., $\sin\phi \approx \phi$, $\sin\theta \approx \theta$), and the products uq and vp are negligible, (6.7) may be represented approximately as

$$V_x = w(\cos\phi\cos\psi)\theta,$$
$$V_y = -w(\cos\psi)\phi, \tag{6.23}$$
$$V_z = w\cos\phi\cos\theta,$$

where, according to (6.2),

$$\dot{w} = Z/m + g\cos\phi\cos\theta. \tag{6.24}$$

Neglecting the fuselage and vertical fin forces along the Z axis in comparison to the main rotor thrust we have that

$$Z \approx -T_{mr},$$

which, along with (6.24), shows that the desired altitude may be reached by appropriate change of the collective pitch angle δ_{col}. Similarly, the desired positions x and y in the horizontal plane may be reached by appropriate deviations of the pitch and roll angles θ and ϕ, respectively. Equation (6.23) suggests to input the following PD regulator actions to the servoactuators:

$$u_z = \frac{K_{p1}(z_{ref} - z) - K_{d1}V_z}{\cos\phi\cos\theta},$$

$$u_x = \frac{K_{p2}(x_{ref} - x) - K_{d2}V_x}{\cos\phi\cos\psi}, \tag{6.25}$$

$$u_y = \frac{K_{p3}(y_{ref} - y) - K_{d3}V_y}{\cos\psi},$$

where $x_{ref}, y_{ref}, z_{ref}$ are the coordinates of the desired position in the 3D space, K_{p1}, K_{p2}, K_{p3} are the proportional, and K_{d1}, K_{d2}, K_{d3}—derivative coefficients of the PD regulators chosen so as to ensure the desired dynamics of the helicopter motion to the prescribed position. The control laws (6.25) allow to decouple approximately the translational motion of the helicopter from its angular motion and to control separately the motions along X, Y, and Z axes.

The control actions u_z, u_x, u_y that ensure the desired position are added to the corresponding outputs of the μ controller that stabilizes the attitude motion as shown in Figure 6.23. The position controller K_{pos} implements the PD regulators (6.25). In the given case, the following regulator coefficients were found appropriate (to three significant digits):

$$K_{p1} = -0.531, \quad K_{p2} = -0.0695, \quad K_{p3} = 0.0225,$$
$$K_{d1} = -0.0327, \quad K_{d2} = -0.102, \quad K_{d3} = 0.0355.$$

These coefficients are found by optimization procedure to provide the desired form of the transient responses and small settling time. Their signs are chosen so as to ensure the stability of the corresponding control loops.

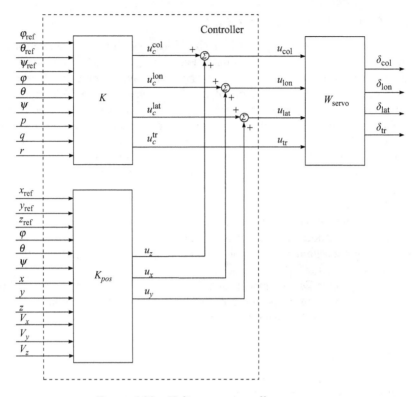

Figure 6.23 Helicopter controller structure

6.3 Hardware-in-the-loop simulation

MATLAB files used in this section

File	Description
sim_heli_double.slx	Simulink model of the nonlinear system in double precision
sim_heli_single.slx	Simulink model of the nonlinear system implementing single precision controller
sim_HIL_SCI_heli.m	HIL simulation of the helicopter control system and visualization of the results
HIL_SCI_heli.slx	Simulink model of the extended helicopter dynamics in double precision executed on the host PC
HIL_SCI_target_heli.slx	Simulink model of the single precision helicopter controller embedded in DSC

6.3.1 Nonlinear system simulation

The transient responses of the nonlinear closed-loop system with discrete-time μ controller are obtained by the Simulink model `sim_heli_double` using the double-precision arithmetic of MATLAB . The control actions produced by the linear controller are added to the trim values of the control actions determined for hovering. To avoid the using of variable step in the integration of the helicopter differential equations, the nonlinear helicopter dynamics is discretized for the sampling interval $T_s = 0.01$ s. This is done by implementing the Bogacki and Shampine method for solving ODE as described in Chapter 2. The helicopter model is realized with MATLAB Embedded Function blocks which allows to compile in C the MATLAB functions included in the Simulink model. This increases significantly the speed of the simulation process.

The wind simulation in the nonlinear model is done using the low-altitude Dryden Wind Turbulence Model as specified, for instance, in [146–149]. The wind model is realized using band-limited white noise with appropriate digital system difference equations.

According to the Dryden model, turbulence is a stochastic process defined by velocity spectra. The power spectral densities for the wind velocities along X, Y, and Z axes are given by the following equations:

$$\Phi_u(\Omega) = \sigma_u^2 \frac{2L_u}{\pi V} \frac{1}{1 + (L_u \Omega)^2},$$

$$\Phi_v(\Omega) = \sigma_v^2 \frac{L_v}{\pi V} \frac{1 + 3(L_v \Omega)^2}{(1 + (L_v \Omega)^2)^2},$$

$$\Phi_w(\Omega) = \sigma_w^2 \frac{L_w}{\pi V} \frac{1 + 3(L_w \Omega)^2}{(1 + (L_w \Omega)^2)^2},$$

where $\sigma_u, \sigma_v, \sigma_w$ are the corresponding turbulence intensities, L_u, L_v, L_w are the gust length scales, and Ω is the spatial frequency. The vertical length scale and turbulence intensity can be assumed to be $L_w = |z|$ and $\sigma_w = 0.1w_{20}$. Here, w_{20} is the given wind speed in knots (1 knot = 0.514444444 m/s) at 20-ft (6-m) altitude. The gust length scales can be found from the following equations:

$$L_w = h$$

$$L_u = \frac{L_w}{(0.177 + 0.000823h)^{1.2}}$$

$$L_v = L_u$$

and the turbulence intensities are computed from

$$\sigma_w = 0.1w_{20}$$

$$\sigma_u = \frac{\sigma_w}{(0.177 + 0.000823h)^{0.4}}$$

$$\sigma_v = \sigma_u.$$

For light turbulence, the wind speed at 20 ft is 15 kn; for moderate turbulence, the wind speed is 30 kn, and for severe turbulence, the wind speed is 45 kn. Further on in the simulations we make use of the value $w_{20} = 30$ kn, corresponding to moderate turbulence.

For a vehicle flying at a speed V through a frozen turbulence field with a spatial frequency of Ω (rad/m), the relationship between Ω and the circular frequency ω (rad/s) is given by $\omega = \Omega V$. This allows to derive difference equations for the wind velocities (in ft/s) in the form

$$u_{\text{wind}}(k+1) = \left(1 - \frac{V}{L_u}T_s\right)u_{\text{wind}}(k) + \sqrt{\frac{2V}{L_u}T_s}\,\sigma_u\eta_u,$$

$$v_{\text{wind}}(k+1) = \left(1 - \frac{V}{L_v}T_s\right)v_{\text{wind}}(k) + \sqrt{\frac{2V}{L_v}T_s}\,\sigma_v\eta_v, \qquad (6.26)$$

$$w_{\text{wind}}(k+1) = \left(1 - \frac{V}{L_w}T_s\right)w_{\text{wind}}(k) + \sqrt{\frac{2V}{L_w}T_s}\,\sigma_w\eta_w,$$

where η_u, η_v, η_w are band limited white noises with unit variances.

The wind velocities are computed by the M-file `wind_model`. The noises η_u, η_v, η_w are generated by the Simulink Band-Limited White Noise block. Since the wind velocity is a vector in the Earth fixed inertial frame, the computed wind velocities are transformed into body coordinates using the direction cosine matrix.

The nonlinear system simulation involves also simulation of the measurement errors which were neglected during the controller design. The modeling of these errors is done under the assumption that the measured Euler angles ϕ, θ, ψ and angular rates p, q, r are obtained by an INS performing Kalman filtering. This means that the errors in the measured variables may be considered as white noises which can be modeled by using the Simulink Band-Limited White Noise block. It is assumed that the angles can be measured to an accuracy of 1 deg and the rates to an accuracy of 0.1 deg/s.

The Simulink model `sim_heli_single` allows to compute the transient responses assuming that the controller works using single precision arithmetic. Simulation results for the nonlinear system are presented in the next section along with the results form hardware-in-the-loop simulation (HIL) simulation of the closed-loop system.

6.3.2 HIL simulation setup

The hardware platform for HIL simulation of helicopter control system consists of a host computer (PC) and eZdsp™ F28335 starter kit equipped with the Texas Instruments *TMS320F28335* Digital Signal Controller (DSC) (the same as the one used for HIL simulation in Chapter 4). The DSC works at 150 MHz and may perform single precision (32-bit) computations by using FPU (Floating-Point Unit). The control algorithm is embedded and runs with frequency 100 Hz on the target DSC. RS-232 interface is used for communication between the host PC and target DSC the interface being configured at 115,200 bps. The serial communication imports from 20 to 30 ms

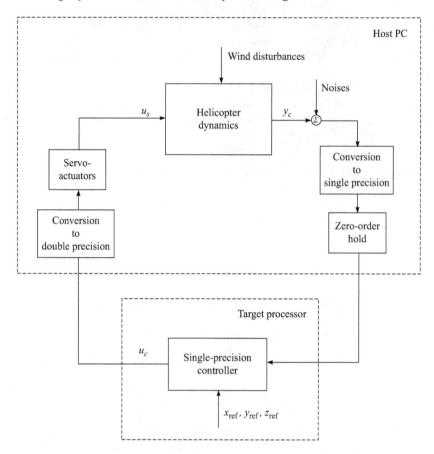

Figure 6.24   HIL simulation of helicopter control system

delay, while the DSC target and host PC compute their output signals much faster. That is why the simulation of the closed-loop system cannot be done in real-time, and it is necessary to use a protocol mode for serial communication. According to the protocol used in Embedded Coder®, a software handshaking between the host and target is implemented. The transmitting side sends "SEND message" indicating that it is ready to transmit. The receiving side sends back "READY message" indicating that it is ready to receive. The transmitting side then sends data and when the transmission is completed it sends a checksum. DSC transmits a frame of data (the control signals) consisting of 10 B to host PC. Once host PC receives data from DSC, it feedbacks a frame of data (the sensor signals) consisting of 26 B. In this way, the controller works exactly as in real-time independently on the presence of communication delays.

The block diagram which represents the hardware-in-the-loop simulation of the closed-loop helicopter control system is shown in Figure 6.24. For more realistic results, the helicopter model is simulated in double-precision floating-point arithmetic, the processor work is simulated in single-precision arithmetic and the signals from INS and to servoactuators are assumed to be 32-b single precision numbers. To reduce the communication load between the host computer and target processor, the reference trajectory is computed online inside the embedded controller.

Thus, three simulation models of the helicopter control system are used, namely,

Model 1 The Simulink model `sim_heli_double` using only double-precision arithmetic in all computations,

Model 2 The Simulink model `sim_heli_single` implementing single-precision arithmetic to simulate the DSC computations, sensor, and servoactuator signals,

Model 3 The Simulink models `HIL_SCI_heli` implementing double-precision arithmetic to simulate helicopter model on PC and `HIL_SCI_target_heli` using single-precision arithmetic to compute control actions on the target DSC.

Models 1 and 2 are implemented only on the host PC, while Model 3 is implemented on PC and on DSC. The host PC and the DSC perform computations with the same sampling frequency, equal to 100 Hz, as assumed in the controller design. All models use fixed step integration of the system difference equations. In what follows we compare the results obtained by Models 1 and 3 in order to assess the effect of using single-precision computations on DSC and in the sensor and actuator signals. Model 2 is used to validate the automatically generated control code, and the results from using this model are very close to the results from HIL simulation. The simulation results are saved on the PC and are visualized in MATLAB.

6.3.3   Results of HIL simulation

The HIL simulation of helicopter control system is done for different desired trajectories in the 3D space including hovering, motion to a desired point of the space, and circular motion at given altitude. The simulation is performed on the full nonlinear model of closed-loop system, including the extended helicopter dynamics, inner loop attitude control, and outer loop position control in the presence of wind disturbances and measurement noises. The variations in position obtained are result of vehicle dynamics, wind disturbances, sensor noises and floating point errors. In this subsection, we present the results for circular motion in the 3D space.

The desired trajectory is set as a circular path of diameter 100 m in the horizontal plane,

$$x(t) = A\sin(2\pi ft), \quad y(t) = A\sin(2\pi ft + \pi/2),$$

with $A = 50$ m, $f = 0.02$ Hz along with a motion in the vertical plane described by

$$z(t) = -2t \text{ m if } t < 10 \text{ s}$$
$$-20 \text{ m if } t \geq 10 \text{ s}.$$

The initial conditions are chosen as

$$x(0) = 0 \text{ m}, \quad y(0) = 50 \text{ m}, \quad z(0) = 0 \text{ m}$$

and the flight time is set equal to 100 s. During this time, the helicopter is performing two circles with average velocity greater than 6 m/s.

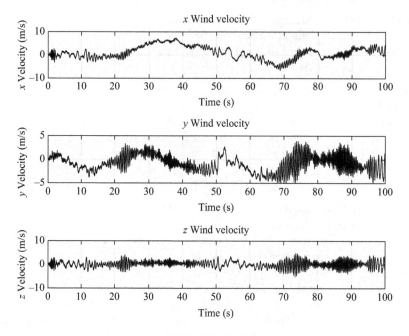

Figure 6.25   Wind velocities

The wind velocity components acting on the helicopter during the flight are shown in Figure 6.25.

The corresponding transient responses along the X, Y, Z axes are shown in Figure 6.26. Independently on the wind disturbances acting on the helicopter, the motion on the desired path is performed with sufficient accuracy.

The transient responses in the Euler angles and in the lateral flapping angle are shown in Figures 6.27–6.30. Significant initial deviation in the pitch angle θ occurs, caused by the aggressive input δ_{lon} to adjust to desired trajectory the initial vehicle motion. There are also some periodical deviations of the Euler angles due to the necessity to ensure the appropriate velocities along the corresponding axes. The deviations of ϕ, θ, and ψ from their trim values during the translational motion change the linearized plant model, but due to the large robust stability margin (310 percent), this change does not affect significantly the helicopter motion.

The control actions that ensure the necessary translations along the X, Y, Z axes are shown in Figures 6.31–6.34. These actions are less than their corresponding maximum allowed values and have some small offsets due to the nonzero mean values of wind disturbances.

To access the simulation accuracy, we computed the mean values and the standard deviations of the differences between the variables, determined by hardware-in-the-loop simulation and the corresponding variables, evaluated by using double-precision arithmetic. The results for the circular motion under consideration are shown in Table 6.5 where the superscripts M_1 and M_3 denote the variables computed by using

3D helicopter motion from HIL simulation

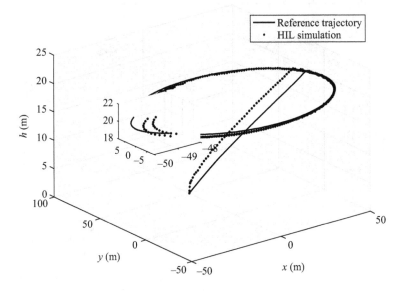

Figure 6.26   3D helicopter motion

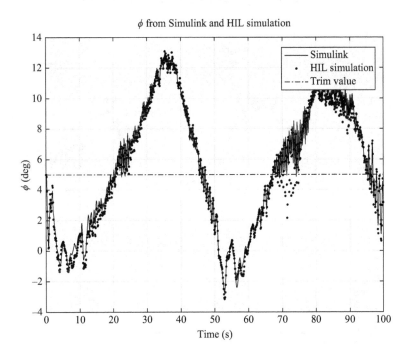

Figure 6.27   Response in φ during circular motion

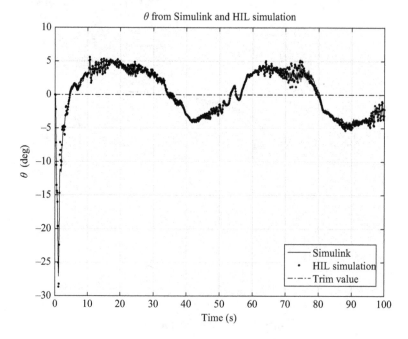

Figure 6.28   Response in θ during position change

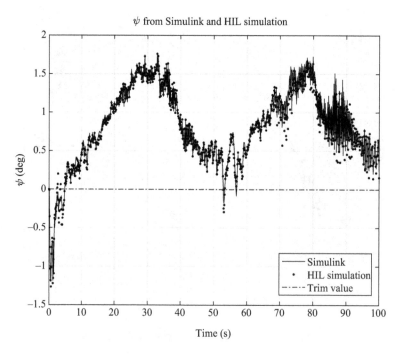

Figure 6.29   Response in ψ during position change

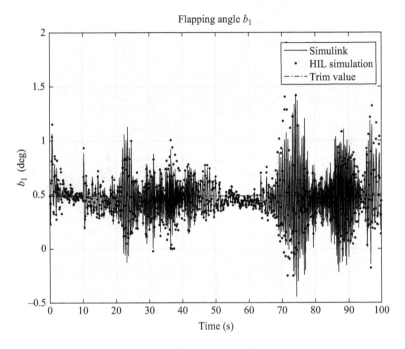

Figure 6.30   Response in lateral flapping angle b_1 during position change

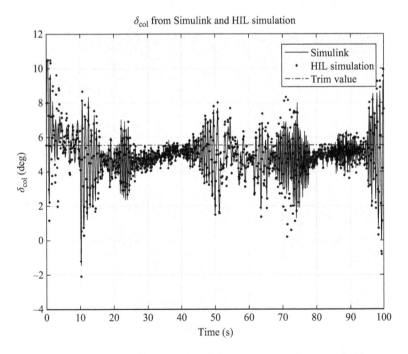

Figure 6.31   Collective pitch angle during position change

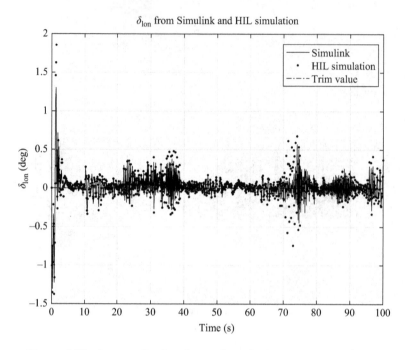

Figure 6.32   Longitudinal cyclic pitch angle during position change

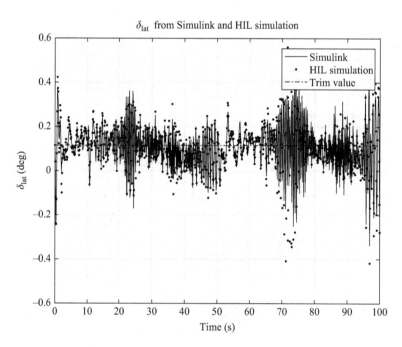

Figure 6.33   Lateral cyclic pitch angle during position change

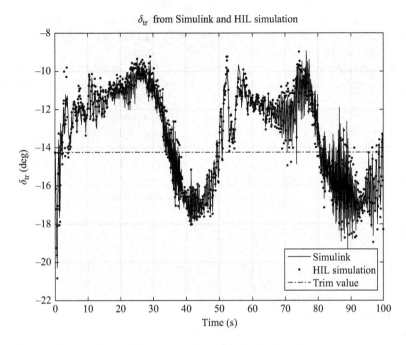

Figure 6.34   Tail rotor pitch angle during position change

Table 6.5   Simulation errors

Quantity	Mean value	Standard deviation	Unit
$x^{M_3} - x^{M_1}$	6.81×10^{-3}	5.79×10^{-2}	m
$y^{M_3} - y^{M_1}$	1.36×10^{-2}	6.08×10^{-2}	m
$z^{M_3} - z^{M_1}$	-8.93×10^{-3}	2.15×10^{-2}	m
$\phi^{M_3} - \phi^{M_1}$	-1.04×10^{-1}	3.11×10^{-1}	deg
$\theta^{M_3} - \theta^{M_1}$	7.92×10^{-2}	3.37×10^{-1}	deg
$\psi^{M_3} - \psi^{M_1}$	-3.11×10^{-3}	5.33×10^{-2}	deg
$a_1^{M_3} - a_1^{M_1}$	3.56×10^{-8}	1.56×10^{-1}	deg
$b_1^{M_3} - b_1^{M_1}$	-5.62×10^{-4}	1.09×10^{-1}	deg
$\delta_{col}^{M_3} - \delta_{col}^{M_1}$	-2.44×10^{-2}	5.63×10^{-1}	deg
$\delta_{lon}^{M_3} - \delta_{lon}^{M_1}$	1.90×10^{-3}	6.34×10^{-2}	deg
$\delta_{lat}^{M_3} - \delta_{lat}^{M_1}$	5.74×10^{-5}	4.16×10^{-2}	deg
$\delta_{tr}^{M_3} - \delta_{tr}^{M_1}$	9.31×10^{-2}	4.19×10^{-1}	deg

Models 1 and 3, respectively. The two models produce results that are sufficiently close which means that the algorithm embedded in the DSC produces control actions that are close to the predicted by the μ synthesis actions.

As noted earlier, the μ controller designed is valid for hovering. According to the experiments, it may be used also to change the helicopter position in sufficiently large area.

The results obtained in this chapter show that it is possible to apply successfully high order robust helicopter controllers by using Digital Signal Processors. The implementation of μ controllers allows to achieve robust stability and robust performance of the linearized closed-loop system which is difficult to achieve with other type of controllers. The μ controller allows to improve the system performance especially in the case of sufficiently large parameter variations and strong disturbances.

6.4   Notes and references

Dynamics models, suitable for control design and simulation of small-scale unmanned helicopters, are derived in many sources, see for instance Castillo, Lozano, and Dzul [150]; Castillo-Effen *et al.* [151]; Mettler [143]; Nonami *et al.* [152]; Raptis and Valavanis [153]; and Sandino, Bejar, and Ollero [154]. These models are based on the first-principles dynamics models developed for full-scale helicopters (see Padfield [141] and Bramwell, Done, and Balmford [155]) accounting for the particular characteristics of miniature helicopters.

The robust helicopter control is based usually on $\mathscr{H}_\infty$ optimization or μ synthesis, see Castillo-Effen *et al.* [151,156]. $\mathscr{H}_\infty$ loop shaping is applied to the hover control of Yamaha R-50 helicopter and validated in real flight as described in La Civita *et al.* [157]. The same approach is used in Boukhnifer, Chaibet, and Larouci [158] to control a 3-DOF miniature helicopter and in Postlethwaite *et al.* [159] and Kureemun *et al.* [160] for robust control of the longitudinal and lateral dynamics of the full-scale Bell 412 helicopter. In Cai *et al.* [161], a state-feedback $\mathscr{H}_\infty$ control law combined with reduced-order observer is designed and successfully implemented to control a small-scale unmanned helicopter. Successful implementation of the μ synthesis in case of Yamaha R-50 helicopter is reported in Shim [162, pp. 126–137], and a similar design is described in Yuan and Katupitiya [163]. In both cases, the linearized helicopter model developed by Mettler [143] is used and an input multiplicative uncertainty of 10 percent is assumed in the helicopter model. The controller designed in Yuan and Katupitiya [163] is of 28th order.

If the helicopter velocity undergoes large variations, it might be impossible to achieve high performance and even closed-loop stability over the entire operating range with a single linear time-invariant controller. In this case, it is appropriate to apply the technique of *gain scheduling*, see Leith and Leithead [164], which is successfully used in the control of uncertain or time-varying systems. This technique involves implementation of a family of controllers designed for different regions of the parameter space so that to guarantee stability and performance in that region. During the system operation, the controllers are changed according to a physical parameter measured in real time which detects in what region the system is working in the corresponding moment of the time.

Chapter 7

Case study 3: robust control of two-wheeled robot

The design of two-wheeled robot control system represents a challenge to the designer. The robot motion in horizontal and vertical planes is described by a nonlinear model whose derivation may be a difficult task. The linearization of this model leads to unstable nonminimum phase plant which should be stabilized in the presence of parameter variations, noises, and disturbances.

This case study presents the design and experimental evaluation of two controllers for vertical stabilization of two-wheeled robot. The first one is a conventional linear quadratic Gaussian (LQG) controller with 17th-order Kalman filter used for state estimation. This controller ensures robust stability of the closed-loop system and good nominal performance. The second one is a μ controller ensuring both robust stability and robust performance. Due to the lack of accurate analytical robot model, the controllers design is based on models derived by closed-loop identification from experimental data. The robot uncertainty is approximated by an input multiplicative uncertainty which leads to a μ controller of order 44, subsequently reduced to 30. The yaw motion is controlled by using a proportional-integral (PI) controller on the basis of yaw angle estimate obtained by a separate second order Kalman filter. A software in MATLAB®/Simulink® environment is developed for generation of control code which is embedded in the Texas Instruments Digital Signal Controller *TMS320F28335*. Results from the simulation of the closed-loop system as well as experimental results obtained during the real-time implementation of the designed controllers are given. The theoretical investigation and experimental results confirm that the closed-loop system achieves robustness in respect to the uncertainties related to the identified robot model.

The chapter is organized as follows. In Section 7.1, we give a brief description of experimental two-wheeled robot. The derivation of robot model using closed-loop identification procedures is considered in Section 7.2. On the basis of model obtained, in Section 7.3, we derive an uncertain plant descriptions with input multiplicative uncertainty representation. The robot nominal model is used in 7.4 to design a LQG controller involving linear quadratic regulator with 17th-order Kalman filter. The uncertain robot model is used in Section 7.4 to design μ controller which ensures robust stability and robust performance of the corresponding closed-loop systems. A comparison of the two controllers in the frequency domain is done in Section 7.6. Results from the experimental evaluation of the two controllers are presented in Section 7.7.

The presentation in this chapter is based on the paper [165].

7.1   Robot description

The general view of the two-wheeled robot in self-balancing mode is shown in Figure 7.1. The robot is statically unstable, and its balancing is achieved by rotating the wheels in appropriate direction. It is assumed that the robot moves on a flat surface.

A schematic diagram of the robot motion is presented in Figure 7.2. The robot motion in the vertical plane is described by the tilt angle ϕ, while the motion in the horizontal plane is characterized by the wheels average angle $\theta = (\theta_L + \theta_R)/2$ where

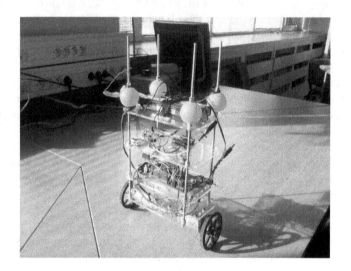

Figure 7.1   Two-wheeled robot

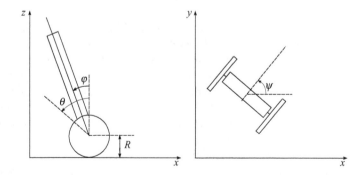

Figure 7.2   Robot motion in vertical and horizontal planes

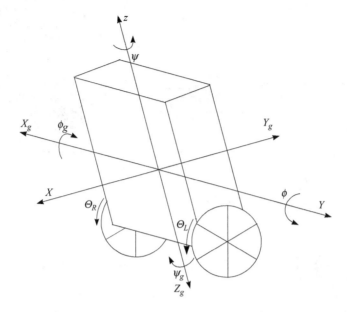

Figure 7.3   Mutual disposition of robot and gyro axes

θ_L and θ_R are the left and right wheel angles, respectively. The motion around the vertical axis is described by the yaw angle ψ.

The robot is constructed from four vertically connected plastic platforms. At the bottom platforms are situated with two 12-V DC brushed drive motors, together with 29:1 gearboxes, wheels, and magnetic quadrature encoders for measuring the wheels angles. Motors are controlled by power amplifier *qik2s12v10* receiving commands from the digital signal controller (DSC). The next platform hosts the Spectrum Digital *eZdspTMF28335* development board , which supports Texas Instruments DSC *TMS320F28335*. The robot controller and the data acquisition (DAQ) system are embedded in the DSC. Stabilization is achieved by using inertial measurement unit *ADIS16405* containing three orthogonal axes microelectromechanical gyroscopes. This unit is mounted on the third platform. The robot is powered by three cell lithium-polymer LiPo 12-V battery situated on the uppermost platform. The input signals to the motors are pulse width modulated (PWM) duty cycles and direction commands sent over RS232 link to the power amplifier forming PWM voltage waveforms for the motors. The signals which are measured in real time are wheels angular rates ($\dot{\theta}_L, \dot{\theta}_R$), body tilt rate ($\dot{\phi}$), and body yaw rate ($\dot{\psi}$). Real-time DAQ system is organized around wireless communication channel based on Bluetooth module.

The mutual disposition of robot and gyro axes is shown in Figure 7.3. ϕ_g, ψ_g denote the tilt angle and yaw angle, measured in the gyro sensors axes. It follows from the figure that $\phi = -\phi_g$ and $\psi = -\psi_g$.

The block diagram of the two-wheeled robot control system is shown in Figure 7.4. The computation of control actions u_L, u_R to the DC drive motors is

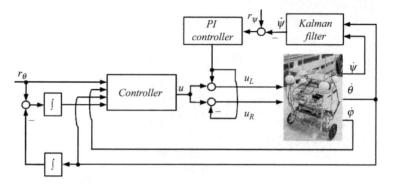

Figure 7.4   Block diagram of robot control system

realized in single precision on the basis of signals from the gyro sensor measuring the angular rate $\dot{\phi}$ (and, after integration, the tilt angle ϕ) and signals measuring the wheel rotation angles θ_L and θ_R. The control of the DC motors is executed by PWM signals. Control signals in the case of vertical stabilization are identical for both motors. The robot turning around the vertical axis (yaw motion) is realized by addition and subtraction of one and the same signal from the control signals to the left and right motors. The signal for robot turn is produced by a separate PI regulator based on feedback from yaw angle ψ which is estimated by a second-order Kalman filter. This approach is based on the assumption that the dynamics of yaw motion is separated from the dynamics of the motion in vertical plane and is validated by the experiments.

The interface between the IMU *ADIS16405* and DSC *TMS320F28335* is presented in Appendix E, and the operation of the rotary encoder is described in Appendix F.

7.2   Closed-loop identification of robot model

MATLAB files used in this section

File	Description
ident_robot.m	M-file for identification of robot dynamics in the vertical plane
ident_psidot_thetadot.m	M-file for identification of robot dynamics around the vertical axis
ident_thetadot_u1.m	M-file for identification of left wheel dynamics
ident_thetadot_u2.m	M-file for identification of right wheel dynamics

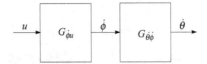

Figure 7.5   Representation of robot dynamics

To determine a mathematical model of the robot, one may apply physical model-ing or identification. The standard assumptions are ideal rigid body dynamics, flat and horizontal ground surface, zero wheel slip, and no friction [166]. Modeling from first principles [167–170] has a lot of advantages related to the determination of linearized description and easy derivation of uncertain model. However, it requires a profound knowledge about physics of the plant and a lot of a priori information. Due to the lack of reliable analytical model, in this study, we prefer to use a numerical model obtained by identification procedure. An additional reason to use an identification model is that such models apart from the robot body dynamics take into account the dynamics of sensors, actuators, and motors which facilitates the plant description. Identification is performed for small deviations of robot from upright position in order to obtain linearized plant model. For this aim, we use the methods described in Appendix D.

Generally, the motions in the vertical and horizontal planes are connected so that the robot plant should be considered as multivariable. However, the experiments show that there is a weak interaction between the dynamics of the robot body in the vertical plane and the dynamics of the rotation around the vertical axis. The influence of the yaw motion on the dynamics in vertical plane may be represented by a relatively slow disturbance which can be suppressed efficiently by a robust controller of the vertical plane motion. Hence, in order to simplify the robot dynamics, one may assume that the motion in vertical plane is independent on the variation of yaw angle which allows to approximate the robot plant by two separate single-input–single-output systems. This assumption is confirmed by the closed-loop experiments presented in Section 7.7 and allows to avoid multivariable plant identification and design of higher-order multivariable controller.

The robot dynamics in the vertical plane and the wheel dynamics is represented by a single-input–single-output plant as shown in Figure 7.5. The subsystem denoted by $G_{\dot{\phi}u}$ corresponds to the dynamics of the vertical plane motion, while the subsystem denoted by $G_{\dot{\theta}\dot{\phi}}$ reflects the wheels dynamics. Note that although the identification problem is simplified in this way, it still may pose difficulties related to the derivation of uncertain plant model.

There are a few reported results about identification of linear discrete-time mod-els of two-wheeled robots. In [171], the dynamics is described by ARX model of fifth order. In [172], identification of third-order state-space model is presented. The advantages of the models for control, obtained by means of identification and the lack of enough results for such models of two-wheeled robots, motivate the determination of uncertain models of such a robot presented in this section.

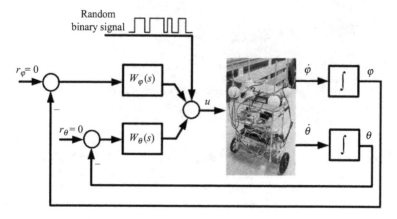

Figure 7.6   Closed-loop identification setup

As mentioned previously, two-wheeled robot is inheritably unstable system which requires the identification experiment to be done in closed-loop setting. In such situation, predictive error method is appropriate if the input–output data are informative and the true system dynamics is in the model set. Usually, the first requirement is guaranteed by adding an external persistently excitation signal to the controller output, and second requirement is guaranteed by firstly estimating the high-order model and then reducing model order by appropriate technique.

Experimental setting intended to identify the plant in Figure 7.5 is shown in Figure 7.6. Stabilizing regulator control signal is disturbed by excitation signal which is random binary signal (RBS). Excitation signal is output of relay fed with Gaussian white noise, generated with MATLAB random generator. The amplitude of RBS is chosen to be ±15 percent, so the control signal stays in linear region and the robot tilts enough without falling.

To determine sufficiently accurate models with direct closed-loop identification, according to diagram presented in Figure 7.6, it is important the stabilizing controller to be "soft." This means that in case of PI controller, the proportional term should be low and the integral term should be sufficient to keep a small error from operating point. Stabilizing controller for the two-wheeled robot is composed of two feedback loops—one for body tilt angle ϕ and one for wheels angle $\theta = (\theta_L + \theta_R)/2$. Transfer functions of the controller are chosen as

$$
\begin{aligned}
W_\phi(s) &= \frac{15(s + 2.5)}{s}, \\
W_\theta(s) &= \frac{0.002(10s + 1)^2(3s + 1)}{s^3}.
\end{aligned}
\tag{7.1}
$$

A PI controller is used for body tilt angle stabilization. The controller for stabilization of wheels angle is composed of PI part followed by two additional integrators.

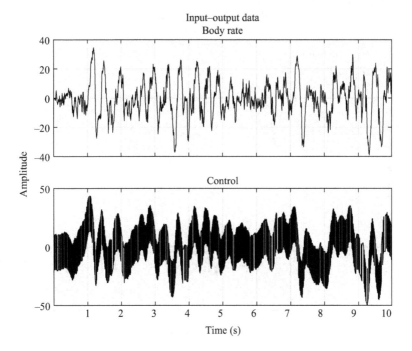

Figure 7.7   Experimental data for $G_{\dot\phi u}$ identification

Also two real zeros are included to appropriately compensate these integrators in frequency domain to preserve the closed-loop stability. More integrators were included to minimize error in wheels angle position. If only standard PI is used, then wheels error is increasing steadily.

Experimental data used in the identification are grouped into two sets. The first set is used for model parameter estimation, and the second one is used for model validation. Sample time is $T_s = 0.005$ s and the number of measurements is $N = 1,000$. A test on the degree of persistence of excitation for the input for both data samples is done using the function `pexcit` from System Identification Toolbox™ [173] of MATLAB. The degree of persistence of excitation for the first data sample is 999, while this for the second data sample is 500.

In Figures 7.7 and 7.8, we show the experimental data used for the robot model identification.

Relationships from u to $\dot\phi$ and from $\dot\phi$ to $\dot\theta$ are assumed linear. Therefore, different types of stochastic polynomial models can be applied. As it follows from Appendix D, the most general discrete-time model is

$$A(q)y(k) = \frac{B(q)}{F(q)}u(k - n_k) + \frac{C(q)}{D(q)}e(k), \tag{7.2}$$

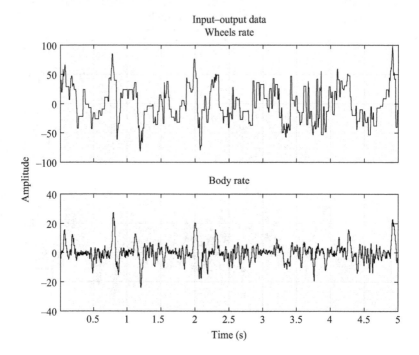

Figure 7.8   *Experimental data for $G_{\dot\theta\dot\phi}$ identification*

where

$$A(q) = 1 + \sum_1^{n_a} a_k q^{-k}, \quad B(q) = \sum_1^{n_b} b_k q^{-k+1}, \quad C(q) = 1 + \sum_1^{n_c} c_k q^{-k},$$

$$D(q) = 1 + \sum_1^{n_d} d_k q^{-k}, \quad F(q) = 1 + \sum_1^{n_f} f_k q^{-k}$$

are polynomials in the delay operator q^{-1}. Model parameters are a_i, b_i, c_i, d_i, f_i, and n_a, n_b, n_c, n_d, n_f are the polynomial orders. The number of delays is n_k. The process $e(k)$ is Gaussian white noise with zero mean and strength σ_e^2.

Further on we make use of the ARX, ARMAX, and BJ models, represented by the following equations:

$$\text{ARX}: A(q)y(k) = B(q)u(k - n_k) + e(k) \tag{7.3}$$

$$\text{ARMAX}: A(q)y(k) = B(q)u(k - n_k) + C(q)e(k) \tag{7.4}$$

$$\text{BJ}: y(k) = B(q)/F(q)u(k - n_k) + C(q)/D(q)e(k). \tag{7.5}$$

Usually, the first estimated model is ARX because of its relative simplicity. In addition, it is estimated by the linear least squares method. If the ARX model does not pass through validation tests, then ARMAX or BJ models are used.

7.2.1 Dynamic models from u to $\dot{\phi}$

Estimation of ARX model (7.3) requires knowledge about structure parameters. For this aim, a set of 500 ARX models is examined using the function `arxstruc` from System Identification Toolbox. The orders of polynomials $A(q)$ and $B(q)$ vary from 1 to 10 and the number of delays vary from 1 to 5. Selection of structure parameters is guided by three criteria—model loss function, Akaike information index, and Rissanen index (see Appendix D). Based on these criteria, only three ARX models from the set of 500 are selected for further identification. Their parameters are statistically estimated, after that four validation tests are applied:

1. Autocorrelation and crosscorrelation test of the model residuals,
2. Test with logarithmic amplitude frequency response of estimated model from input signal to model residuals,
3. Akaike predictive error calculated from model residuals,

$$FPE = \frac{1}{N} \sum_{k=1}^{N} e^2(k) \frac{1 + d/N}{1 - d/N},$$

where d is the number of estimated parameters.

4. Comparison of model output and measured output by the index

$$FIT = 100 \times \left[1 - \frac{\|\hat{y} - y\|_2}{\|y - \bar{y}\|_2} \right] \%,$$

where $\hat{y}$ is the model output, y is the measured output, and $\bar{y}$ is the mean of y.

In all cases, the experimental data set used for validation is different from that used for parameter estimation. Tests 1 and 2 decide whether the model is valid. From various valid models, the best model is the model which has smallest FPE and largest FIT.

In our case, the three selected ARX models do not pass test 1 because the autocorrelation function is crossing the confidence interval for lag 10. Therefore, the selected ARX models must be rejected.

The next step is ARMAX model (7.4) estimation which is done by the function `armax` from System Identification Toolbox. Selected structure parameters are $n_a = 10, n_b = 10, n_c = 10, n_k = 3$. Figure 7.9 shows zeros and poles of that model together with their 99 percent confidence intervals. There is an intersection of some zeros confidence intervals with poles confidence intervals which is a motivation for model order reduction. Hence, an ARMAX model with structure parameters $n_a = 7, n_b = 7, n_c = 7, n_k = 3$ is estimated.

The performance of two estimated ARMAX models are compared in Table 7.1. From the two ARMAX models, the model with structure $n_a = 7, n_b = 7, n_c = 7, n_k = 3$ is chosen based on Tests 1, 2, 3, and 4. This model is appropriate for controller design.

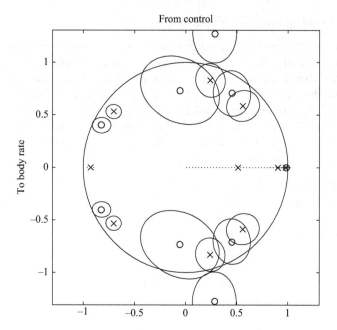

Figure 7.9   Poles and zeros of ARMAX model with $n_a = 10, n_b = 10$,
$n_c = 10, n_k = 3$

Table 7.1   Performance of the subsystem model from u to $\dot{\phi}$

Model	FIT	FPE
ARMAX $n_a = 10, n_b = 10, n_c = 10, n_k = 3$	46.36	7.85
ARMAX $n_a = 7, n_b = 7, n_c = 7, n_k = 3$	49.78	7.57

Some validation results for the model with $n_a = 7, n_b = 7, n_c = 7, n_k = 3$ are given in Figures 7.10 and 7.11. The model passes through the test of autocorrelation and crosscorrelation function of model residuals. Therefore, the parameter estimates are unbiased, the models of robot dynamics and $\dot{\phi}$ noise dynamics are adequate.

In Figure 7.12, we show the logarithmic response of the model between residuals and the input signal u. The response is inside the confidence regions which supports the claim that the model chosen is adequate.

ARMAX model, resulting from identification, is assumed as nominal and has the form

$$\dot{\phi}(z) = G_{\dot{\phi}u,\text{nom}}(z)u(z) + v_{\dot{\phi}}(z), \tag{7.6}$$

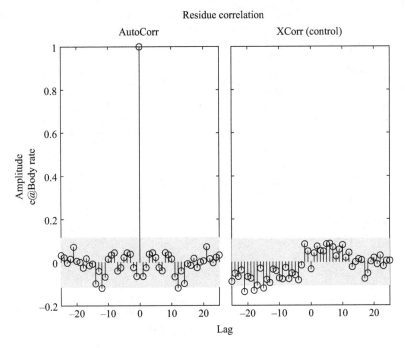

Figure 7.10 Correlation test of residuals of ARMAX model with $n_a = 7, n_b = 7$,
$n_c = 7, n_k = 3$

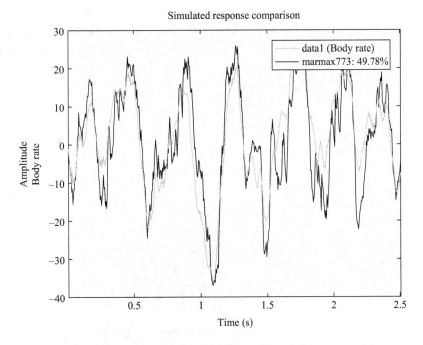

Figure 7.11 Responses of the ARMAX model and the measured data

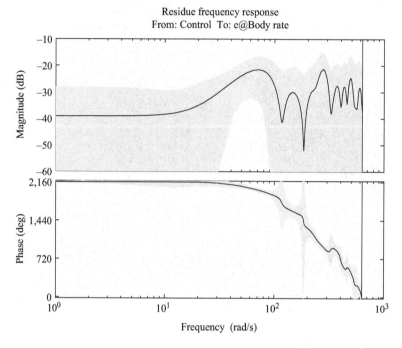

Figure 7.12   Logarithmic response of the residuals for ARMAX model

where $G_{\dot{\phi}u,\mathrm{nom}}(z)$ is the nominal transfer function from u to $\dot{\phi}$, the noise $v_{\dot{\phi}}$ is obtained during the identification procedure and reflects the uncertainty in the model found. In the given case, we have that

$$G_{\dot{\phi}u,\mathrm{nom}}(z) = \frac{0.0306z^6 + 0.003992z^5 + 0.01837z^4 + 0.02122z^3 -}{z^2(z^7 - 0.7768z^6 - 1.502z^5 + 0.8929z^4 +} \cdots \rightarrow$$

$$\leftarrow \cdots \frac{-0.02372z^2 - 0.04351z - 0.002676}{+1.218z^3 - 0.6273z^2 - 0.5815z + 0.3823)}. \tag{7.7}$$

The noise $v_{\dot{\phi}}$ is expressed as

$$v_{\dot{\phi}}(z) = N_{\dot{\phi}}(z)n_{\dot{\phi}}(z), \tag{7.8}$$

where

$$N_{\dot{\phi}}(z) = \frac{z^7 + 0.315z^6 - 1.236z^5 - 0.6616z^4 + 0.4729z^3 +}{z^7 - 0.7768z^6 - 1.502z^5 + 0.8929z^4 +} \cdots \rightarrow$$

$$\leftarrow \cdots \frac{+0.3178z^2 - 0.0696z - 0.02562}{+1.218z^3 - 0.6273z^2 - 0.5815z + 0.3823}$$

is shaping filter determined so that $n_{\dot{\phi}}$ is a white noise.

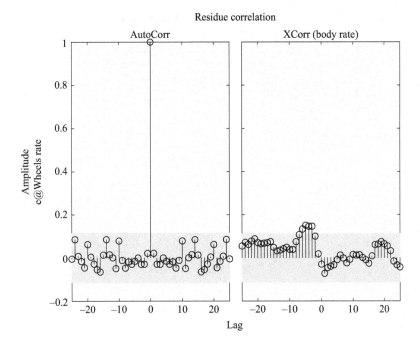

Figure 7.13   Correlation test of residuals of BJ model

7.2.2   Dynamic models from $\dot{\phi}$ to $\dot{\theta}$

In this case, none of the examined ARX or ARMAX models pass through the residuals autocorrelation test. Hence, all ARX and ARMAX models were rejected and BJ model (7.5) is estimated as being more general. A BJ model with structure parameters $n_b = 3, n_f = 3, n_c = 3, n_d = 3, n_k = 1$ is estimated using the function bj from System Identification Toolbox.

Results of validation test for this model are represented in Figures 7.13 and 7.14.

In Figure 7.15, we present the logarithmic response of the model between the residuals and body rate ϕ. As it is seen, this response is inside the confidence regions.

The subsystem from $\dot{\phi}$ to $\dot{\theta}$ is described by the following equation:

$$\dot{\theta}(z) = G_{\dot{\theta}\dot{\phi},\text{nom}}(z)\dot{\phi}(z) + v_{\dot{\theta}}(z), \qquad (7.9)$$

where

$$G_{\dot{\theta}\dot{\phi},\text{nom}}(z) = \frac{0.6775z^2 - 0.7007z - 0.1773}{z^3 - 1.047z^2 - 0.5094z + 0.5561} \qquad (7.10)$$

is the nominal transfer function and $v_{\dot{\theta}}$ is the noise reflecting the model uncertainty. This noise is represented as

$$v_{\dot{\theta}}(z) = N_{\dot{\theta}}(z)n_{\dot{\theta}}(z), \qquad (7.11)$$

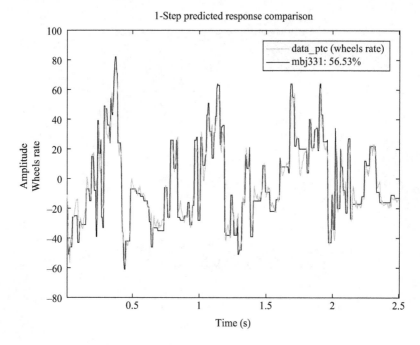

Figure 7.14   *Responses of the BJ model and the measured data*

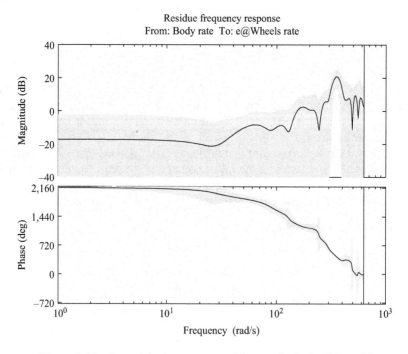

Figure 7.15   *Logarithmic response of the residuals for BJ model*

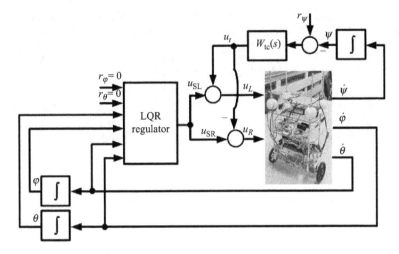

Figure 7.16 Yaw dynamics identification setup

where the shaping filter

$$N_{\dot{\theta}}(z) = \frac{z^3 - 1.516z^2 + 0.8294z - 0.1448}{z^3 - 2.344z^2 + 2.081z - 0.7329}$$

is determined so that $n_{\dot{\theta}}$ is a white noise. Note that both transfer functions $G_{\dot{\phi}u,\text{nom}}(z)$ and $G_{\dot{\theta}\dot{\phi},\text{nom}}(z)$ are nonminimum phase which is unfavorable in respect to the controller design.

It is necessary to point out that the uncertainty models obtained are not unique. If data sets with different measurements are used, then it is possible to obtain models with different shaping filters which is due to the properties of the identification methods implemented.

7.2.3 Dynamic model of the yaw motion

As described previously, the dynamics of the yaw motion is considered as separate from the dynamics of the vertical plane motion. Model of robot yaw motion is assessed by optimization procedure using the transient response of a tunable parametric model. This requires a closed-loop experiment where robot is vertically stabilized and yaw motion is initiated.

Experimental setting is shown in Figure 7.16.

Vertical and longitudinal stabilization in this experiment is achieved by LQR regulator based on estimated ARMAX and BJ models. Control signal is identical for both motors, i.e., $u_{SL} = u_{SR} = u_s$. In addition, a PI regulator controls yaw motion based on feedback from ψ angle, with transfer function

$$W_{tc}(s) = \frac{2s + 0.01}{s}.$$

The output of $W_{tc}(s)$ is signal u_t which is added to control signal u_L for the left motor and subtracted from control signal u_R of the right motor. The effect of this is turning of the robot around the vertical axis. Thus, for the two control signals, one obtains

$$u_L = u_s - u_t, u_R = u_s + u_t.$$

It is assumed that the motion around the vertical axis is described by the linear relationship

$$\psi(s) = G_{\psi\dot{\theta}_L}(s)G_{\dot{\theta}_L u_L}(s)u_L(s) + G_{\psi\dot{\theta}_R}(s)G_{\dot{\theta}_R u_R}(s)u_R(s).$$

First-order models are determined, describing the dynamics between u_L and $\dot{\theta}_L$; between u_R and $\dot{\theta}_R$; and between $\dot{\theta}_L, \dot{\theta}_R$, and $\dot{\psi}$. The quadratic costs used in the optimization are

$$J_{\dot{\theta}_L u_L} = \frac{1}{M} \sum_{i=1}^{M} (\dot{\theta}_{mL}(i) - \dot{\theta}_L(i))^2,$$

$$J_{\dot{\theta}_R u_R} = \frac{1}{M} \sum_{i=1}^{M} (\dot{\theta}_{mR}(i) - \dot{\theta}_R(i))^2, \quad \quad (7.12)$$

$$J_{\dot{\psi}} = \frac{1}{M} \sum_{i=1}^{M} (\dot{\psi}_m(i) - \dot{\psi}(i))^2$$

As a result, the following yaw model transfer functions are found:

$$G_{\dot{\theta}_L u_L}(s) = \frac{8.5867}{0.1328s + 1}, \quad G_{\dot{\theta}_R u_R}(s) = \frac{8.7270}{0.1186s + 1},$$

$$G_{\psi\dot{\theta}_L}(s) = -\frac{2.004}{0.0119s + 1}, \quad G_{\psi\dot{\theta}_R}(s) = \frac{2.009}{0.0119s + 1}.$$

In Figure 7.17, we show the two control signals u_L and u_R used in the optimization. In Figures 7.18 and 7.19, we compare the measured signals and the model output signals. The minimum values of the quadratic costs (7.12) are

$$J_{\dot{\theta}_L u_L} = 297.877, \quad J_{\dot{\theta}_R u_R} = 87.184, \quad J_{\dot{\psi}} = 3.896.$$

It is seen from the figures that the models obtained describe with sufficient accuracy the rotation dynamics around the vertical axis (Figure 7.20).

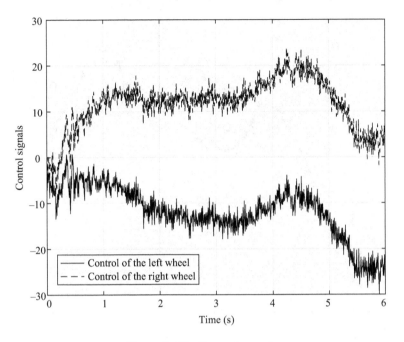

Figure 7.17   Control signals

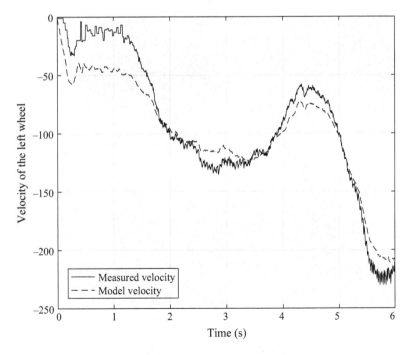

Figure 7.18   Left wheel velocity

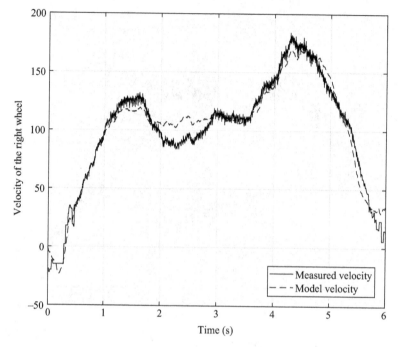

Figure 7.19   Right wheel velocity

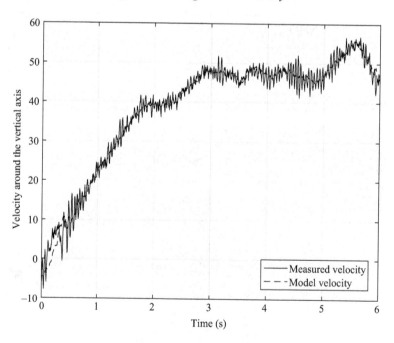

Figure 7.20   Velocity around vertical axis

7.3 Derivation of uncertain models

MATLAB files used in this section

File	Description
unc_model_opt.m	Finds the more accurate uncertainty model ($G_{\dot\phi u}$ and $G_{\dot\theta\dot\phi}$ are of orders 12 and 10, respectively)
unc_modelṁ	Finds reduced-order uncertainty model ($G_{\dot\phi u}$ and $G_{\dot\theta\dot\phi}$ are of orders 9 and 5, respectively)
ident_unc.m	Determines the nominal model and 3-sigma confidence intervals of the uncertain model frequency responses

7.3.1 Signal-based uncertainty representation

Based on (7.6) and (7.9), the uncertain robot model may be represented as shown in Figure 7.21. We will call this model as *model with signal-based uncertainty representation* because the uncertainty is represented by the noise signals. Further on we assume that $n_{\dot\phi}$ and $n_{\dot\theta}$ are white noises with unity variances. The model is obtained directly from the identification procedure without additional computations.

7.3.2 Input multiplicative uncertainty representation

Condition of unbiased parameter estimates obtained by the identification guarantees that the exact parameter values are contained in the confidence intervals of parameter estimates with probability close to 1. This allows to derive parametric uncertainty model with scalar uncertainties. The number of these uncertainties is equal to the number of estimated parameters. In the case of ARMAX model with $n_a = 7$, $n_b = 7$, $n_c = 7$, $n_k = 3$, this number is equal to 21 which makes the implementation of such model unpractical. That is why it is preferred to derive input multiplicative uncertainty models of the transfer functions $G_{\dot\phi u}$ and $G_{\dot\theta\dot\phi}$ which contain unstructured (complex) uncertainties. Based on parameter confidence intervals, maximum relative deviations from nominal models in the frequency domain are obtained. In order to derive models with multiplicative uncertainties, these deviations are approximated, through

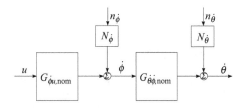

Figure 7.21 Robot model with signal-based uncertainty representation

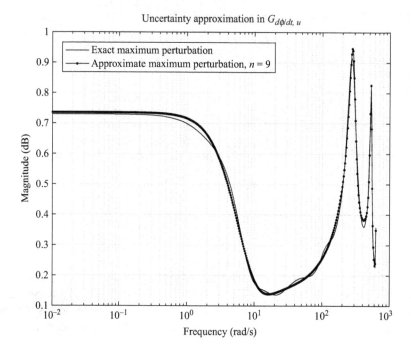

Figure 7.22 Ninth-order approximation of the relative uncertainty in $G_{\dot\phi u}$

optimization procedure in the M-file `unc_model_opt`, with shaping filters which are represented by transfer function of 12th and 10th order, respectively. However, in the design of the μ controller in Section 7.4, we use a reduced-order uncertainty model in which the transfer functions $G_{\dot\phi u}$ and $G_{\dot\theta\phi}$ are of orders 9 and 5, respectively. The higher order uncertainty model, implemented by the M-file `unc_opt_model`, is used in the robustness analysis of the LQG and μ controllers.

The maximum relative uncertainties in $G_{\dot\phi u}$ and in $G_{\dot\theta\phi}$ along with their low order approximations are represented in Figures 7.22 and 7.23, respectively. Note that the maxima along the frequency of the uncertainties are in the higher frequency range. The approximations are used to determine the weighting shaping filters in the corresponding input multiplicative uncertainty representations. Resulting uncertain models for the two plant subsystems are

$$G_{\dot\phi u}(z) = G_{\dot\phi u,\text{nom}}(z)(1 + W_{\dot\phi}(z)\Delta_{\dot\phi}), \qquad (7.13)$$

$$G_{\dot\theta\phi}(z) = G_{\dot\theta\phi,\text{nom}}(z)(1 + W_{\dot\theta}(z)\Delta_{\dot\theta}), \qquad (7.14)$$

where $W_{\dot\phi}(z)$, $W_{\dot\theta}(z)$ are the shaping filters and $\Delta_{\dot\phi}$, $\Delta_{\dot\theta}$ are scalar uncertainties which satisfy

$$|\Delta_{\dot\phi}| < 1, \quad |\Delta_{\dot\theta}| < 1.$$

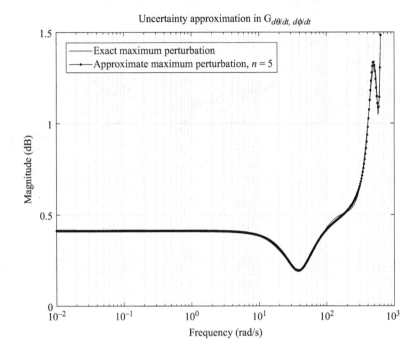

Figure 7.23   Fifth-order approximation of the relative uncertainty in $G_{\dot{\theta}\dot{\phi}}$

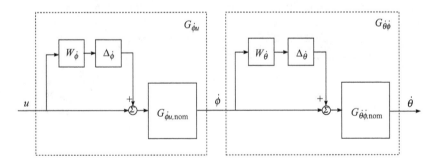

Figure 7.24   Robot model with input multiplicative uncertainty representation

The robot model with input multiplicative uncertainty is shown in Figure 7.24.

The Bode plots of the uncertain models $G_{\dot{\phi}u}$ and $G_{\dot{\theta}\dot{\phi}}$ are shown in Figures 7.25 and 7.26, respectively.

The model given in Figure 7.24 is used for μ robot controller design described in Section 7.5.

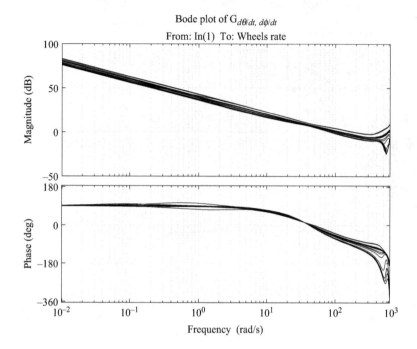

Figure 7.25   Bode plot of $G_{\dot\phi u}$

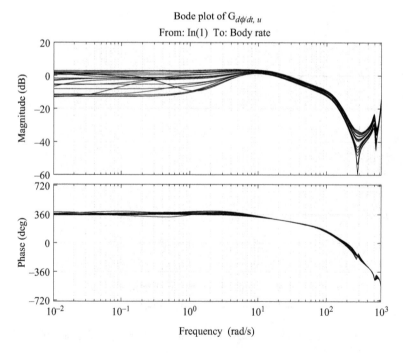

Figure 7.26   Bode plot of $G_{\dot\theta\dot\phi}$

7.4 LQG controller design

MATLAB files used in this section

File	Description
`LQG_kalman_phi.m`	Design of LQG controller
`dfrs_LQG`	Frequency responses of the LQG controller
`kalman_filter_phi.slx`	Simulation of tilt angle Kalman filter
`kalman_psi_sim.slx`	Simulation of yaw angle Kalman filter

Consider first the design of a LQG robot controller using the model with signal-based uncertainty representation shown in Figure 7.21.

The block diagram of the robot control system with LQG regulator is shown in Figure 7.27. The system state is estimated by the Kalman filter *KF* and the control action is generated by the linear quadratic regulator *LQR*. The estimate $\hat{\dot{\psi}}$ of the yaw rate $\dot{\psi}$ is obtained by the Kalman filter $KF_{\dot{\psi}}$. The PI controller PI_{ψ} is used to ensure the desired performance of the yaw dynamics.

The control actions to the plant are realized by a DSC in real time with sampling frequency $f_s = 200$ Hz. For this reason, the LQG design is implemented to determine a discrete-time controller at this sampling frequency.

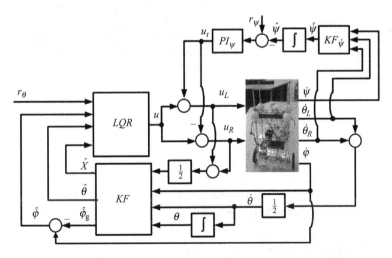

Figure 7.27 Block diagram of the closed-loop system with LQG regulator

In the LQR design, the ARMAX and BJ models of $G_{\dot{\phi}u,\text{nom}}(z)$ and $G_{\dot{\theta}\dot{\phi},\text{nom}}(z)$ are represented by the following state equations:

$$x_{\dot{\phi}}(k+1) = A_{\dot{\phi}}x_{\dot{\phi}}(k) + B_{\dot{\phi}}u(k) + J_{\dot{\phi}}v_{\dot{\phi}}(k), \tag{7.15}$$

$$\dot{\phi}(k) = C_{\dot{\phi}}x_{\dot{\phi}}(k) + H_{\dot{\phi}}v_{\dot{\phi}}(k),$$

$$x_{\dot{\theta}}(k+1) = A_{\dot{\theta}}x_{\dot{\theta}}(k) + B_{\dot{\theta}}u(k) + J_{\dot{\theta}}v_{\dot{\theta}}(k), \tag{7.16}$$

$$\dot{\Theta}(k) = C_{\dot{\theta}}x_{\dot{\theta}}(k) + H_{\dot{\theta}}v_{\dot{\theta}}(k)$$

where $x_{\dot{\phi}}$, $x_{\dot{\theta}}$ are state vectors with dimensions 9 and 6, respectively, $v_{\dot{\phi}}$, $v_{\dot{\theta}}$ are discrete-time white Gaussian noises with unit variances, and $A_{\dot{\phi}}$, $B_{\dot{\phi}}$, $J_{\dot{\phi}}$, $C_{\dot{\phi}}$, $H_{\dot{\phi}}$, $A_{\dot{\theta}}$, $B_{\dot{\theta}}$, $J_{\dot{\theta}}$, $C_{\dot{\theta}}$, $H_{\dot{\theta}}$ are constant matrices with corresponding dimensions containing the model parameters. The noises $v_{\dot{\phi}}$, $v_{\dot{\theta}}$ are obtained during the identification procedure and reflect the uncertainty in the model found. These noises are also represented as models with input multiplicative uncertainty which are used in the analysis in frequency domain.

The nonmeasurable wheels angle is described by the first-order difference equation

$$\theta(k+1) = \theta(k) + T_s\dot{\theta}(k), \tag{7.17}$$

where T_s is the sample time. The following state equations are also included in the plant description:

$$x_{\dot{\phi}_i}(k+1) = x_{\dot{\phi}_i}(k) - T_s\dot{\phi}(k), \tag{7.18}$$

$$x_i(k+1) = x_i(k) + T_s(r_\theta(k) - \theta(k)), \tag{7.19}$$

where r_θ is the wheels reference angle. These equations allow to compute approximations of the discrete-time integrals of $\dot{\phi}$ and of tracking error $r_\theta - \theta$, respectively. Both integrals are used in the design of linear-quadratic regulator in order to ensures efficient stabilization in the vertical plane and zero steady-state tracking error. In this way, one obtains the full plant equations of 18th order in the form

$$x(k+1) = Ax(k) + Bu(k) + Jv(k),$$
$$y(k) = Cx(k) + Hv(k), \tag{7.20}$$

where

$$x = [x_{\dot{\phi}}^T \; x_{\dot{\theta}}^T \; \Theta \; x_{\dot{\phi}_i} \; x_i]^T, \; y = [\dot{\phi} \; \dot{\Theta} \; \Theta \; \phi]^T, \; v = [v_{\dot{\phi}} \; v_{\dot{\theta}}]^T.$$

and the matrices A, B, J, C, H are obtained combining (7.15)–(7.19).

The aim of the controller design is to minimize the quadratic performance index

$$J(u) = \sum_{k=0}^{\infty} [x(k)^T Qx(k) + u(k)^T Ru(k)], \tag{7.21}$$

where Q and R are positive definite matrices chosen to ensure acceptable transient response of the closed-loop system.

As already shown in Chapter 4, the optimal control which minimizes (7.21) in respect to the system (7.20) is given by

$$u(k) = -Kx(k), \tag{7.22}$$

where the optimal feedback matrix K is determined by

$$K = (R + B^T PB)^{-1} B^T PA \tag{7.23}$$

and the matrix P is the positive definite solution of the discrete-time matrix algebraic Riccati equation

$$A^T PA - P - A^T PB(R + B^T PB)^{-1} B^T PA + Q = 0. \tag{7.24}$$

The solution of the Riccati equation (7.24) and determination of the gain matrix K (7.23) is done by the function $\texttt{dlqr}$ of MATLAB.

Let the matrix K is partitioned according to the dimensions of $x_{\dot\phi}, x_{\dot\Theta}, \Theta, x_{\phi_i}, x_i$ as

$$K = [K_{\dot\phi}\ K_{\dot\Theta}\ K_\Theta\ K_{\phi_i}\ K_{x_i}].$$

Since the state $x(k)$ of system (7.20) is not accessible, the optimal control (7.22) is implemented as

$$u(k) = -K_{\dot\phi}\hat x_{\dot\phi}(k) - K_{\dot\Theta}\hat x_{\dot\Theta}(k) - K_\Theta\hat\Theta(k) - K_{\phi_i}\hat x_{\phi_i}(k) - K_{x_i}\hat x_i(k), \tag{7.25}$$

where $\hat x_{\dot\phi}(k), \hat x_{\dot\Theta}(k)$ are estimates of $x_{\dot\phi}(k)$ and $x_{\dot\Theta}(k)$, respectively, and

$$\hat x_{\phi_i}(k+1) = \hat x_{\phi_i}(k) - T_s\hat{\dot\phi}(k), \tag{7.26}$$

$$\hat x_i(k+1) = \hat x_i(k) + T_s(r_\Theta(k) - \hat\Theta(k)) \tag{7.27}$$

are estimates of $x_{\phi_i}(k)$ and $x_i(k)$, respectively. The estimates $\hat{\dot\phi}(k)$ and $\hat\Theta(k)$ are obtained by the aid of a Kalman filter. In particular, the quantity $\hat{\dot\phi}(k)$ is obtained from

$$\hat{\dot\phi}(k) = \dot\phi(k) - \hat{\dot\phi}_g(k), \tag{7.28}$$

where $\dot\phi(k)$ is the measured tilt angular rate and $\hat{\dot\phi}_g(k)$ is the estimate of the rate gyroscope bias. The gyro bias $\dot\phi_g$ is modeled by the additional equation

$$\dot\phi_g(k+1) = \dot\phi_g(k) + J_g v_{\dot\phi g}(k), \tag{7.29}$$

where $v_{\dot\phi g}$ is a white Gaussian noise with unit variance and the coefficient J_g is determined experimentally as $J_g = 10^{-4}$ to obtain a good estimate of $\dot\phi$. Combining (7.15)–(7.17) and (7.29), the Kalman filter is designed for the 17th-order system:

$$x_f(k+1) = A_f x_f(k) + B_f u(k) + J_f v_f(k),$$

$$y_f(k) = C_f x_f(k) + H_f v_f(k), \tag{7.30}$$

where

$$x_f = [x_{\dot\phi}^T\ x_{\dot\Theta}^T\ \Theta\ \phi_g]^T, y_f = [\dot\phi\ \dot\Theta\ \Theta\ \phi_g]^T, v_f = [v_{\dot\phi}\ v_{\dot\Theta}\ v_g]^T$$

and A_f, B_f, J_f, C_f, H_f are matrices of corresponding dimensions.

The discrete-time Kalman filter for the system (7.30) is obtained as

$$\hat{x}_f(k+1) = A_f\hat{x}_f(k) + B_fu(k) + K_f(y_f(k+1) - C_fA_f\hat{x}_f(k) - C_fB_fu(k)),$$
$$\hat{y}_f(k) = C_f\hat{x}_f(k) \tag{7.31}$$

The filter matrix K_f is determined as

$$K_f = D_fC_f^T(C_fD_fC_f^T + 10^{-5})^{-1}, \tag{7.32}$$

where the matrix D_f is obtained as the positive semidefinite solution of the discrete-time matrix algebraic Riccati equation

$$A_fD_fA_f^T - D_f - A_fD_fC_f^T(C_fD_fC_f^T + 10^{-5})^{-1}C_fD_fA_f^T$$
$$+ J_fD_{v_f}J_f^T = 0 \tag{7.33}$$

and the matrix $D_{v_f} = I_3$ is the variance of the noise v_f. Note that in (7.32) and (7.33), the variance of the zero output noise in (7.30) is taken equal to 10^{-5} to avoid singularity of the corresponding matrix. The matrix K_f is computed by the function `kalman` of MATLAB.

The quantity $\hat{\dot{\phi}}(k)$, which is used in computation of the estimate (7.26), is obtained as shown in (7.28) where $\hat{\dot{\phi}}_g(k)$ is the last element of the state estimate $\hat{x}_f$.

The magnitude responses of the LQG controller obtained in respect to the inputs r, $\dot{\phi}$, and $\dot{\theta}$ are shown in Figure 7.28.

As mentioned previously, a PI controller of the yaw motion is also designed. The yaw angular velocity $\dot{\psi}$ is measured by a gyroscope of the same type as the gyro used to measure the tilt rate $\dot{\phi}$. This gyroscope contains a noise ψ_g which is modeled by the additional equation

$$\dot{\psi}_g(k+1) = \dot{\psi}_g(k) + J_gv_{\dot{\psi}g}(k), \tag{7.34}$$

where $v_{\dot{\psi}g}$ is a white Gaussian noise with unit variance and the coefficient J_g is determined as described above. A second-order Kalman filter is designed to produce sufficiently accurate estimate $\hat{\dot{\psi}}$ of the yaw angle.

In Figure 7.29, we show the closed-loop system for yaw motion control. The control actions u_L and u_R to the left and right motor, respectively, are produced by a PI regulator with coefficients $K_P = 2$ and $K_I = 0.01$. These coefficients ensure sufficiently fast and accurate dynamics of the yaw motion. The control signal u_{LQG} from the LQG controller is added to the two motor signals.

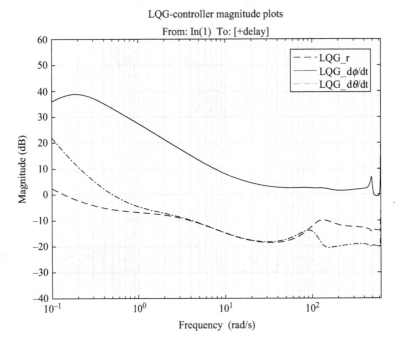

Figure 7.28   *Frequency responses of the LQG controller*

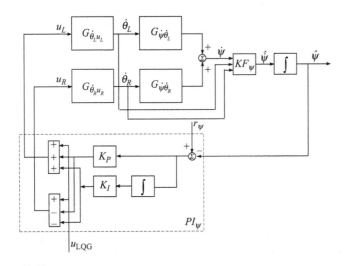

Figure 7.29   *Block diagram of the yaw motion control*

7.5   μ Controller design

MATLAB files used in this section

File	Description
olp_robot_2dof.m	Determines the open-loop interconnection for the μ synthesis
unc_model.m	Determines reduced order uncertainty model
dms_robot_2dof.m	Performs D–K iterations
dfrs_mu.m	Obtains the frequency responses of the closed-loop system with μ controller
clp_mu_sys.slx	Simulink model of the closed-loop system

The μ controller is designed on the basis of the robot model with input multiplicative uncertainty representation, shown in Figure 7.24.

To obtain good performance of the closed-loop system, we shall implement a two-degree-of-freedom controller. The block diagram of the continuous-time closed-loop system in that case is shown in Figure 7.30 where r_θ is the wheels reference angle and e_p, e_u are the weighted closed-loop system outputs. The aim of the controller K, determined by μ synthesis, is to ensure robust stability and robust performance of the closed-loop system in the presence of input multiplicative uncertainty. In order to obtain better position accuracy, a feedback from the integrated tracking error err $= r_\theta - \theta$ is introduced to the controller.

Let

$$y_c = \left[\dot{\phi}, \dot{\theta}, \int (r_\theta - \theta) \right]^T$$

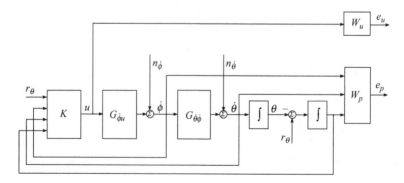

Figure 7.30   Block diagram of the closed-loop system in case of μ synthesis

be the output feedback vector where $\int (r_\theta - \theta)$ denotes discrete-time integration of the difference $r_\theta - \theta$ using the forward Euler method. Assume that the controller transfer function matrix K is represented as

$$K = [K_r \ K_y],$$

where K_r is the controller prefilter transfer function and

$$K_y = [K_{\dot\phi} \ K_{\dot\theta} \ K_{\int \text{err}}]$$

is the output feedback transfer function matrix. The matrix K_y is partitioned with respect to the dimensions of $\dot\phi$, $\dot\theta$, $\int (r_\theta - \theta)$. Then, the control actions are calculated from the expression

$$u = K \begin{bmatrix} r_\theta \\ y_c \end{bmatrix}$$

(7.35)

$$= K_r r_\theta + K_{\dot\phi}\dot\phi + K_{\dot\theta}\dot\theta + K_{\int \text{err}} \int (r_\theta - \theta).$$

Using the notation

$$G_{\text{int}}(z) = \frac{T_s}{z-1},$$

$$G_u(z) = \begin{bmatrix} 1 \\ G_{\dot\theta\dot\phi}(z) \\ -G_{\text{int}}(z)^2 G_{\dot\theta\dot\phi}(z) \end{bmatrix} G_{\dot\phi u}(z), \quad G_{n_{\dot\phi}}(z) = \begin{bmatrix} 1 \\ G_{\dot\theta\dot\phi}(z) \\ -G_{\text{int}}(z)^2 G_{\dot\theta\dot\phi}(z) \end{bmatrix},$$

$$G_{n_{\dot\theta}}(z) = \begin{bmatrix} 0 \\ 1 \\ -G_{\text{int}}(z)^2 \end{bmatrix}, \quad G_{r_\theta}(z) = \begin{bmatrix} 0 \\ 0 \\ G_{\text{int}}(z) \end{bmatrix},$$

where $T_s = 0.005$ s is the sampling period, the closed-loop system is described by the following equations:

$$y_c = G_{r_\theta} r_\theta + G_u u + G_{n_{\dot\phi}} n_{\dot\phi} + G_{n_{\dot\theta}} n_{\dot\theta},$$

(7.36)

$$u = K_r r_\theta + K_y y_c.$$

(7.37)

The weighted closed-loop system errors e_p and e_u satisfy the following equation:

$$\begin{bmatrix} e_p \\ e_u \end{bmatrix} = \begin{bmatrix} W_p S_o(G_{r_\theta} + G_u K_r) & W_p S_o G_{n_{\dot\phi}} & W_p S_o G_{n_{\dot\theta}} \\ W_u S_i(K_r + K_y G_{r_\theta}) & W_u S_i K_y G_{n_{\dot\phi}} & W_u S_i K_y G_{n_{\dot\theta}} \end{bmatrix} \begin{bmatrix} r_\theta \\ n_{\dot\phi} \\ n_{\dot\theta} \end{bmatrix},$$

(7.38)

where $S_i = (1 - K_y G_u)^{-1}$ is the input sensitivity transfer function and $S_o = (I - G_u K_y)^{-1}$ is the output sensitivity transfer function matrix.

As usual, the performance criteria used in controller design require the transfer function matrix from the exogenous input signals r_θ, $n_{\dot\phi}$, and $n_{\dot\theta}$ to the output signals e_p and e_u to be small in the sense of $\| \cdot \|_\infty$, for all possible uncertain plant models. This leads to small weighted signals $\dot\phi$, $\dot\theta$, and $\int (r_\theta - \theta)$ and small control action u. The transfer function matrices W_p and W_u reflect the relative importance of the different frequency ranges for which the performance requirements should be fulfilled.

The μ synthesis is applied for several choices of performance weighting functions that ensure a good balance between system performance and robustness. After several trials, the performance weighting function (in the continuous-time case) is chosen as low pass filter as

$$
W_p(s) = \begin{bmatrix} 52.2\dfrac{0.05s+1}{s+1} & 0 & 0 \\[2ex] 0 & 0.029\dfrac{0.6s+1}{0.007s+1} & 0 \\[2ex] 0 & 0 & 4.7\dfrac{4s+1}{125s+1} \end{bmatrix},
$$

and the control weighting function as a high-pass filter as

$$
W_u(s) = \frac{s/0.07 + 1}{s/200 + 1}/30{,}000.
$$

Since in the given case the plant uncertainty is taken into account by the corresponding input multiplicative uncertainty, the transfer functions $N_{\dot\phi}$, $N_{\dot\theta}$ are set equal to one and $n_{\dot\phi} = n_{\dot\theta} = 0$. However, in order to obtain better convergence of the D–K iterations, two small noises with 2-norms equal to 10^{-4} are added to the angular velocities $\dot\phi$ and $\dot\theta$, respectively. The μ-synthesis is done using the function dksyn from Robust Control Toolbox™. After the third iteration, the maximum value of μ is decreased to 0.736 and the final controller obtained is of 44th order. The controller order is reduced to 30th by using the command reduce of Robust Control Toolbox without deterioration of closed-loop performance.

The controller magnitude plots in respect to the inputs r, $\dot\phi$, $\dot\theta$, and $\int err$ are shown in Figure 7.31.

It should be noted that the robust control law presented is designed in MATLAB using double precision (64 b) floating-point arithmetic. In our case, this control law is embedded in a processor which use single precision (32 b) arithmetic. This circumstance may affect the behavior of the discrete-time closed-loop system. For instance, some controllers become unstable when implemented in single precision, which is undesirable in practice. For this reason, the controller stability in single precision should be checked after the design.

The μ controller incorporates the same PI regulator of the yaw motion as the one used with LQG controller.

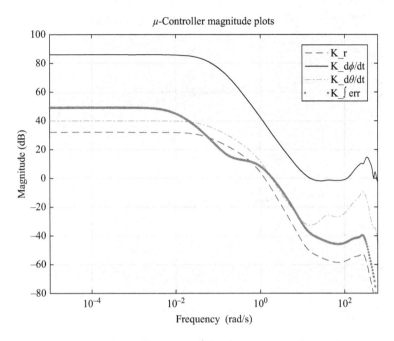

Figure 7.31 Controller magnitude plots in case of μ synthesis

7.6 Comparison of designed controllers

MATLAB files used in this section

File	Description
dfrs_mu_uncertain.m	Determines the frequency responses of LQG and μ controllers
robust_comparison.m	Determines robust stability and robust performance for LQG and μ controllers
comparison_experiments.m	Compares the experimental results for both controllers

In this section, we compare the frequency domain properties of the two controllers designed. In the case of μ controller we use a 30th-order approximation obtained by the function reduce.

The Bode plots of the closed-loop system is shown in Figure 7.32. It is seen that for both controllers, the closed-loop bandwidth is approximately 1 rad/s, the LQG controller having the larger bandwidth.

In Figures 7.33 and 7.34, we show the influence of the measurement noises in $\dot{\phi}$ and $\dot{\theta}$, respectively, on the control action u. It is seen that the effect of noise in $\dot{\phi}$ is much stronger than the effect of noise in $\dot{\theta}$.

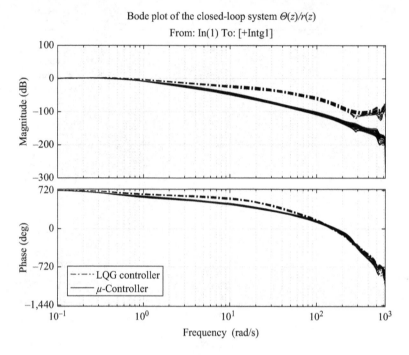

Figure 7.32   Bode plot of the closed-loop system

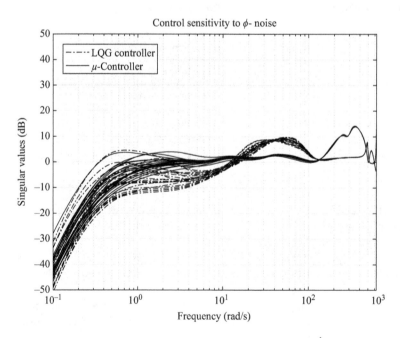

Figure 7.33   Sensitivity of control to noise in $\dot{\phi}$

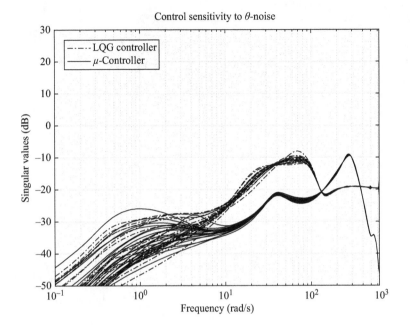

Figure 7.34   Sensitivity of control to noise in θ

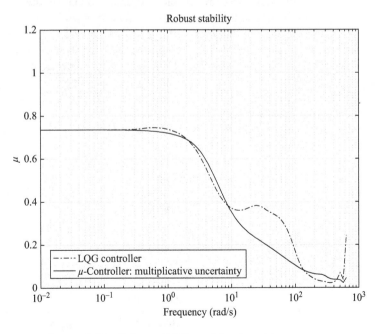

Figure 7.35   Robust stability of the closed-loop system

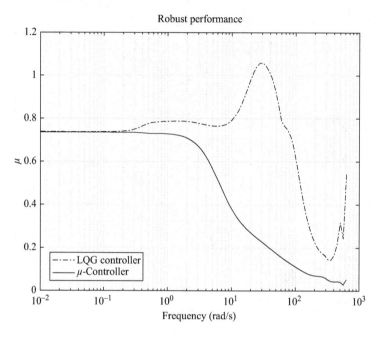

Figure 7.36   Robust performance of the closed-loop system

In Figures 7.35 and 7.36, we show the plots of upper bounds on the structured singular value corresponding to the robust stability and robust performance, respectively, of the closed-loop system for both controllers. In both cases, the plant model with input multiplicative uncertainty is used in the analysis. For both controllers, the system achieves robust stability in respect to the uncertainties corresponding to the identification of robot model. For both controllers, the robust performance analysis is done for the same weighting functions used in the μ synthesis. For this reason, better performance margin is obtained in the case of the μ controller. The LQR regulator with Kalman filter does not achieve robust performance in presence of uncertainties corresponding to the identified robot model.

As disadvantages of the μ synthesis, one may point out the difficult choice of the weighting functions as well as the possibility that the D–K iteration does not converge to a stabilizing controller. The μ controllers usually are of high order which may be reduced significantly without performance deterioration.

7.7   Experimental evaluation

MATLAB files used in this section

File	Description
DSP_controller_ver2lqrc.slx	Simulink LQG controller model
DSP_controller_ver21.slx	Simulink μ controller model
DSP_OP_ver2.slx	Interface program for data collection

Table 7.2   Wheels angle reference trajectory

t (s)	θ_{ref} (deg)
$0 \leq t < 10$	0
$10 \leq t < 10 + 1{,}000/\dot{\theta}_{ref}$	$\dot{\theta}_{ref}(t - 10)$
$10 + 1{,}000/\dot{\theta}_{ref} \leq t < 140$	1,000
$140 \leq t < 140 + 2{,}000/\dot{\theta}_{ref}$	$1{,}000 - \dot{\theta}_{ref}(t - 140)$
$140 + 2{,}000/\dot{\theta}_{ref} \leq t < 320$	$-1{,}000$
$320 \leq t < 320 + 1{,}000/\dot{\theta}_{ref}$	$-1{,}000 + \dot{\theta}_{ref}(t - 320)$
$320 \leq t \leq 400$	0

Table 7.3   Yaw angle reference

t (s)	ψ_{ref} (deg)
$0 \leq t < 60$	0
$60 \leq t < 80$	$5(t - 60)$
$80 \leq t < 100$	100
$100 \leq t < 120$	$100 - 5(t - 100)$
$120 \leq t < 400$	0

A simulation scheme of the control system and a specialized software in MATLAB/Simulink environment is developed to implement the robot control code for the two designed controllers. With the aid of Simulink Coder™ [9], Embedded Coder™ [34], and Code Composer Studio™, a code is generated from this software which is embedded in the Texas Instruments *TMS320F28335* Digital Signal Controller [174].

Several experiments with the two controllers designed are performed, and obtained results are compared. In this section, we present results related to tracking wheels trajectories which are predefined as in Table 7.2. Each experiment has a duration of 400 s. The reference wheel velocity is constant during a specific trajectory and takes the values $\dot{\theta}_{ref} = 25$, 35, and 50 rad/s for different experiments. The experiments are done on a laboratory floor covered by ceramic tails which leads to additional small disturbances due to the small gaps between tails.

During each experiment, the yaw angle reference changes as shown in Table 7.3.

Consider first the experimental results with LQG controller for wheel velocity equal to 25 deg/s. It is seen from Figures 7.37 and 7.38 that the wheels track accurately the reference, and the robots keeps well its vertical position. The usage of $\hat{\psi}(k)$ instead of $\psi(k)$ ensures exact rotation of the robot around the vertical axis. It should be noted that there is an increasing with the time bias in the measured value of $\psi(k)$ due to the integration of the gyro noise (Figure 7.39).

Let us compare now the transient responses of system with LQG and μ controller.

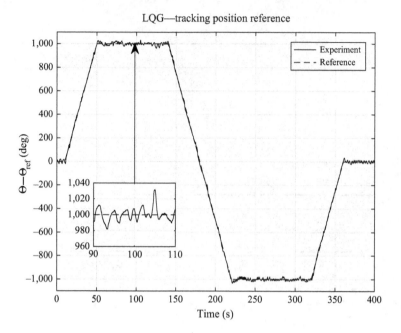

Figure 7.37   *LQG—tracking position reference*

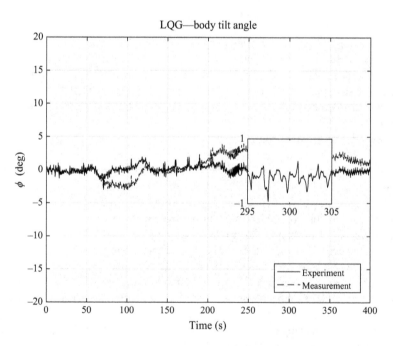

Figure 7.38   *LQG—body tilt angle*

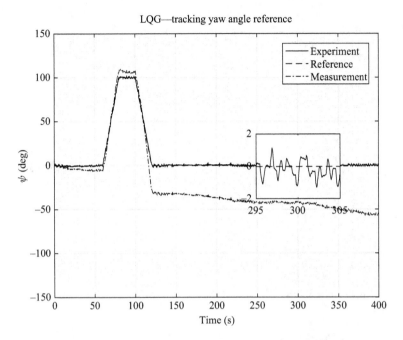

Figure 7.39 LQG—tracking yaw angle reference

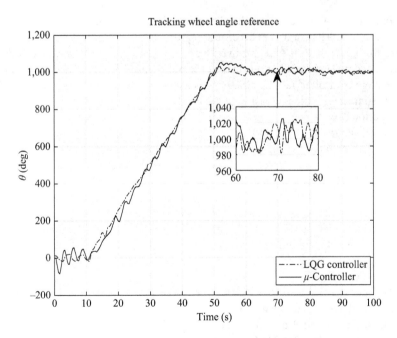

Figure 7.40 Tracking wheel angle reference

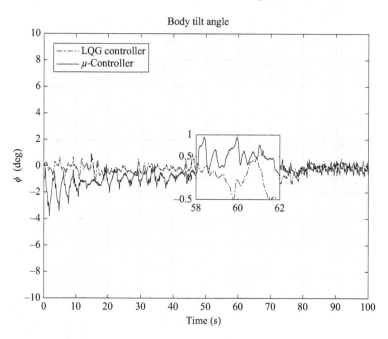

Figure 7.41   Tilt body angle

In Figure 7.40, we compare the change of wheel angle for $\dot{\theta}_{ref} = 25$ rad/s for the two controllers. (Here and after for clearer representation, the experimental results are shown for $t \leq 100$ s.) It is seen that in both cases, the tracking error is relatively small. There are some small oscillations in the case of μ controller which may be explained by the large controller gain and the presence of dead zones in the gear boxes. Also, there is some negligible increasing of the oscillations magnitude for both controllers in the period between $t = 60$ s and $t = 80$ s which is due to the influence of yaw angle reference changes in this period.

The variations of the tilt body angle during the same experiments are compared in Figure 7.41. After the transient response due to the nonzero initial conditions dies, the tilt angle remains less than 1 deg for both controllers. The measured angle ϕ contains a bias which is removed offline (for the μ controllers) and online (for the LQG controller) by using a 17th-order Kalman filter. It is interesting to note that this bias does not affect the vertical stabilization of the robot in case of μ controller which is due to the good filtration properties of the closed-loop system with such controller. That is why the inclusion of Kalman filter in the μ controller does not improve the closed-loop system performance in the given case.

The tracking error during the yaw motion also remains small as shown in Figure 7.42 due to the usage of Kalman filter.

Finally, the control action for both controllers does not exceed 25 U (1 U of control corresponds to 0.094 V to motors) with maximum allowable value equal to 50 (Figure 7.43).

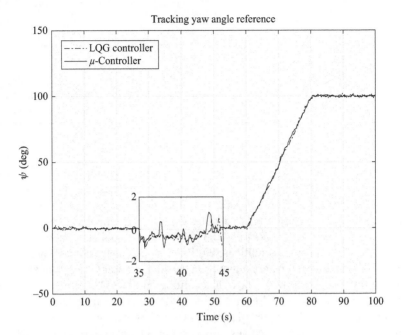

Figure 7.42 Tracking yaw angle reference

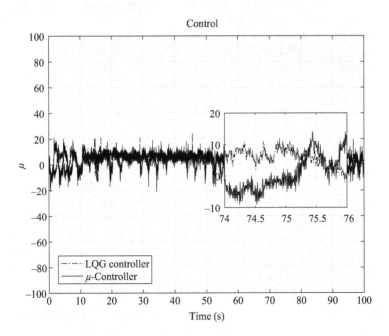

Figure 7.43 Control action

Table 7.4 Controller performance for different speeds

Controller	$\sigma(\theta - \theta_{\text{ref}})$ (deg)	$\sigma(\Delta X)$ (m)	$\sigma(\phi)$ (deg)	$\sigma(\psi - \psi_{\text{ref}})$ (deg)	$\sigma(u)$
LQG	1.19×10^1	8.27×10^{-3}	5.48×10^{-1}	9.88×10^{-1}	6.57×10^0
μ	1.97×10^1	1.38×10^{-2}	1.10×10^0	6.35×10^{-1}	5.89×10^0

In Table 7.4, we present the mean square values of variables characterizing the robot motion along the reference trajectory using experimental data for wheels velocity $\dot{\theta}_{\text{ref}} = 25$ deg/s and for duration equal to 400 s which corresponds to 80,000 sample measurements. The variable $\sigma(\Delta X)$ denotes the square mean value of robot position error computed as $\sigma(\Delta X) = R\sigma(\theta - \theta_{\text{ref}})\pi/180$ for $R = 0.04$ m. To avoid the influence of the nonzero initial conditions, the first 2,000 samples corresponding to the first 10 s of each experiment are removed in the computation of mean square values. Both controllers perform well. The LQG controller produces better results in respect to the tracking of wheels and tilt angle but gives worse accuracy in respect to the yaw motion. The better accuracy of the LQG controller is at the expense of its poorer robustness. Also, the better accuracy of the μ controller in comparison to LQG controller in the yaw motion is explained with stronger suppression of the influence from the vertical stabilization channel. Note, however, that these conclusions are derived on the basis of few experiments for three reference trajectories with restricted duration and may not be valid in more general case.

Further improvement of the closed-loop performance may be achieved by using a multivariable plant model comprising of the vertical stabilization and the yaw motion control. Determining such model may require implementation of identification methods intended for multivariable systems and may increase the controller order.

7.8 Notes and references

Two-wheeled robots have several peculiarities which make them challenging both from theoretical and practical point of view. The most popular commercial product, built on the idea of self-balancing two-wheeled robot, is the SegwayPersonal Transporter (PT), produced by Segway Inc., USA [175]. Some of the SegwayPTs have maximum speed of 20 km/h and can travel as far as 39 km on a single battery charge. Another robot, the self-balancing two-wheeled robot NXTway-GS [176], built on the basis of the LEGO® Mindstorms NXT developer kit, is widely used in education. Also, the telepresence and video conferencing two-wheeled robot Double® [177] is an interesting example of this type of robots.

The two-wheeled robots have dynamics which is similar to the inverted pendulum dynamics so that they are inherently unstable and should be stabilized around the vertical position using a control system. A detailed review of several results in the area of two-wheeled robot control is done in Chan, Stol, and Halkyard [166]. Different kinds of proportional-integral-derivative (see Lee and Jung [178]; Ren, Chen, and Chen [179]; Qiu and Huang [180]), linear-quadratic (see Fang [181], Lupián and Avila [182], Sun, Lu, and Yuan [183]), or pole placement control laws (see Muralidharan and Mahindrakar [184]; Nawawi, Ahmad, and Osman [185]) are usually implemented in order to achieve vertical stabilization and desired position of the robot in horizontal plane. Model predictive control for this purpose is considered in Azimi and Koofigar [186], and several fuzzy logic controllers are proposed, see for instance Sayidmarie *et al.* [187]; Wong *et al.* [188]; Xu, Guo, and Lee [188]. Implementation of nonlinear control methods is discussed in Li, Zhang, and Yang [189]; Zhuang, Hu, and Yao [170]; Raffo *et al.* [190]. However, with all these controllers, it is difficult or even impossible to ensure robust stability and robust performance of the closed-loop system in presence of uncertainties and disturbances. There are a few papers devoted to the robust control of two-wheeled robots. Robustness is achieved by using $\mathscr{H}_\infty$ control, see Hu and Tsai [167], Ruan and Chen [191], Shimada and Hatakeyama [192], disturbance observer, see Dinale *et al.* [193], or sliding mode control, see Wu, Liang, and Wang [194]; Yau *et al.* [195]. As it is well known, the μ controllers have the advantage that they may ensure both robust stability and robust performance of the closed-loop system, see Skogestad and Postlethwaite [87]; Zhou, Doyle, and Glover [82]. An obstacle in the implementation of high-order μ control laws for stabilization of two-wheeled robots is the difficulties related to the development, testing, and verification of the necessary real-time software which is highly dependent on the type of digital controller platform used. These difficulties are reduced significantly using the technologies for automatic code generation and embedding implemented in MATLAB/Simulink program environment [9].

Appendix A
Elements of matrix analysis

In this appendix, we summarize some standard results of matrix analysis, used in the book.

MATLAB® functions used in this appendix

Function	Description
`inv`	Inverse of a nonsingular matrix
`eig`	Matrix eigenvalues
`svd`	Singular value decomposition
`rank`	Rank of a matrix
`norm`	Norm of vector, matrix, or system

A.1 Vectors and matrices

A complex *column vector* v with n elements is written as

$$v = \begin{bmatrix} v_1 \\ v_2 \\ \vdots \\ v_n \end{bmatrix},$$

where $v_i, i = 1, \ldots, n$ are complex scalars. Formally, we write $v \in \mathbb{C}^n$, if v is a complex vector or $v \in \mathcal{R}^n$, if v is a real vector. The *row vector* $v^T = [v_1, v_2, \ldots, v_n]$ denotes the transposed of the vector v.

Consider a complex $n \times m$ matrix A with elements $a_{ij} = \mathrm{Re}\, a_{ij} + j\mathrm{Im}\, a_{ij}$. Here, n is the number of rows (number of "outputs," when A is a transfer function matrix), and m is the number of columns ("inputs"). Formally, we write $A \in \mathbb{C}^{n \times m}$, if A is a complex matrix or $A \in \mathcal{R}^{n \times m}$, if A is a real matrix. If a matrix A is square ($n = m$), we say that A is nth order matrix.

Matrix with all elements equal to zero is called *zero matrix* and is denoted by 0 or by $0_{n \times m}$ when it is necessary to indicate its size. A matrix A with elements a_{ij} is *diagonal* if $a_{ij} = 0$ for $i \neq j$. The square diagonal matrix with diagonal elements equal to 1 is called *unit matrix* and is denoted by I (or by I_n when it is necessary to indicate its order).

The transposed of the matrix A is denoted by A^T (with elements a_{ji} in position (i,j)), the conjugate of A by $\bar{A}$ (with elements Re $a_{ij} - j\text{Im } a_{ij}$), the conjugate transpose (Hermitian conjugate) of A by $A^H := \bar{A}^T$ (with elements Re $a_{ji} - j\text{Im } a_{ji}$). The trace Tr($A$) of a matrix A is the sum of its diagonal elements and det A is the determinant of A.

The *Kronecker product* of the $n \times m$ matrix A and the matrix B is defined as

$$A \otimes B = \begin{bmatrix} a_{11}B & a_{12}B & \dots & a_{1m}B \\ a_{21}B & a_{22}B & \dots & a_{2m}B \\ \vdots & \vdots & & \vdots \\ a_{n1}B & a_{n2}B & \dots & a_{nm}B \end{bmatrix}.$$

A square matrix $A \in \mathbb{C}^{n \times n}$ is *nonsingular*, if there is a matrix $A^{-1} \in \mathbb{C}^{n \times n}$ (the *inverse* of A) such that $AA^{-1} = I$. In such case, one also has that $A^{-1}A = I$. A^{-1} is unique whenever it exists. If A^{-1} does not exist, then A is *singular* matrix. If $A \in \mathbb{C}^{n \times n}$ is nonsingular and $b \in \mathbb{C}^n$, then $x = A^{-1}b$ is the unique solution of the linear system $Ax = b$.

A square matrix Q is *symmetric*, if $Q^T = Q$, and *Hermitian*, if $Q^H = Q$. A Hermitian matrix is *positive definite*, if $x^H Q x > 0$ for all nonzero vectors $x \in C$. The positive definite matrices are nonsingular. A Hermitian matrix Q is *positive semidefinite*, if $x^H Q x \geq 0$ for all $x \in \mathbb{C}$. The matrix Q is called a *square root* of the positive semidefinite matrix P if $Q^2 = P$. This is denoted as $Q = \sqrt{P}$. The complex matrix U is *unitary*, if

$$U^H U = I. \tag{A.1}$$

If the matrix U is unitary, then $UU^H = I$ and $U^{-1} = U^H$.

The real matrix U is *orthogonal*, if

$$U^T U = I. \tag{A.2}$$

If the matrix U is orthogonal, then $UU^T = I$ and $U^{-1} = U^T$.

A.2 Eigenvalues and eigenvectors

Eigenvalues and eigenvectors. *Let A is a complex $n \times n$ matrix. The eigenvalues λ_i, $i = 1, 2, \dots, n$, are the n roots of the n-order characteristic equation*

$$\det(\lambda I - A) = 0 \tag{A.3}$$

of the matrix A. The (right) eigenvector v_j, corresponding to the eigenvalue λ_j, is the nontrivial solution ($v_j \neq 0$) of

$$(A - \lambda_i)v_j = 0 \Leftrightarrow Av_j = \lambda_j v_j. \tag{A.4}$$

The corresponding left eigenvectors w_i satisfy

$$w_j^H(A - \lambda_j I) = 0 \Leftrightarrow w_j^H A = \lambda_j w_j^H. \tag{A.5}$$

Note that the left eigenvectors of A are the (right) eigenvectors of A^H.

The eigenvalues are sometimes called *characteristic values*. The set of eigenvalues of A counting their algebraic multiplicity as roots of (A.3) is called *spectrum* of A and is denoted by $\lambda(A)$. The largest of the modules of the eigenvalues of A is the *spectral radius* of A, $\rho(A) := \max_i |\lambda_i(A)|$. This is the radius of the smallest central circle in $\mathbb{C}$, which contains $\lambda(A)$.

The positive definite matrices have real positive eigenvalues and the positive semidefinite matrices have nonnegative eigenvalues (some of the eigenvalues are equal to zero).

Eigenvector corresponding to distinct eigenvalues are linearly independent. In case of multiple eigenvalues, an nth order matrix may have less than n linearly independent eigenvectors. Such matrix is called *defective*.

If the $n \times n$ matrix A is nondefective, i.e., it has n linearly independent eigenvectors, it is possible to reduce it to diagonal form. If

$$\Lambda = \text{diag}\,\{\lambda_1, \ldots, \lambda_n\}. \tag{A.6}$$

is a diagonal matrix which contains the eigenvalues of A and

$$V = [v_1, v_2, \ldots, v_n]; \quad \Lambda = \text{diag}\,\{\lambda_1, \ldots, \lambda_n\}$$

is matrix with the corresponding eigenvectors, then

$$AV = V\Lambda. \tag{A.7}$$

and

$$\Lambda = V^{-1}AV. \tag{A.8}$$

Such matrices are called *diagonalizable*. Note that a matrix may have multiple eigenvalues, but nevertheless, it may be possible to reduce it to diagonal form. The symmetric nth order matrices always have n orthogonal eigenvectors and hence are diagonalizable. Furthermore, the positive definite and positive semidefinite matrices also are diagonalizable. In particular, if a real matrix Q is positive (semi)definite it may be represented as

$$U^T Q U = \Lambda, \quad U^T U = I. \tag{A.9}$$

Therefore, the square root of a positive (semi)definite matrix may be computed as

$$\sqrt{Q} = U\sqrt{\Lambda}U^T = U \text{diag}\left\{\sqrt{\lambda_1}, \sqrt{\lambda_2}, \ldots, \sqrt{\lambda_n}\right\} U^T.$$

A.3   Singular value decomposition

Singular value decomposition (SVD). *Each complex $n \times m$ matrix A can be represented by the singular value decomposition*

$$A = U\Sigma V^H, \tag{A.10}$$

where the $n \times n$ matrix U and the $m \times m$ matrix V are unitary, and $n \times m$ matrix Σ is diagonal:

$$\Sigma = \begin{bmatrix} \Sigma_1 \\ 0 \end{bmatrix}; \quad n \geq m, \tag{A.11}$$

or

$$\Sigma = \begin{bmatrix} \Sigma_1 & 0 \end{bmatrix}; \quad n \leq m, \tag{A.12}$$

where

$$\Sigma_1 = \text{diag} \{\sigma_1, \sigma_2, \dots, \sigma_k\}; \quad k = \min (n, m), \tag{A.13}$$

and

$$\bar{\sigma} := \sigma_1 \geq \sigma_2 \geq \dots \geq \sigma_k := \underline{\sigma} \tag{A.14}$$

The numbers $\sigma_i \geq 0$ are called singular values of the matrix A.

The columns of V, denoted by v_i, are called *right* or *input* singular vectors, and the columns of U, denoted by u_i, are called *left* or *output* singular vectors. We define $\bar{u} := u_1, \bar{v} := v_1; \underline{u} := u_k$ and $\underline{v} := v_k$.

The decomposition (A.10) is not unique, since the matrices U and V are not unique. However, the singular values σ_i are unique.

The rank, rank(A), of an $n \times m$ matrix A is the number of its positive singular values. Let rank(A) $= r$. Then, the matrix A is called rank deficient if $r < k = \min(n, m)$. In this case $\sigma_i = 0$ for $i = r + 1, \dots, k$. A rank deficient square matrix is a singular matrix.

It follows from the SVD definition that

$$\bar{\sigma}(A^H) = \bar{\sigma}(A) \text{ and } \bar{\sigma}(A^T) = \bar{\sigma}(A). \tag{A.15}$$

An important property of the singular values is that

$$\bar{\sigma}(AB) \leq \bar{\sigma}(A)\bar{\sigma}(B). \tag{A.16}$$

For a nonsingular matrix A (or B), we have the inequalities

$$\underline{\sigma}(A)\bar{\sigma}(B) \leq \bar{\sigma}(AB) \text{ or } \bar{\sigma}(A)\underline{\sigma}(B) \leq \bar{\sigma}(AB). \tag{A.17}$$

The following inequality also holds

$$\underline{\sigma}(A)\underline{\sigma}(B) \leq \underline{\sigma}(AB). \tag{A.18}$$

For a nonsingular square matrix A, we have that

$$\bar{\sigma}(A^{-1}) = \frac{1}{\underline{\sigma}(A)} \text{ and } \bar{\sigma}(A) = \frac{1}{\underline{\sigma}(A^{-1})}. \tag{A.19}$$

Other useful relationships for the singular values are (*Fan's inequalities*)

$$\underline{\sigma}(A) - \bar{\sigma}(B) \leq \underline{\sigma}(A + B) \leq \underline{\sigma}(A) + \bar{\sigma}(B), \tag{A.20}$$

and

$$\underline{\sigma}(A) - 1 \leq \frac{1}{\bar{\sigma}(I + A)^{-1}} \leq \underline{\sigma}(A) + 1, \tag{A.21}$$

provided that $I + A$ is nonsingular.

A.4 Vector and matrix norms

In many cases, it is useful to have a single number which characterizes the size of a vector, matrix, signal, or system. For this purpose, we use functions which are called norms.

A norm of the element v (vector or matrix) is a real number, denoted by $\|v\|$, *which satisfies the following properties:*

1. *Nonnegativity:* $\|v\| \geq 0$, *as* $\|v\| = 0$ *exactly when* $v = 0$.
2. *Homogeneity:* $\|\alpha v\| = |\alpha| \, \|v\|$ *for all complex scalars* α.
3. *Triangle inequality:*

$$\|v + w\| \leq \|v\| + \|w\|. \tag{A.22}$$

A.4.1 Vector norms

Consider a column vector v with n real or complex elements v_i. We shall consider three norms which are special cases of the vector p-norm

$$\|v\|_p = \left(\sum |v_i|^p\right)^{1/p}, \tag{A.23}$$

where it is necessary to have $p \geq 1$ to satisfy the triangle inequality (property 3 of norm). Here, $|v_i|$ is the module of the complex scalar v_i. To illustrate the different norms, we shall compute each of them for the vector

$$z = \begin{bmatrix} z_1 \\ z_2 \\ z_3 \end{bmatrix} = \begin{bmatrix} 1 \\ 2i \\ -3 \end{bmatrix}.$$

Vector 1-norm (or norm–sum). We have that

$$\|v\|_1 := \sum_i |v_i| \quad (\|z\|_1 = 1 + 2 + 3 = 6). \tag{A.24}$$

Vector 2-norm (Euclidean norm). This is the most frequently used vector norm which is interpreted as a vector length. We have that

$$\|v\|_2 := \sqrt{\sum_i^n |a_i|^2} \quad \left(\|z\|_2 = \sqrt{1 + 4 + 9} = 3.7417\right). \tag{A.25}$$

The Euclidean norm satisfies the property

$$v^H v = \|v\|_2^2, \tag{A.26}$$

where v^H is the Hermitian conjugate of the vector v.

Vector norm ∞. This the maximum in module vector element. We use the notion $\|v\|_{\max}$, such that

$$\|v\|_{\max} \equiv \|v\|_\infty := \max_i |v_i| \quad (\|z\|_{\max} = |-3| = 3). \tag{A.27}$$

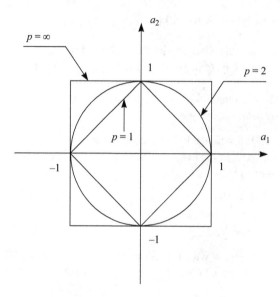

Figure A.1   Contours for the vector p-norm, $\|a\|_p = 1$ for $p = 1, 2, \infty$

For all $p, q \geq 1$, there exists a constant c_{pq}, such that $\|a\|_p \leq c_{pq}\|a\|_q$ for each a. In this sense, the p-norms frequently are called *equivalent*. For instance, for a vector v with n elements, it is fulfilled that

$$\|v\|_{\max} \leq \|v\|_2 \leq \sqrt{n}\,\|v\|_{\max}, \tag{A.28}$$

$$\|v\|_2 \leq \|v\|_1 \leq \sqrt{n}\,\|v\|_2. \tag{A.29}$$

Figure A.1 illustrates the difference between the vector norms by plotting the line $\|v\|_p = 1$ in $\mathscr{R}^2$.

A.4.2   Matrix norms

Consider the constant matrices $A, B, \ldots$. The $n \times m$ complex matrix A may represent, for example, the value for particular ω of the frequency response $G(j\omega)$ of the system $G(s)$ with m inputs and n outputs.

*The quantity $\|A\|$ is a **matrix norm**, if in addition to the three properties of norms, it also satisfies the multiplicative property*

$$\|AB\| \leq \|A\|\,\|B\|. \tag{A.30}$$

The condition (A.30) is important in considering the properties of system connections.

For a numerical illustration, we shall use the 2×2 matrix

$$M = \begin{bmatrix} -5 & 1 \\ 4 & -7 \end{bmatrix}.$$

Figure A.2   Representation of (A.32)

One of the most frequently used matrix norms is the **Frobenius matrix norm** (or Euclidean norm). It is defined as the square root of the sum of squares of the modules of matrix elements,

$$\|A\|_F = \sqrt{\sum_{i,j} |a_{i,j}|^2} = \sqrt{\operatorname{Tr}(A^H A)}. \quad (\|M\|_F = \sqrt{91} = 9.5394). \tag{A.31}$$

The most important matrix norms are the so-called induced matrix norms which are closely related to the amplification of signals in systems. Consider the following equation illustrated in Figure A.2:

$$z = Aw. \tag{A.32}$$

We may consider w as an input vector, and z as an output vector and to consider "amplification" of the matrix A, defined by the ratio $\|z\|/\|w\|$. The maximum amplification over the all possible input directions, i.e., all possible directions of the vector w, represents a special interest. It is given by the *induced norm*, which is defined as

$$\|A\|_p := \max_{w \neq 0} \frac{\|Aw\|_p}{\|w\|_p}, \tag{A.33}$$

where $\|w\|_p = (\sum_i |w_i|^p)^{1/p}$ denotes the vector p-norm. In other words, we look for a direction of the vector w, such that the ratio $\|z\|_p/\|w\|_p$ is maximized. In this way, the induced norm yields the maximum possible "amplification" of a matrix. The following equivalent definition is also used:

$$\|A\|_p = \max_{\|w\|_p \leq 1} \|Aw\|_p = \max_{\|w\|_p = 1} \|Aw\|_p. \tag{A.34}$$

The geometric interpretation of the induced norm is that it is the maximum norm of all vectors z, obtained after transformation by the matrix A of all vectors w, having unity norm. This is illustrated in Figure A.3 for vectors in $\mathscr{R}^2$ with 2-norm.

The induced 1, 2, and ∞ matrix norms can be determined from

$$\|A\|_1 = \max_j \left(\sum_i |a_{ij}| \right) \quad \text{"maximum column sum"} \tag{A.35}$$

$$\|A\|_\infty = \max_i \left(\sum_j |a_{ij}| \right) \quad \text{"maximum row sum"} \tag{A.36}$$

$$\|A\|_2 = \bar{\sigma}(A) = \sqrt{\rho(A^H A)} \quad \text{"spectral norm"} \tag{A.37}$$

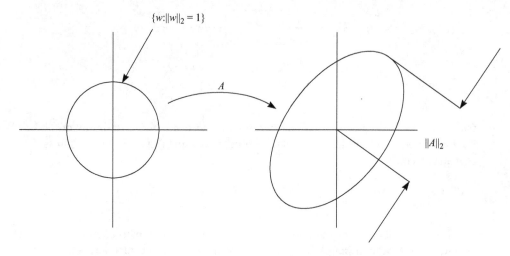

Figure A.3   Geometric interpretation of induced 2-norm

where the *spectral radius* $\rho(M) = \max_i |\lambda_i(A)|$ is the largest eigenvalue of the matrix A. The induced 2-norm of a matrix is equal to its maximum singular value and is called *spectral norm*. For the example matrix we obtain

$$\|M\|_1 = 9; \quad \|M\|_\infty = 11; \quad \|M\|_2 = \bar{\sigma}(M) = 8.8772.$$

All induced norms $\|A\|_p$ are matrix norms and satisfy the multiplicative property

$$\|AB\|_p \leq \|A\|_p \|B\|_p. \tag{A.38}$$

A.4.3   Relationships between matrix norms

Let $A \in \mathbb{C}^{n \times m}$, then

$$\bar{\sigma}(A) \leq \|A\|_F \leq \sqrt{\min(n, m)}\, \bar{\sigma}(A), \tag{A.39}$$

$$\bar{\sigma}(A) \leq \sqrt{\|A\|_1 \|A\|_\infty}, \tag{A.40}$$

$$\frac{1}{\sqrt{m}} \|A\|_\infty \leq \bar{\sigma}(A) \leq \sqrt{l}\, \|A\|_\infty, \tag{A.41}$$

$$\frac{1}{\sqrt{n}} \|A\|_1 \leq \bar{\sigma}(A) \leq \sqrt{m}\, \|A\|_1. \tag{A.42}$$

All these norms satisfy (A.30).

A very important property of the Frobenius norm and the induced norm is that they are invariant to unitary transformations, i.e., for unitary matrices $U_i, i = 1, 2$, satisfying $U_i U_i^H = I$, we have that

$$\|U_1 A U_2\|_F = \|A\|_F, \tag{A.43}$$

$$\bar{\sigma}(U_1 A U_2) = \bar{\sigma}(A). \tag{A.44}$$

From the singular value decomposition of the matrix $A = U\Sigma V^H$ and (A.44), one obtains a relationship between the Frobenius norm and the singular values $\sigma_i(A)$,

$$\|A\|_F = \sqrt{\sum_i \sigma_i^2(A)}. \tag{A.45}$$

It follows from this equation that $\|A\|_F = \|A\|_2$ exactly when the rank r of A is equal to 1. Also it is seen that

$$\sqrt{r}\,\sigma_r \leq \|A\|_F \leq \sqrt{r}\,\sigma_1. \tag{A.46}$$

According to the Perron–Frobenius theorem, which applies to a square matrix A, it follows that

$$\min_D \|DAD^{-1}\|_1 = \min_D \|DAD^{-1}\|_\infty = \rho(|A|), \tag{A.47}$$

where D is a diagonal scaling matrix, $|A|$ is a matrix whose elements are the modules of the elements of A and $\rho(|A|) = \max_i |\lambda_i(|A|)|$ is the Perron root (Perron–Frobenius eigenvalue).

A.5 Notes and references

The material presented in this appendix is standard. A profound and rigorous treatment of matrix analysis is presented in Horn and Johnson [196]. Useful facts from matrix analysis in connection with linear systems theory can be found in Bernstein [197] and Laub [198].

Appendix B

Elements of linear system theory

In this appendix, we present without proofs some basic results from the theory of Linear Control Systems.

MATLAB® functions used in this appendix

Function	Description
ss	Constructs state-space model
tf	Derives transfer function matrix
ctrb	Computes the controllability matrix
obsv	Computes the observability matrix
minreal	Determines minimal realization
lyap	Solves the Lyapunov equation
dlyap	Solves the discrete Lyapunov equation
pole	Computes poles of a linear system
tzero	Computes invariants zeros of a linear system

B.1   Description

A *linear continuous time-invariant control system* (*continuous LTI system*) is described by the set of differential and algebraic equations of the form

$$\dot{x}(t) = Ax(t) + Bu(t), \quad x(0) = x_0, \tag{B.1}$$

$$y(t) = Cx(t) + Du(t), \tag{B.2}$$

where $x(t) \in R^n$ is the *state vector*, $u(t) \in \mathscr{R}^m$ is the *input vector*, and $y(t) \in \mathscr{R}^r$ is the *output vector*; the matrices A, B, C, and D are real constant matrices with appropriate dimensions.

A linear system with single input ($m = 1$) and single output ($r = 1$) is called a SISO (single-input–single-output) system; otherwise, it is called a MIMO (multiple-input–multiple-output) system.

The *transfer function matrix* of the system (B.1), (B.2) from u to y is defined by

$$Y(s) = G(s)U(s), \tag{B.3}$$

where $U(s)$ and $Y(s)$ are the Laplace transforms of $u(t)$ and $y(t)$, respectively, with zero initial conditions. It is easy to show that

$$G(s) = C(sI - A)^{-1}B + D. \tag{B.4}$$

The transfer matrix $G(s)$ is *proper* if $D \neq 0$ and *strictly proper*, if $D = 0$.

The following notation is frequently used

$$\left[\begin{array}{c|c} A & B \\ \hline C & D \end{array}\right] := C(sI - A)^{-1}B + D.$$

A *linear discrete time-invariant control system* (*discrete-time LTI system*) is described by the set of difference and algebraic equations of the form

$$x(k + 1) = Ax(k) + Bu(k), \tag{B.5}$$

$$y(k) = Cx(k) + Du(k), \tag{B.6}$$

where $x(k)$, $u(k)$, and $y(k)$ are the state vector, input (control) vector, and output vector, respectively, at the time $t = k\Delta t$ and where A, B, C, and D are constant matrices.

The transfer function matrix of the discrete-time system (B.5), (B.6) from u to y is defined as

$$Y(z) = G(z)U(z), \tag{B.7}$$

where $U(z)$ and $Y(z)$ are the z transforms of $u(k)$ and $y(k)$, respectively. One has that

$$G(z) = C(zI - A)^{-1}B + D. \tag{B.8}$$

The transfer function matrix is invariant under nonsingular state transformations.

B.2 Stability

An unforced continuous-time linear system

$$\dot{x}(t) = Ax(t)$$

is said to be *stable* if all the eigenvalues of A are in the open left half plane, i.e., $\mathrm{Re}\lambda(A) < 0$. A matrix with such a property is said to be *stable*.

An unforced discrete-time linear system

$$x(k + 1) = Ax(k)$$

is said to be *stable* if all the eigenvalues of A are inside the unit circle, i.e., $|\lambda(A)| < 1$. A matrix with such a property is said to be stable or *convergent*.

B.3 Controllability and observability

The linear system (B.1) or the pair (A, B) is said to be *controllable* if one of the following equivalent conditions hold:

1. Given any x_0, $t_1 > 0$ and x_1, there exists an input $u(.)$ such that the solution of (B.1) satisfies $x(t_1) = x_1$.

2. The controllability matrix

$$M_c = [B \; AB \; A^2B \ldots A^{n-1}B]$$

has full row rank, i.e., rank $M_c = n$.

3. (Popov–Belevitch–Hautus test for controllability). The matrix $[A - \lambda I, B]$ has full row rank for all $\lambda \in \mathscr{C}$.

4. Let λ and w be any eigenvalue and the corresponding left eigenvector of A (i.e., $w^T A = \lambda w^T$); then $w^T B \neq [0 \; 0 \ldots 0]$.

5. There exist an $m \times n$ matrix K such that the eigenvalues of $A + BK$ can take on arbitrary values.

If the above conditions are not fulfilled, then the system or the pair (A, B) is said to be *uncontrollable*.

If the pair (A, B) is uncontrollable, then there exists a nonsingular T such that

$$TAT^{-1} = \begin{bmatrix} A_{11} & A_{12} \\ 0 & A_{22} \end{bmatrix}, \quad TB = \begin{bmatrix} B_1 \\ 0 \end{bmatrix}, \quad CT^{-1} = [C_1 \; C_2] \tag{B.9}$$

with (A_{11}, B_1) controllable. In this case, the transfer function matrix of the system is given by

$$G(s) = C_1(sI - A_{11})^{-1}B_1 + D,$$

which shows that the transfer function represents only the controllable part of the system.

The linear system (B.1) or the pair (A, B) is said to be *stabilizable* if there exists a state feedback $u(t) = Kx(t)$ such that $A + BK$ is stable (i.e., $\mathrm{Re}\lambda_i(A + BK) < 0$). If the system is stabilizable and the pair (A_{11}, B_1) in representation (B.9) is controllable, then the matrix A_{22} is stable.

The linear system described by (B.1) and (B.2) or the pair (C, A) is said to be *observable* if one of the following equivalent conditions hold:

1. For any $t_1 > 0$, the initial state $x(0) = x_0$ can be determined from the time history of the input $u(t)$ and the output $y(t)$ in the time interval $[0, t_1]$.

2. The observability matrix

$$M_o = \begin{bmatrix} C \\ CA \\ CA^2 \\ \vdots \\ CA^{n-1} \end{bmatrix}$$

has full column rank, i.e., rank $M_o = n$.

3. The matrix

$$\begin{bmatrix} A - \lambda I \\ C \end{bmatrix}$$

has full column rank for all $\lambda \in \mathscr{C}$.

4. Let λ and v be any eigenvalue and the corresponding right eigenvector of A (i.e., $Av = \lambda v$); then $Cv \neq 0$.
5. There exist an $n \times r$ matrix L such that the eigenvalues of $A + LC$ can be freely assigned.
6. The pair (A^T, C^T) is controllable.

If the above conditions are not fulfilled, then the system or the pair (C, A) is said to be *unobservable*.

If the pair (C, A) is unobservable, then there exists a nonsingular T_o such that

$$CT^{-1} = [C_1 \ 0], \quad TAT^{-1} = \begin{bmatrix} A_{11} & 0 \\ A_{21} & A_{22} \end{bmatrix}, \quad TB = \begin{bmatrix} B_1 \\ B_1 \end{bmatrix} \tag{B.10}$$

with (C_1, A_{11}) observable. In this case, the transfer function matrix of the system is given by

$$G(s) = C_1(sI - A_{11})^{-1}B_1 + D,$$

which shows that the transfer function represents only the observable part of the system.

The linear system described by (B.1), (B.2) or the pair (C, A) is said to be *detectable* if there exists an output injection matrix L such that $A + LC$ is stable. If the system is detectable and the pair (C_1, A_{11}) in representation (B.10) is observable, then the matrix A_{22} is stable.

The linear discrete-time system (B.5) or the pair (A, B) is said to be *reachable* if one of the following equivalent conditions hold.

1. Given $x_0 = 0$, there exists an input $\{u(k)\}$ for $k \in [0, n-1]$ such that x_n takes an arbitrary value.
2. The reachability matrix

$$M_c = [B \ AB \ A^2B \dots A^{n-1}B]$$

    has full row rank, i.e., rank $M_c = n$.
3. The matrix $[A - \lambda I, B]$ has full row rank for all $\lambda \in \mathscr{C}$.
4. Let λ and w be any eigenvalue and the corresponding left eigenvector of A (i.e., $w^T A = \lambda w^T$); then $w^T B \neq [0 \ 0 \dots 0]$.
5. There exist an $m \times n$ matrix K such that the eigenvalues of $A + BK$ can take on arbitrary values.

If the above conditions are not fulfilled, then the system or the pair (A, B) is said to be *unreachable*.

The discrete-time system (B.5) or the pair (A, B) is said to be *stabilizable* if there exists a state feedback $u(k) = Kx(k)$ such that the eigenvalues of $A + BK$ are inside the unit circle, i.e., $|\lambda(A + BK)| < 1$. If the system is stabilizable and the pair (A_{11}, B_1) in representation

$$TAT^{-1} = \begin{bmatrix} A_{11} & A_{12} \\ 0 & A_{22} \end{bmatrix}, \quad TB = \begin{bmatrix} B_1 \\ 0 \end{bmatrix}$$

is reachable, then $|\lambda_i(A_{22})| < 1$.

The discrete-time system described by (B.5) and (B.6) or the pair (C, A) is said to be *observable* if one of the following equivalent conditions hold:

1. Knowledge of $\{u(k)\}$ and $\{y(k)\}$ for $k \in [0, n-1]$ suffices to determine x_0.
2. The observability matrix

$$M_o = \begin{bmatrix} C \\ CA \\ CA^2 \\ \vdots \\ CA^{n-1} \end{bmatrix}$$

has full column rank, i.e., rank $M_o = n$.
3. The matrix

$$\begin{bmatrix} A - \lambda I \\ C \end{bmatrix}$$

has full column rank for all $\lambda \in \mathscr{C}$.
4. Let λ and v be any eigenvalue and the corresponding left eigenvector of A (i.e., $Av = \lambda v$); then $Cv \neq 0$.
5. There exists an $n \times r$ matrix L such that the eigenvalues of $A + LC$ can be freely assigned.
6. The pair (A^T, C^T) is reachable.

The discrete-time system described by (B.5) and (B.6) or the pair (C, A) is said to be *detectable* if there exists a matrix L such that the eigenvalues of $A + LC$ are inside the unit circle, i.e., $|\lambda(A + CA)| < 1$. If the system is detectable and the pair (C_1, A_{11}) in representation

$$CT^{-1} = [C_1 0], \quad TAT^{-1} = \begin{bmatrix} A_{11} & 0 \\ A_{21} & A_{22} \end{bmatrix}$$

is observable, then $|\lambda_i(A_{22})| < 1$.

Assume that $G(s)$ is a real rational transfer function matrix that is proper. Then, the state-space model (A, B, C, D) that satisfies

$$\left[\begin{array}{c|c} A & B \\ \hline C & D \end{array} \right] = G(s)$$

is called a *realization* of $G(s)$. The matrix A is of minimal order if and only if the pair (A, B) is controllable and the pair (C, A) is observable. In such case, the realization (A, B, C, D) is called *minimal realization* of $G(s)$.

B.4 Lyapunov equations

The Lyapunov equations play fundamental role in the stability, controllability and observability analysis, as well as in the optimization of deterministic and stochastic linear systems.

Consider first the continuous Lyapunov equation

$$A^T P + PA + Q = 0 \tag{B.11}$$

with given $n \times n$ real matrices A and Q, $Q^T = Q$. This equation has a unique solution for P if and only if $\lambda_i(A) + \lambda_j(A) \neq 0$.

Assume that A is stable, then $P > 0$ if $Q > 0$ and $P \geq 0$ if $Q \geq 0$. If A is stable and $Q \geq 0$, then $(\sqrt{Q}, A)$ is observable if $P > 0$. Hence for a stable A, the pair (C, A) is observable if and only if the solution of the Lyapunov equation

$$A^T L_o + L_o A + C^T C = 0$$

is positive definite. The solution L_o is called the *observability Grammian* and is expressed as

$$L_o = \int_0^\infty e^{A^T t} C^T C e^{At} dt.$$

The Lyapunov stability criterion for continuous LTI systems can be formulated as follows.

Suppose P is the solution of Lyapunov (B.11). Then A is stable if $P \geq 0$, $Q \geq 0$, and $(\sqrt{Q}, A)$ is detectable. If $(\sqrt{Q}, A)$ is observable, then actually $P > 0$.

In state estimation problems the Lyapunov equation appears in the dual form

$$AP + PA^T + Q = 0. \tag{B.12}$$

If $P \geq 0$, $Q \geq 0$ and $(A, \sqrt{Q})$ is stabilizable, then A is stable. Furthermore, if $(A, \sqrt{Q})$ is controllable, then P is positive definite. For a stable A, the pair (A, B) is controllable if and only if the solution of the Lyapunov equation

$$AL_c + L_c A^T + BB^T = 0$$

is positive definite. The solution L_c is called the *controllability Grammian* and may be expressed as

$$L_c = \int_0^\infty e^{At} BB^T e^{A^T t} dt.$$

Consider now the *discrete Lyapunov equation* (*Stein equation*)

$$A^T PA - P + Q = 0 \tag{B.13}$$

with given $n \times n$ real matrices A and Q, $Q^T = Q$. This equation has a unique solution for P if and only if $\lambda_i(A)\lambda_j(A) \neq 1$.

Assume that A is stable, then $P > 0$ if $Q > 0$ and $P \geq 0$ if $Q \geq 0$. For a stable A, the pair (C, A) is observable if and only if the solution of the equation

$$A^T L_o A - L_o + C^T C = 0$$

is positive definite. The observability Grammian L_o is expressed as

$$L_o = \sum_{k=0}^\infty A^{T k} C^T C A^k.$$

The Lyapunov stability criterion for discrete-time systems can be formulated as follows. Suppose P is a solution of (B.13). Then A is stable, i.e., $|\lambda_i(A)| < 1$, if $P \geq 0$, $Q \geq 0$ and $(\sqrt{Q}, A)$ is detectable. If $(\sqrt{Q}, A)$ is observable, then actually $P > 0$.

Consider the dual discrete-time Lyapunov equation

$$APA^T - P + Q = 0. \tag{B.14}$$

If $P \geq 0$, $Q \geq 0$, $(A, \sqrt{Q})$ is stabilizable, then A is stable. Furthermore, if $(A, \sqrt{Q})$ is controllable, then P is positive definite.

In the discrete-time case, the controllability Grammian is a solution of

$$AL_c A^T - L_c + BB^T = 0$$

and is expressed as

$$L_c = \sum_{k=0}^{\infty} A^k BB^T A^{Tk}.$$

The Lyapunov equation (B.12) is solved by the MATLAB function `lyap` and the discrete Lyapunov equation (B.14)—by the function `dlyap`.

B.5 Poles and zeros

Let

$$\left[\begin{array}{c|c} A & B \\ \hline C & D \end{array} \right]$$

be the minimal realization of the transfer function matrix $G(s)$. Then, the eigenvalues of A are called *poles* of $G(s)$.

A complex number z_0 is called an *invariant zero* of the system realization if there exists a vector $\begin{bmatrix} x \\ u \end{bmatrix} \neq 0$ such that

$$\begin{bmatrix} A - z_0 I & B \\ C & D \end{bmatrix} \begin{bmatrix} x \\ u \end{bmatrix} = 0.$$

Let $G(s)$ be a real rational proper transfer function matrix and (A, B, C, D) is its minimal realization. Then, a complex number z_0 is a *transmission zero* of $G(s)$ if and only if it is an invariant zero of the minimal realization.

Consider a SISO system with strictly proper transfer function $G(s)$ and let (A, B, C) be a realization of $G(s)$. If the matrix A is nondefective, i.e., has a full set of linearly independent eigenvectors, then $G(s)$ may be represented as

$$G(s) = C(sI - A)^{-1} B = \sum_{i=1}^{n} \frac{C v_i w_i^T B}{s - \lambda_i},$$

where λ_i is an eigenvalue of A and v_i, w_i are the corresponding right and left eigenvectors. The pole λ_i will appear in the transfer function only if the products $C v_i$ and

$w_i^T B$ are both nonzero, i.e., only if the realization is controllable and observable. If for some i it is fulfilled that Cv_i or/and $w_i^T B = 0$ then the pole λ_i is canceled by zero. This shows that if the realization is uncontrollable or unobservable, then zero/pole cancelation will occur in the transfer function.

In contrast to the SISO systems, a MIMO system may have pole and zero at the same location, see for an example [82, p. 83].

B.6   Notes and references

The linear system theory is presented in several good books, among them Antsaklis and Michel [199], Callier and Desoer [98], Fairman [200], Kailath [201], and Wonham [202].

Stochastic processes

In this appendix, we present some notation and basic facts pertaining to the theory of stochastic processes. For a systematic and rigorous treatment of this subject, the reader should consult the references given at the end of the appendix.

C.1 Random variables

Let v be a real, scalar random variable (RV). The probability that the RV v is less than the real number V is denoted by

$$P(v < V).$$

The *distribution function* of the RV v is defined as

$$F_v(V) = P(v < V).$$

The derivative of the distribution function is the *density function* of the RV v denoted by

$$p_v(V) = \frac{dF_v}{dV}(V).$$

The density function represents the probability that the RV v assumes a value in the infinitesimal interval $(V, V + \Delta V)$. Since the range of V is divided into elements of length dV and $p_v(V)dV$ is the probability of falling into one of these elements, the density $p_v(V)$ is a function which shows the relative probability of v falling into the interval dV at V.

The outcome of a particular experiment is called a *realization*. Note that the realizations of an RV are not equal to the RV itself and RV exists independently of any of its realizations.

The *expected value* (also called the *expectation*, the *mean*, or the *average*) of an RV is defined as its average value over a large (theoretically infinite) number of experiments. Specifically, the mean of v is defined as

$$m_v = E\{v\} = \int_{-\infty}^{\infty} V p_v(V) dV. \tag{C.1}$$

The *variance* of a scalar valued RV v is defined by

$$\sigma_v^2 = E\{(v - m_v)^2\} = \int_{-\infty}^{\infty} (V - m_v)^2 p_v(V) dV. \tag{C.2}$$

The variance of an RV is a measure of how much we expect the RV to vary from its mean.

The *standard deviation* of an RV is σ, which is the square root of the variance. It is possible to show that

$$\sigma_v^2 = E\{v^2\} - m_v^2 \tag{C.3}$$

where $E\{v^2\}$ is the *mean squared value* of v. The quantity

$$\sqrt{E\{v^2\}}$$

is called *root mean squared value* of v.

In general, the ith *moment of an RV* v is the expected value of the ith power of v. The ith *central moment of an RV* v is the expected value of the ith power of v minus its mean:

> ith moment of $v = E\{v^i\}$,
> ith central moment of $v = E\{(v - m_v)^i\}$.

For example, the first moment of an RV is equal to its mean, the first central moment is equal to 0 and the second central moment is equal to the variance.

An RV x has the *Gaussian* or *Normal* distribution if its density function is described by

$$p_x(X) = \frac{1}{\sqrt{2\pi}\sigma} \exp\left(-\frac{(X - m)^2}{2\sigma^2}\right). \tag{C.4}$$

Since

$$\int_{-\infty}^{\infty} Xp_x(X)dX = m \text{ and}$$

$$\int_{-\infty}^{\infty} (X - m)^2 p_x(X)dX = \sigma^2,$$

it follows that m and σ are the expected value and standard deviation of the RV x. The notation $x \sim N(m, \sigma^2)$ is used to indicate that the scalar RV x has a Gaussian distribution with mean m and variance σ^2. Hence, the Gaussian distribution is completely specified only by these two parameters.

The probability distribution function for a normally distributed RV is given by

$$F_x(X) = \frac{1}{\sqrt{2\pi}\sigma} \int_{-\infty}^{\infty} \exp\left(-\frac{(Z - m)^2}{2\sigma^2}\right) dZ.$$

The probability that a normally distributed variable $x(t)$ will be between $m - a$ and $m + a$ is given by

$$P((m - a) \le x(t) \le (m + a)) = erf\left\{\frac{a}{\sqrt{2}}\right\}, \tag{C.5}$$

where the function $erf(.)$ is defined as $erf(x) = 2F(x) - 1$ and is available in tabulated form. Equation (C.5) leads to the following relationships:

$$P(|x(t) - m| \leq \sigma) = 0.6827,$$
$$P(|x(t) - m| \leq 2\sigma) = 0.9545, \qquad \text{(C.6)}$$
$$P(|x(t) - m| \leq 3\sigma) = 0.9973.$$

The expressions (C.6) show that a Gaussian RV must remain within the bounds $\pm 3\sigma$ of its mean 99.73% of the time for sufficiently large number of experiments. This is very useful in practice when an estimate of the standard deviation σ is available.

For the vector RV $x = [x_1, x_2, \ldots, x_n]^T$, the notation $x \sim N(m, P)$ is used to indicate that x has the multivariate Gaussian or Normal density function described by

$$p_x(X) = \frac{1}{\sqrt{(2\pi)^n \det(P)}} \exp\left(-\frac{1}{2}(X - m)^T P^{-1}(X - m)\right).$$

Similarly to the scalar case, the density of the vector Gaussian RV is completely determined by the mean value $m = [m_1, m_2, \ldots, m_n]^T$ and the $n \times n$ covariance matrix P.

According to the *Central Limit Theorem*, if n RVs are independent and identically distributed, then the distribution of their sum will converge to a normal distribution as n becomes large. This is a fundamental result which explains the wide application of the Gaussian distribution in the analysis of random processes.

A *discrete RV* is one that can only assume discrete values. The distribution and the density functions can be defined for discrete RVs in the same way as for continuous variables except that the integration must be replaced by summation.

If the values of the discrete RV v are denoted by $V_i, i = 1, \ldots, N$, then the expectation or average value of this variable is given by

$$E\{v\} = \sum_{i=1}^{N} V_i P(v = V_i),$$

where N is infinite in some cases.

C.2 Stochastic processes

A *stochastic process*, also called a *random process*, $x(t)$, is an RV X that changes with time. The process $x(t)$ can be either a scalar or a vector. The outcome of a particular random experiment is called a *realization*.

For a vector stochastic process $x(t) = [x_1(t), x_2(t), \ldots, x_n(t)]^T$, the mean of the process is a vector of the means of process components,

$$m_x(t) = E\{x(t)\} = [E\{x_1(t)\}, E\{x_2(t)\}, \ldots, E\{x_n(t)\}]^T, \qquad \text{(C.7)}$$

$E\{.\}$ denotes mathematical expectation.

The vector stochastic processes are characterized by the following second moments.

- *Crosscorrelation function* between two stochastic processes $x(t)$ and $y(t)$,

$$R_{xy}(t_1, t_2) = E\left\{x(t_1)y(t_2)^T\right\}.\tag{C.8}$$

- *Autocorrelation function* of a stochastic process $x(t)$,

$$R_x(t_1, t_2) = E\left\{x(t_1)x(t_2)^T\right\}.\tag{C.9}$$

- *Crosscovariance function* between two stochastic processes $x(t)$ and $y(t)$,

$$C_{xy}(t_1, t_2) = E\left\{[x(t_1) - m_x(t_1)][y(t_2) - m_y(t_2)]^T\right\}.\tag{C.10}$$

- *Autocovariance function* of a stochastic process $x(t)$,

$$C_x(t_1, t_2) = E\left\{[x(t_1) - m_x(t_1)][x(t_2) - m_x(t_2)]^T\right\}.\tag{C.11}$$

The matrices $R_x(t_1, t_2)$ and $C_x(t_1, t_2)$ are symmetric. For instance, the autocorrelation matrix is given by

$$R_x(t_1, t_2) = E\left\{x(t_1)x(t_2)^T\right\}$$

$$= \begin{bmatrix} x_1(t_1)x_1(t_2) & x_1(t_1)x_2(t_2) & \cdots & x_1(t_1)x_n(t_2) \\ \vdots & \vdots & \ddots & \vdots \\ x_n(t_1)x_1(t_2) & x_n(t_1)x_2(t_2) & \cdots & x_n(t_1)x_n(t_2) \end{bmatrix}.$$

The *covariance matrix of a random process* is obtained from the autocovariance function for $t_1 = t_2 = t$,

$$P_x(t) = E\left\{[x(t) - m_x(t)][x(t) - m_x(t)]^T\right\}.\tag{C.12}$$

Note that $P_x(t)$ is an $n \times n$ matrix which is positive semidefinite for all t.

The autocovariance function is equal to the autocorrelation function when the random process has zero mean. In general, the autocovariance function can be expressed in terms of the autocorrelation function and the mean as

$$C_x(t_1, t_2) = R_x(t_1, t_2) - m_x(t_1)m_x^T(t_2).$$

If the crosscorrelation function

$$R_{xy}(t_1, t_2) = E\left\{x(t_1)y(t_2)^T\right\}$$

between the Gaussian processes $x(t)$ and $y(t)$ is equal to zero for all t_1, t_2, then these processes are called *independent*.

Two random processes $x(t)$ and $y(t)$ are *uncorrelated* with each other if, for all t_1, t_2,

$$E\left\{x(t_1)y^T(t_2)\right\} = m_x(t_1)m_y^T(t_2)$$

or, equivalently, if

$$E\left\{[x(t_1) - m_x(t_1)][y(t_2) - m_y(t_2)]^T\right\} = 0.$$

If two stochastic processes are independent, then they are also uncorrelated.

A stochastic process $x(t)$ is *stationary* if its distribution is independent of time. A stochastic process is *wide sense stationary (WSS)* if the mean and the variance of the process are independent of the time. A WSS process must have a constant mean and its correlation and covariance can only depend on the time difference $\tau = t_1 - t_2$ between the occurrence of the two RVs, i.e.,

$$E\left\{x(t_1)x(t_2)^T\right\} = R_x(\tau).$$

Stationary processes are WSS but the WSS does not imply strict-sense stationary. However, in the special case of Gaussian processes, a WSS process is strict-sense stationary as well.

A stochastic process is said to be *ergodic* if a certain moment calculated by averaging over the members of an ensemble at a fixed moment of the time has the same value as calculated by time averaging over any particular realization, given sufficient time. The ergodic processes are stationary processes, but not every stationary process is ergodic one. The ergodicity assumption is very useful in practice, since it is often quite easy to obtain a sufficiently long time sequence, but may be quite difficult to obtain an ensemble of time sequences. If this assumption holds, the statistics of the ensemble of a stochastic process may be obtained as the corresponding temporal statistics computed from appropriate data sequence.

The *power spectral density (PSD)* of a wide sense stationary random process $x(t)$ is defined as the Fourier transform of its autocorrelation function $R_x(\tau)$,

$$S_x(\omega) = \int_{-\infty}^{\infty} R_x(\tau)e^{-j\omega\tau}\,d\tau. \tag{C.13}$$

From the Fourier inversion formula, it follows that

$$R_x(\tau) = \frac{1}{2\pi}\int_{-\infty}^{\infty} S_x(\omega)e^{j\omega\tau}\,d\omega. \tag{C.14}$$

The PSD is a quantity, which is proportional to the power of the random process in the interval from ω to $\omega + d\omega$. Note that $S_x(\omega)$ is always positive. If $x(t)$ is a real process, then $R_x(\tau)$ is real and even; hence, $S_x(\omega)$ is also real and even. The function $S_x(\omega)$ is frequently called *two-sided PSD*.

The PSD is typically specified in units of (quantity)^2/hertz.

An important property of the PSD is that its integration over all frequencies from $-\infty$ to ∞ gives the average power of the function $x(t)$, i.e.,

$$\frac{1}{2\pi}\int_{-\infty}^{\infty} S_x(\omega)d\omega = R_x(0) = E\{x(t)x^T(t)\}. \tag{C.15}$$

In case of real signals, it is possible to consider the spectral density only for positive values of the frequency. For a real signal, one has that

$$\frac{1}{2\pi}\int_0^{\infty} S_x^o(\omega)d\omega = \frac{1}{2\pi}\int_{-\infty}^{\infty} S_x(\omega)d\omega = E\{x(t)x^T(t)\},$$

where $S_x^o(\omega) = 2S_x(\omega)$ is the spectral density for positive frequency values. The function $S_x^o(\omega)$ is called the *one-sided PSD*. The one-sided PSD contains the total power of the random signal and is used frequently in the graphical representation of the PSD which is plotted as a function of f on a log–log plot.

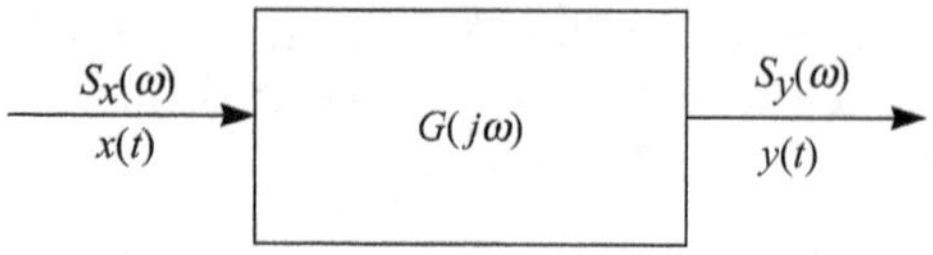

Figure C.1   Input and output spectral densities

Consider a linear system with transfer function matrix $G(s)$ (Figure C.1). The power spectral densities of the system input and output are related by

$$S_y(\omega) = G(j\omega)S_x(\omega)G^T(-j\omega). \tag{C.16}$$

In the single-input–single-output case where y and x are both scalars, (C.16) reduces to

$$S_y(\omega) = |G(j\omega)|^2 S_x(j\omega). \tag{C.17}$$

According to (C.17), in the scalar case, the mean squared value of the system output is given by

$$E\{y^2\} = \frac{1}{2\pi} \int_{-\infty}^{\infty} S_y(\omega)d\omega = \frac{1}{2\pi} \int_{-\infty}^{\infty} |G(j\omega)|^2 S_x(j\omega)d\omega. \tag{C.18}$$

A *discrete-time stochastic process* is a sequence of RVs. Discrete-time random processes have characteristics which are analogous to those of continuous processes and are described in a similar way.

The mean or expectation value of the stochastic process $x(k)$ is defined as

$$m_x(k) = E\{x(k)\}. \tag{C.19}$$

The autocorrelation function of the process is

$$R_x(k,j) = E\left\{x(k)x(j)^T\right\}, \tag{C.20}$$

while the covariance matrix is

$$C_x(k,j) = E\left\{[x(k) - m_x(k)][x(j) - m_x(j)]^T\right\}. \tag{C.21}$$

The covariance matrix of the process is

$$P_x(k) = E\left\{[x(k) - m_x(k)][x(k) - m_x(k)]^T\right\}. \tag{C.22}$$

A WSS discrete-time process fulfills the conditions:

- $m_x(k) = const$ (in its components),
- $R_x(k,k) = E\left\{x(k)x(k)^T\right\} < \infty$ (in its elements),
- $C_x(k,j) = E\left\{[x(k) - m_x(k)][x(j) - m_x(j)]^T\right\} = C_x(k-j)$.

The PSD $S(\omega)$ of a WSS discrete-time process $x(k)$ is defined as

$$S_x(\omega) = \sum_{k=-\infty}^{k=\infty} R_x(k)e^{-jk\omega} \tag{C.23}$$

which is valid for $-\pi \leq \omega \leq \pi$. One has that

$$R_x(k) = \frac{1}{2\pi} \int_{-\pi}^{\pi} S_x(\omega) e^{jk\omega} d\omega \tag{C.24}$$

and hence

$$R_x(0) = E\left\{x(k)x(k)^T\right\} = \frac{1}{2\pi} \int_{-\pi}^{\pi} S_x(\omega) d\omega \tag{C.25}$$

Let an asymptotically stable time invariant linear discrete-time system, described by the transfer function matrix $H(z)$, has as an input $x(k)$ a WSS discrete-time stochastic process with PSD $S_x(\omega)$. Then, the output $y(k)$ of this system is a realization of a discrete-time stochastic process with PSD given by

$$S_y(\omega) = H(e^{j\omega}) S_x(\omega) H^T(e^{-j\omega}). \tag{C.26}$$

C.3 White noise

A scalar continuous-time random process $v(t)$ is referred to as a *white noise* process if its PSD is constant over the whole frequency range, i.e.,

$$S_v(\omega) = \sigma_v^2, \tag{C.27}$$

where σ_v^2 is the noise intensity (or strength). The adjective "white" indicates that this noise has constant power at all frequencies. Any process that does not have equal power over the frequency interval is referred to as *colored*.

The autocorrelation function of the white noise is determined as

$$R_v(\tau) = E\{v(t)v(t+\tau)\} = \sigma_v^2 \delta(\tau) \tag{C.28}$$

where δ is the Dirac delta function: $\delta(\tau) = \lim_{\varepsilon \to 0} \delta_\varepsilon(\tau)$ and

$$\delta_\varepsilon(\tau) = \begin{cases} \dfrac{1}{2\varepsilon} & \text{if } |\tau| < \varepsilon \\ 0 & \text{otherwise} \end{cases}$$

which has units of $1/s$. In the derivation of (C.28), one uses the relationship

$$\frac{1}{2\pi} \int_{-\infty}^{\infty} e^{j\omega\tau} d\omega = \delta(\tau).$$

White noise is completely unpredictable, as can be seen from (C.28) because it is uncorrelated for any $\tau \neq 0$, while it has finite variance and standard deviation at $\tau = 0$. According to (C.15), such process is physically unrealizable since it implies infinite average power. However, the white noise idealization is convenient to find approximate error models. Specifically, white noise is used to drive appropriate filters (called *shaping filters*) which produce finite power output over a particular frequency range. Thus, within the bandwidth of the shaping filter, the output of the filter is approximately equal to the noise of the system being modeled.

A white noise can have any probability distribution but the Gaussian distribution is often assumed. The notation $v \sim N(0, \sigma^2)$ is used to describe a continuous-time, Gaussian, white noise v. In this case, σ^2 is interpreted as the PSD, not variance.

The units of the symbol σ_v^2 may be determined in two different ways [85]. First, from the PSD perspective, σ_v^2 represents the power per unit frequency interval. Therefore, the units of σ_v^2 are the units of v squared divided by the unit of frequency Hz. Second, from the perspective of correlation, R_v has the units of v squared and the Dirac delta function $\delta(\tau)$ has units of Hz $= 1/s$. Therefore, σ_v^2 has dimension corresponding to the square of the units of v times s. For example, if v is an angle in degrees, then σ_v^2 has dimension $\deg^2 s = \deg^2/(1/s) = \deg^2/Hz$. The units of σ_v are then either $\deg\sqrt{s}$ or $\deg/\sqrt{Hz}$.

A discrete-time stochastic process $v(k)$ is called white noise if

$$S(j\omega) = \sigma_v^2 \text{ for all } \omega \in [-\pi, \pi] \tag{C.29}$$

According to the discrete Fourier transform, this implies that

$$R_v(k,j) = E\left\{v(k)v(j)^T\right\} = \sigma_v^2 \delta_{j-k}, \tag{C.30}$$

where

$$\delta(k) = \frac{1}{2\pi} \int_{-\pi}^{\pi} e^{jk\theta} d\theta$$

is the Kronecker delta function, defined as

$$\delta(k) = \begin{cases} 1 \text{ if } k = 0 \\ 0 \text{ otherwise.} \end{cases}$$

The Kronecker delta function $\delta(k)$ is dimensionless. The dimension of σ_v is the same as the dimension of $v(k)$.

Contrary to the continuous-time case, the discrete-time noise is physically realizable. Its power is finite, because the integral of the PSD is only over the discrete frequency range $\omega \in [-\pi, \pi]$.

C.4   Gauss–Markov processes

If the input $v(t)$ of a linear system

$$\dot{x}(t) = Ax(t) + Gv(t),$$
$$y(t) = Cx(t)$$

is a Gaussian stochastic process, then this system represents a *Gauss–Markov process*. Since any linear operation performed on a Gaussian RV results in a Gaussian RV, the system state $x(t)$ and output $y(t)$ will be Gaussian stochastic processes. In what follows we describe two Gauss–Markov processes which are frequently used in stochastic modeling.

One of the basic Gauss–Markov processes is the so-called random walk. The random walk is an output of an integrator driven by zero mean white noise. It is described by the following equation:

$$\frac{dx(t)}{dt} = w(t), \tag{C.31}$$

where $E\{w(t)\} = 0, E\{w(t)w(t+\tau)\} = \sigma_w^2 \delta(\tau)$. It is possible to prove that

$$E\{x^2(t)\} = \sigma_w^2 t$$

which shows that the variance of the random walk is growing linearly with the time. This process is suitable to model errors which grow without bound or are slowly varying. Since the transfer function of the integrator is

$$G(s) = \frac{1}{s},$$

the spectral density of the random walk is given by

$$S_x(\omega) = |G(j\omega)|^2 \sigma_w^2 = \frac{\sigma_w^2}{\omega^2}.$$

As with white noise, the units of the random walk process may be determined in two different ways. For instance, if x is angular rate measured in deg/s, then $E\{x^2(t)\}$ has units of deg^2/s^2. Therefore, the units of σ_w are $(\text{deg/s})/\sqrt{s}$ or $(\text{deg/s}^2)/\sqrt{\text{Hz}}$.

Another typical random process used in stochastic modeling is a first-order Gauss–Markov process described by the differential equation

$$\frac{dx(t)}{dt} = -\frac{1}{T}x(t) + w(t), \tag{C.32}$$

where $w(t)$ is a white noise with strength σ_w^2. Clearly, the first-order Gauss–Markov process represents the output of a first-order lag

$$G(s) = \frac{1}{Ts+1}$$

with time constant T, driven by a white noise. Using (C.18), one obtains that the spectral density of the first-order Gauss–Markov process is given by

$$S_x(\omega) = \frac{1}{1 + \omega^2 T^2} \sigma_w^2.$$

Using (C.13), it is possible to show that

$$\sigma_x^2 = \frac{1}{2\pi} \int_{-\infty}^{\infty} \frac{1}{1 + \omega^2 T^2} \sigma_w^2 d\omega = \frac{\sigma_w^2}{2T}.$$

Using this relationship, the spectral density $S(\omega)$ is represented as

$$S_x(\omega) = 2\sigma_x^2 \frac{T}{1 + \omega^2 T^2}.$$

In this way, the autocorrelation function of $x(t)$ is given by

$$R(\tau) = \sigma_x^2 e^{-\omega_0 |\tau|},$$

where the parameter $\omega_0 = 1/T$ is called *bandwidth* and the time-constant T is called *correlation time*.

Since the units of σ_x^2 are the units of x squared, it follows that σ_w^2 has dimension corresponding to the units of x squared times s. If, for example, x has units of deg/s, then the units of σ_w are (deg/s)$\sqrt{s}$ or (deg/s)/$\sqrt{Hz}$.

C.5   Generation of white noise in MATLAB®

The generation of random sequences with normal distribution is done in MATLAB by the function `randn`. These sequences, after appropriate scaling, may be used to simulate approximately white noises with desired strengths.

In Simulink®, one may use the Band-Limited White Noise block to produce discrete signals which approximate the effect of continuous white noise in a given frequency range. Specifically, for a given noise power (strength) σ_c^2 and sample time T_s, the Band-Limited White Noise block generates a noise with two-sided spectral density S equal to σ_c^2 and correlation time equal to T_s. The correlation time determines the frequency range in which the generated signal has constant PSD and hence may be considered as a white noise. Theoretically, the continuous-time white noise has correlation time equal to 0. In practice, it is recommended that the correlation time is sufficiently small and satisfies

$$T_s \le \frac{1}{c}\frac{2\pi}{\omega_B}$$

where ω_B is the bandwidth of the simulated system in rad/s and $c \ge 100$. Note that to produce the correct intensity of the noise, the covariance of the noise is scaled to reflect the implicit conversion from a continuous PSD to a discrete noise covariance. (The strengths of the discrete-time and continuous time noises, σ_d^2 and σ_c^2, respectively, are related by $\sigma_d^2 = \sigma_c^2/T_s$.) This scaling ensures that the response of the simulated continuous system to the approximate white noise has the same covariance as the system would have to true white noise. Because of this scaling, the covariance of the signal from the Band-Limited White Noise block is not the same as the Noise power (strength) parameter. This parameter is actually the value of the two-sided PSD S of the white noise. This block approximates the covariance of white noise as σ_c^2/T_s.

Example C.1. Generation of Band-Limited White Noise
Assume that it is necessary to produce a Band-Limited White Noise with one-sided PSD equal to $S_o = 0.001$ and correlation time 0.005 s. Since $S_o = 2S$, this means that the noise power parameter and the sampling time of the Band-Limited White Noise block should be set as $S = S_o/2$ and $T_s = 0.005$ s, respectively. The simulation is done in Simulink for 1,000 s the noise being saved in the variable x. The one-sided spectral density of the noise is computed by the command lines

```
window = 500;
fs = 1/T_s;
[Pxx,f] = pwelch(x.data,window,[],[],fs);
semilogy(f,Pxx), grid
```

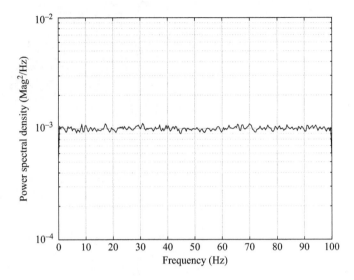

Figure C.2   Spectral density of the band-limited white noise

The one-sided spectral density of x is shown in Figure C.2. Clearly, the PSD of x is close to the prescribed value $S_o = 0.001$ in the frequency range [0–100] Hz. The covariance of x is close to $\sigma_x^2 = S/T_s = 0.1$.   ☐

To produce a noise signal with desired spectral content, one can use Band-Limited White Noise and an appropriate shaping filter.

C.6   Notes and references

The theory of RVs and stochastic processes is presented in depth in Papoulis [203]; Maybeck [72]; Farrell [85, Chapter 4]; Hendricks, Jannerup, and Sørensen [45, Chapter 6].

Appendix D
Identification of linear models

An alternative approach for model building is plant identification. *Identification* means obtaining of system model based on experimental (measurement) input/output data. A model can be identified by two approaches named "black box" and "gray box," respectively. The black-box approach requires only measurements of system inputs and outputs. The black-box models do not reveal explicitly the physical structure and parameters of the system, but they may be simple and appropriate for controller design. The gray-box approach except input/output measurements requires information about model structure. This approach is useful when we can derive model structure from physical principles but do not know the values of parameters, like mass, moment of inertia, friction coefficients. The main advantage of gray-box models is that their variables and parameters have physical interpretation and reveal explicitly physical relations between variables.

The detailed systematic representation of various types of model structures and methods for their parameters estimation is out of scope of this appendix. Our aim is to introduce the reader into the practical aspects of identification process and to describe how to perform identification using the capabilities of MATLAB® and Simulink®. First, we shall briefly present identification of black-box models based on the most often used in practice model structures and methods for their parameters estimation. Next, we describe shortly the gray-box identification approach based on the linear state-space model. In Chapter 2, the black-box and gray-box approaches are implemented on the cart–pendulum system described in Example 2.1. In Chapter 7, the black-box approach is applied for identification of a real system—two-wheeled robot.

D.1   Identification of linear black-box model

The identification process of linear black-box models includes following four stages:

- Experiment design and input/output data acquisition
- Model structure selection
- Estimation of model parameters
- Validation of estimated model to decide whether the model is adequate for the aims of controller design.

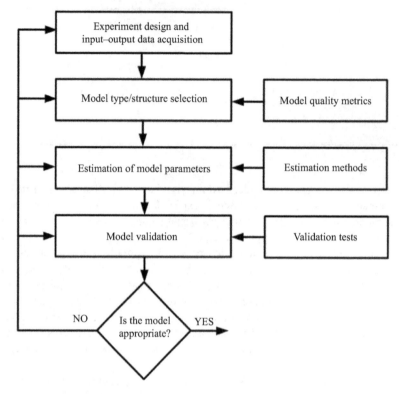

Figure D.1   Black-box model identification procedure

As there does not exists unique experiment design technique, unique model type, unique model structure, and unique parameter estimation methods which can guarantee that the estimated model will pass validation tests, the identification procedure can be regarded as an iterative (trial and error) procedure illustrated in Figure D.1.

If the model obtained does not pass the validation tests, then the experiment conditions, selected model type and structure, and/or estimation algorithm should be reconsidered. Now we briefly present the essential components of each identification stage.

D.1.1 Experiment design and input/output data acquisition

The aim of this stage is to perform an experiment and to collect the input/output data. Good experiment design means that the collected data contains information about significant plant dynamics, measured variables are the right ones, and measurements are performed with sufficient accuracy for enough long time. In general in experiment we should

- use input signals that excite enough plant dynamics,
- measure data with appropriate sample time,

- measure data long enough to collect information about important time constant,
- set up data acquisition system to have sufficient large signal-to-noise ratio.

Input signals must have sufficiently rich spectrum which cover the bandwidth of identified system and in the same time must be small since in practice the magnitude of input signals is always constrained. For example, the step signal seldom excites system enough. Moreover, it deviates output signal from its steady state value for long time which is unacceptable in many practical cases. Good input signal used in identification experiment must be persistently exciting [48, Section 13.2].

Definition D.1 A quasistationary signal $u(t)$ with spectrum $\Phi_u(\omega)$ is said to be *persistently exciting* of order n if for all filters of the form

$$M_n(z^{-1}) = m_1 z^{-1} + m_2 z^{-2} + \cdots + m_n z^{-n},$$ (D.1)

the relation $|M_n(e^{j\omega})|^2 \Phi_u(\omega) \equiv 0$ implies that $M_n(e^{j\omega}) \equiv 0$.

We also should note that $|M_n(e^{j\omega})|^2 \Phi_u(\omega)$ is the spectrum of the signal $v(t) = M_n(z^{-1})u(t)$. Hence, a signal that is persistently exciting of order n cannot be filtered to zero by $(n-1)$th-order moving-average filter. Another definition is given in terms of the autocorrelation function of the input signal $R_u(n)$.

Definition D.2 The quasistationary input signal $u(t)$ is persistently exciting of order n if the matrix

$$\bar{R}_n = \begin{vmatrix} R_u(0) & R_u(1) \ldots & R_u(n-1) \\ R_u(1) & R_u(0) \ldots & R_u(n-2) \\ \vdots & \vdots & \vdots \\ R_u(n-1) & R_u(n-2) & R_u(0) \end{vmatrix}$$ (D.2)

is nonsingular. Note that the order of persistent excitation of input signal equals to the number of parameters which one can estimate from dataset.

Example D.1. Persistently exciting signal
Let us consider the sinusoidal signal $u(t) = \sin(\omega t)$. It is easy to show that

$$R_u(m) = \frac{1}{2} \cos(\omega m)$$ (D.3)

Then, we evaluate

$$\det(\bar{R}_1) = \det\left(\frac{1}{2}\cos(0)\right) = \frac{1}{2} \neq 0$$ (D.4)

$$\det(\bar{R}_2) = \frac{1}{2}\det\begin{bmatrix} 1 & \cos\omega \\ \cos\omega & 1 \end{bmatrix} \neq 0$$ (D.5)

$$\det(\bar{R}_3) = \frac{1}{2}\det\begin{bmatrix} 1 & \cos\omega & \cos 2\omega \\ \cos\omega & 1 & \cos\omega \\ \cos 2\omega & \cos\omega & 1 \end{bmatrix} = 0.$$ (D.6)

Therefore, the signal $u(t) = \sin(\omega t)$ is persistently exciting of order 2, which means that if a sinusoidal input is used, we can estimate model with two parameters.

The degree of input signal persistence of excitation in MATLAB environment is determined by function `pexcit` from System Identification Toolbox™. It calculates the smallest rank of matrix (D.2).   □

The primary aim of experiment design is to provide appropriate data for identification which means that the experiment should be informative enough. We say that the identification experiment is informative enough if it generates dataset that is informative enough. This means that by dataset we can distinguish between two different models in intended model set [48, see Section 13.2].

Theorem D.1 Consider the set of SISO models S given by the rational transfer functions

$$W(z^{-1}) = \frac{b_1 z^{-1} + b_2 z^{-2} + \cdots + b_{nb} z^{-nb}}{1 + a_1 z^{-1} + a_2 z^{-2} + \cdots + a_{na} z^{-na}} = \frac{B(z^{-1})}{A(z^{-1})}.$$

Then the open-loop experiment with an input that persistently exciting of order $na + nb$ is informative with respect to S.

Remark D.1 An open-loop experiment is informative if the input signal is persistently exciting. The necessary order of persistent excitation is equal to the number of parameters to be estimated.

Remark D.2 Persistently excitation of m-dimensional signal $u(t)$ is defined analogously to Definition D.1. Let it be defined by (D.1) with m_i as $1 \times m$ row matrices. Then, one says that $u(t)$ is persistently exciting of order n if

$$M_n(e^{j\omega})\Phi_u(\omega)M_n^T(e^{j\omega}) \equiv 0,$$

implies that $M_n(e^{j\omega}) \equiv 0$.

Note that Definition D.2 has a counterpart for the multivariable case too.

All mentioned above impose requirements to the input signal. In case of linear model identification, the choice of input signal is governed by three basic facts (see [48, Section 13.3], [17, Section 3.2.2]):

- The bias and variance of the estimate depend only on the input spectrum and not on the actual waveform of the input.
- The input must have limited amplitude.
- Periodic inputs may have certain advantages.

It is well known that the estimates covariance matrix is inversely proportional to the input signal power. Thus, we would like to have as much power in the input as possible but in practice we have amplitude limitations. The desired property of input signal wave form is defined in terms of the *crest factor*, which for zero mean signal is defined as

$$C_r^2 = \frac{\max_k u^2(k)}{\lim_{N \to \infty} 1/N \sum_{k=1}^{N} u^2(k)}. \tag{D.7}$$

The good signal waveform is one that has small crest factor. The theoretic lower bound of C_r^2 is 1, which is achieved for symmetric binary signal $u(k) = \pm \bar{u}$. This gives advantage of binary signal. Indeed, in practice this type of input signal is often used. However, it should be mentioned that binary input will not allow validation against nonlinearities.

In practice, it is suitable to decide upon a frequency band to identify the system in question and then select a signal with more or less flat spectrum over this band. A simple choice is to generate input signal as white Gaussian noise. This signal is theoretically unbounded, so it has to be saturated at certain amplitude. For example, picking it to three standard deviations gives a crest factor of 3, and in the same time, only average 1% of the points will be affected—this leads to minor distortion of the spectrum.

Another commonly used choice of input signal is a random binary signal (RBS), which is obtained by the sign of the filtered white Gaussian noise. The level of this signal can be adjusted to any desired one so that the amplitude limitations of input signal to be satisfied. The crest factor will be equal to the ideal value 1. The problem is that taking the sign of the filtered Gaussian noise will change its spectrum. Therefore, we do not have full control of shaping the spectrum, but in the case of off-line identification, we can check the spectrum of RBS before using it as input to the identified system. Third commonly used input signal is pseudo-RBS (PRBS), which is deterministic periodic signal with like white noise properties. It is generated by the following equation:

$$u(k) = \text{modulo} \left(\sum_{i=1}^{n} a_k u(k-i), 2 \right), \tag{D.8}$$

where a_k are constants and function modulo $(x, 2)$ is the reminder as x is divided by 2, thus $u(k)$ only taking values 0 and 1. The vector of past inputs $[u(k-1), u(k-2), \ldots, u(k-n)]$ can assume 2^n different values. The sequence $u(k)$ therefore will be periodic with period of at most 2^n. In fact since n consecutive zeros will make further u's zero, we can eliminate that state and the maximum period length is $M = 2^n - 1$. The actual length of period depends on the choice of a_k. It can be shown that for each n exists choice of a_k which give maximum period length and the corresponding signals are called the maximum length PRBS.

The interest in maximum length PRBS follows from the fact that any maximum length PRBS taking levels $\pm \bar{u}$ has the so-called first- and second-order properties

$$\frac{1}{M} \sum_{k=1}^{M} u(k) = \frac{\bar{u}}{M},$$

$$R_u(n) = \frac{1}{M} \sum_{k=1}^{M} u(k)u(k+n), \tag{D.9}$$

which leads to

$$R_u(n) = \begin{cases} \bar{u}^2, & \text{for } n = 0, \pm M, \pm 2M, \ldots, \\ -\bar{u}^2/M & \text{otherwise} \end{cases}$$

where M is the maximum length of the period. The spectrum of PRBS is given by

$$\Phi_u(\omega) = \frac{2\pi \bar{u}^2}{M} \sum_{t=1}^{M-1} \delta(\omega - 2\pi k/M),$$

where $0 \leq \omega \leq 2\pi$ and δ is the Kronecker operator. This property shows that the maximum length PRBS behaves like "periodic white noise" and is persistently exciting of order $M - 1$. Like RBS PRBS has crest factor equal to 1. The advantages and disadvantages of PRBS compared to the binary random noise can be summarized as follows:

- As deterministic signal PRBS has its second-order properties (second equation in (D.9)), when is evaluated over whole periods (for random signals must rely on the law of large numbers to have a good second order properties for finite samples),
- There is essentially only one PBRS for each choice of a_k. Different initial values when generating (D.8) only correspond to shifting the sequence. It is therefore not straightforward to generate mutually uncorrelated sequences of PRBS,
- In order to have good properties of PRBS, an integer number of periods has to be considered, which limits the choice of experiment length,
- The algorithm generating PRBS can be easily implemented by microcontroller.

In MATLAB, the identification input signal can be created by the function `idinput`, which generates random Gaussian signal or RBS (by default), or PRBS, or sum of sinusoids. As a conclusion, it is always advisable after generation of input signal to study its properties offline before using it in identification experiment.

Another important issue of experiment design is choosing of sample time for measuring data. As mentioned in Chapter 1, sampling the data leads to losses of information and it is important to select sampling instances so that these losses are negligible. Detail description of appropriate choosing of sample time in case of identification is given in [48, Section 13.7]. Here, we will present some practical advices for choosing of sample time. The model which is identified with small sample time fits data in high frequencies, where the measurement noise influence on the data may be significant. Moreover very small sample time, which is near to the smallest time constant, leads to the numerical problems according to that the model poles are near to the unit circle. Often such models have zeros outside of the unit circle. Then, if the estimated plant has time delay, it is modeled by large number of samples. Each of these effects causes problems in control design task. Large sample time leads to the losses of information about significant plant dynamics and the variance of estimated parameters increases drastically near to the smallest time constant. Again these effects cause problems in control design task. It is obvious that the choice of sample time is trade-off between the reduction of high frequency noise influence and capturing the plant dynamics. A good practical advice is to choose sampling frequency about ten times the bandwidth of the system under consideration. Another useful advice is first to record step response and then choose sample time 30–40 times smaller than rising time.

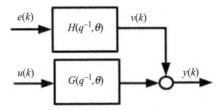

Figure D.2   Block diagram of SISO system model

D.1.2   Model structure selection and parameters estimation

The selection of appropriate model structure is crucial for a successful identification. To perform this choice, first we must estimate set of models and next evaluate in some manner the quality of each model. The model performance is assessed with various quality metrics, which provide a tool to pick the best model from given set. From control view of point, the best model means possibly simplest low-order model that explains input/output data "reasonably well." The difference between the obtained model and actual plant output can be regarded as the response of the uncertain part of the plant. It can be used for determination of uncertainty model. Such models are used in controller design in the framework of robust control which is described in Chapter 4. There are various types of linear and nonlinear models, parametric and nonparametric models, discrete-time and continuous-time models. Here, we first will briefly present most often used in control practice linear time-invariant polynomial and state-space models and general methods for their parameters estimation. Second, we introduce some of quality metrics for model performance evaluation.

The general description of SISO system in the time domain is presented in Figure D.2. It is a linear discrete-time invariant model with additive disturbance

$$y(k) = G(q^{-1}, \theta)u(k) + v(k),$$
$$v(k) = H(q^{-1}, \theta)e(k),$$

(D.10)

where $G(q^{-1}, \theta)$ is a proper rational transfer function between output $y(k)$ and input $u(k)$, $v(k)$—additive output disturbance, $H(q^{-1}, \theta)$—a proper rational transfer function which is used to modeling the noise, $e(k)$—zero mean white noise with covariance λ, and θ—vector with unknown model parameters, which must be estimated from experimental data. These estimates should be evaluated in such manner that the model be "good." The model will be "good" if it predicts "well" the output, which means that it produces small prediction error

$$e(k, \theta) = y(k) - \hat{y}(k/k - 1, \theta),$$

(D.11)

where $\hat{y}(k/k - 1, \theta)$ is the predicted output value. This value is calculated based on input/output data to the moment $k - 1$. The question is how to qualify what is "small" prediction error. Various metrics can be applied, but most often used cost function in practice and in MATLAB System Identification Toolbox is

$$J(\theta) = \frac{1}{N} \sum_{k=1}^{N} e^2(k, \theta),$$ (D.12)

where N is the number of input/output measurements. The estimate $\hat{\theta}_N$ of θ is defined as

$$\hat{\theta}_N = \arg\min_{\theta} (J(\theta)).$$ (D.13)

The estimates $\hat{\theta}_N$ can be determined after calculation of one-step-ahead predicted output $\hat{y}(k/k - 1, \theta)$ and after parametrization of transfer functions in (D.10). The one-step-ahead predictor of output signal which minimizes mean square of prediction error is [48, Section 3.2]

$$\hat{y}(k/k - 1, \theta) = H^{-1}(q^{-1}, \theta)G(q^{-1}, \theta)u(k) + [I - H^{-1}(q^{-1}, \theta)]y(k).$$ (D.14)

Equations (D.10)–(D.14) define the Prediction Error Method which is implemented in many functions of System Identification Toolbox. After parametrization of (D.10) with rational transfer functions, the parameters will be the coefficients in nominators and denominators of these transfer functions. Such models are also known as black-box polynomial models.

One of the simplest relationships between output and input is given by

$$y(k) + a_1 y(k - 1) + a_2 y(k - 2) + \cdots + a_{na} y(k - na) =$$
$$b_1 u(k - 1) + b_2 u(k - 2) + \cdots + b_{nb} u(k - nb) + e(k)$$ (D.15)

Here, the white noise $e(k)$ enters directly in the difference equation. After introducing the vector with unknown parameters θ, polynomials $A(q^{-1})$, and $B(q^{-1})$ as

$$\theta = [a_1, a_2, \ldots, a_{na}, b_1, b_2, \ldots, b_{nb}]^T,$$
$$A(q^{-1}) = 1 + a_1 q^{-1} + a_2 q^{-2} + \cdots + a_{na} q^{-na},$$ (D.16)
$$B(q^{-1}) = b_1 q^{-1} + b_2 q^{-2} + \cdots + b_{nb} q^{-nb},$$

(D.15) takes the form

$$y(k) = \frac{B(q^{-1})}{A(q^{-1})} u(k) + \frac{1}{A(q^{-1})} e(k).$$ (D.17)

Equation (D.17) is a special case of (D.10) if

$$G(q^{-1}, \theta) = \frac{B(q^{-1})}{A(q^{-1})}, \quad H(q^{-1}, \theta) = \frac{1}{A(q^{-1})}.$$ (D.18)

The model (D.17) depicted in Figure D.3 is the so-called ARX model, because it comprises the autoregressive part (AR) $A(q^{-1})y(k)$ and exogenous part (X) $B(q^{-1})u(k)$. In the special case of $na = 0$, the model (D.17) is transformed to the finite impulse response (FIR) model. The main disadvantage of ARX model is that the output disturbance is obtained from white noise $e(k)$ which goes through the

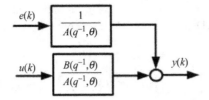

Figure D.3   Structure diagram of ARX model

denominator dynamics $A(q^{-1})$ of system. The main advantage is that the predictor (D.14) defines linear regression. Indeed, let we introduce the regression vector $\varphi(k)$ as

$$\varphi(k) = [-y(k-1), -y(k-2), \dots, -y(k-na),$$
$$u(k-1), u(k-2), \dots, u(k-nb)]^{T}. \tag{D.19}$$

After substitution of (D.18) into (D.14) and taking into account (D.16), (D.19), the predictor (D.14) based on ARX model takes the form

$$\hat{y}(k/k-1,\theta) = \varphi^{T}(k)\theta. \tag{D.20}$$

The predictor output is formed by scalar product of the known regression vector and the parameter vector. From (D.20) and (D.11), we obtain

$$y(k) = \varphi^{T}(k)\theta + e(k). \tag{D.21}$$

Equation (D.21) defines linear regression model whose parameters are determined by simple estimation method, such as linear least square. According to (D.20) and (D.21), the prediction error (D.11) becomes

$$e(k,\theta) = y(k) - \varphi^{T}(k)\theta. \tag{D.22}$$

Thus, the cost function (D.12) for linear regression model (D.21) is

$$J(\theta) = \frac{1}{N} \sum_{k=1}^{N} (y(k) - \varphi^{T}(k)\theta)^{2}, \tag{D.23}$$

which is known as *least square criterion* for linear regression model. The cost function (D.23) is a quadratic function in θ and can be minimized analytically, which gives the least square estimate

$$\hat{\theta} = \arg\min J(\theta) = \left[\frac{1}{N} \sum_{k=1}^{N} \varphi(k)\varphi^{T}(k) \right]^{-1} \frac{1}{N} \sum_{k=1}^{N} \varphi(k)y(k). \tag{D.24}$$

It is important to note that the estimate has unique solution if the matrix

$$R_{\hat{\theta}}(N) = \frac{1}{N} \sum_{k=1}^{N} \varphi(k)\varphi^{T}(k) \tag{D.25}$$

is nonsingular, which is provided by the informative identification experiment. Substituting (D.21) in (D.24), the bias $M\{\hat{\theta}\} - \theta$ of least square estimate is calculated from

$$
\begin{aligned}
M\{\hat{\theta}\} - \theta &= M\left\{\left[\frac{1}{N}\sum_{k=1}^{N}\varphi(k)\varphi^T(k)\right]^{-1}\frac{1}{N}\sum_{k=1}^{N}\varphi(k)(\varphi^T(k)\theta + e(k))\right\} - \theta \\
&= M\left\{\left[\frac{1}{N}\sum_{k=1}^{N}\varphi(k)\varphi^T(k)\right]^{-1}\frac{1}{N}\sum_{k=1}^{N}\varphi(k)e(k)\right\} \\
&= M\left\{R_{\hat{\theta}}^{-1}(N)\frac{1}{N}\sum_{k=1}^{N}\varphi(k)e(k)\right\}.
\end{aligned}
\tag{D.26}
$$

The estimate $\hat{\theta}$ will be unbiased if (D.26) is equal to zero. It can be shown that if the process $e(k)$ is a sequence of independent random variables with zero mean (white noise), than the least square estimate will be unbiased. Now, let consider constraints for output disturbance imposed by the requirement for unbiased estimation in case of ARX model. Assume that the actual plant output is formed by

$$
y(k) = \frac{B(q^{-1})}{A(q^{-1})}u(k) + v(k),
\tag{D.27}
$$

and we model plant dynamic by ARX model (D.17). Then after substituting (D.27) in (D.17), one obtains

$$
e(k) = A(q^{-1})v(k).
\tag{D.28}
$$

It is seen from (D.28) that the process $e(k)$ will be white noise if the plant output disturbance $v(k)$ is formed by

$$
v(k) = \frac{1}{A(q^{-1})}\eta(k),
\tag{D.29}
$$

where $\eta(k)$ is a zero mean white noise. Equation (D.29) imposed requirement for output disturbance which is not realistic from practical point of view.

The least square method applied to the linear regression model has many advantages, such as simplicity of ARX model and achievement of global minima of (D.23). Its main shortcoming is the restriction (D.29). If the plant output disturbance cannot be modeled in accordance with (D.29), then the estimated coefficients will not be equal to the true ones. Nevertheless, the least square method for estimation of ARX model is one of the most used in control practice identification method because of its simplicity and due to the fact that in many cases the bias of estimates is small provided of small noise power.

The main disadvantage of the ARX model (D.15) is the lack of adequate freedom in describing the output disturbance in (D.10). We can introduce more flexibility to that by describing the output disturbance as a moving average of white noise. This gives the so-called ARMAX model

$$y(k) + a_1 y(k-1) + a_2 y(k-2) + \cdots + a_{na} y(k-na)$$
$$= b_1 u(k-1) + b_2 u(k-2) + \cdots + b_{nb} u(k-nb) + e(k)$$
$$+ c_1 e(k-1) + \cdots + c_{nc} e(k-nc). \tag{D.30}$$

Equation (D.30) can be rewritten as

$$A(q^{-1})y(k) = B(q^{-1})u(k) + C(q^{-1})e(k), \tag{D.31}$$

where $C(q^{-1}) = 1 + c_1 q^{-1} + c_2 q^{-2} + \cdots + c_{nc} q^{-nc}$. It is clear that (D.31) corresponds to the description (D.10) with

$$G(q^{-1}, \theta) = \frac{B(q^{-1})}{A(q^{-1})}, H(q^{-1}, \theta) = \frac{C(q^{-1})}{A(q^{-1})}. \tag{D.32}$$

The vector with unknown parameters will be

$$\theta = [a_1, a_2, \ldots, a_{na}, b_1, b_2, \ldots, b_{nb}, c_1, c_2, \ldots, c_{nc}]^T.$$

The predictor for ARMAX model is obtained by inserting (D.32) in (D.14),

$$\hat{y}(k/k-1, \theta) = \frac{B(q^{-1})}{C(q^{-1})} u(k) + \left[I - \frac{A(q^{-1})}{C(q^{-1})} \right] y(k),$$

or

$$C(q^{-1})\hat{y}(k/k-1, \theta) = B(q^{-1})u(k) + [C(q^{-1}) - A(q^{-1})]y(k). \tag{D.33}$$

Adding $(1 - C(q^{-1}))\hat{y}(k/k-1, \theta)$ to the both sides of (D.33), we obtain

$$\hat{y}(k/k-1, \theta) = B(q^{-1})u(k) + [1 - A(q^{-1})]y(k)$$
$$+ [C(q^{-1}) - 1][y(k) - \hat{y}(k/k-1, \theta)]. \tag{D.34}$$

Introducing the vector

$$\varphi(k, \theta) = [-y(k-1), -y(k-2), \ldots, -y(k-na), u(k-1),$$
$$u(k-2), \ldots, u(k-nb), e(k-1, \theta), e(k-2, \theta), \ldots, e(k-nc, \theta)]^T$$

and taking into account (D.22), the predictor can be rewritten as

$$\hat{y}(k/k-1, \theta) = \varphi^T(k, \theta)\theta. \tag{D.35}$$

From (D.11) and (D.35), we obtain

$$y(k) = \varphi^T(k, \theta)\theta + e(k, \theta). \tag{D.36}$$

Equation (D.36) is similar to the linear regression model (D.21), but it is not linear in θ due to the dependence of vector φ on parameters θ. The model (D.36) is known as *pseudolinear regression*. The ARX and ARMAX models are special cases of general polynomial model

$$A(q^{-1})y(k) = \frac{B(q^{-1})}{F(q^{-1})} u(k) + \frac{C(q^{-1})}{D(q^{-1})} e(k). \tag{D.37}$$

This model may give rise to 32 different model sets, depending on which of five polynomials A, B, C, D, and F are used. Very often in practice the dynamics from input u to output y has delay of nk samples, so some of the leading coefficients of $B(q^{-1})$ are zero, that is

$$B(q^{-1}) = b_{nk}q^{-nk} + b_{nk+1}q^{-nk-1} + \cdots + b_{nk+nb-1}q^{-nk-nb+1} = q^{-nk}\bar{B}(q^{-1}).$$

Then, the general model can be rewritten as

$$A(q^{-1})y(k) = \frac{\bar{B}(q^{-1})}{F(q^{-1})}q^{-nk}u(k) + \frac{C(q^{-1})}{D(q^{-1})}e(k). \tag{D.38}$$

The predictor for model (D.38) is

$$\hat{y}(k/k-1,\theta) = \frac{D(q^{-1})\bar{B}(q^{-1})}{C(q^{-1})F(q^{-1})}q^{-nk}u(k) + \left[1 - \frac{D(q^{-1})A(q^{-1})}{C(q^{-1})}\right]y(k). \tag{D.39}$$

If we introduce the vectors

$$\theta = [a_1,\ldots,a_{na},b_{nk},\ldots,b_{nk+nb-1},f_1,\ldots,f_{nf},c_1,\ldots,c_{nc},d_1,\ldots,d_{nd},]^T,$$

$$\varphi(k,\theta) = [-y(k-1),\ldots,-y(k-na),u(k-nk),\ldots,u(k-nk+nb+1),$$

$$-w(k-1,\theta),\ldots,-w(k-nf,\theta),e(k-1,\theta),\ldots,e(k-nc,\theta),$$

$$-v(k-1,\theta),\ldots,-v(k-nd,\theta)]^T, \tag{D.40}$$

where

$$w(k,\theta) = \frac{\bar{B}(q^{-1})q^{-nk}}{F(q^{-1})}u(k),\ v(k,\theta) = A(q^{-1})y(k) - w(k,\theta),\ e(k,\theta) = \frac{D(q^{-1})}{C(q^{-1})}v(k,\theta),$$

it can be shown that the predictor (D.39) takes the form (D.35). The pseudolinear regression in the general case of model (D.38) has the form

$$y(k) = \varphi^T(k,\theta)\theta + e(k,\theta), \tag{D.41}$$

where θ and $\varphi^T(k,\theta)$ are formed by (D.40). The structure (D.38) is too general for most practical purposes. It is simplified by fixing to unity of one or several of polynomials A, B, C, D, and F. Some of the commonly used black-box models, which are special case of (D.38), are summarized in Table D.1.

In case of pseudolinear regression (D.41), the cost function (D.12) is not linear in θ and cannot be minimized by analytical methods. The solution should be found

Table D.1   Some common black-box models

Polynomials used	Name of the model structure
$B(q^{-1})$	FIR (finite impulse response)
$A(q^{-1}), B(q^{-1})$	ARX
$A(q^{-1}), B(q^{-1}), C(q^{-1})$	ARMAX
$B(q^{-1}), F(q^{-1})$	OE (output error)
$B(q^{-1}), F(q^{-1}), C(q^{-1}), D(q^{-1})$	BJ (Box–Jenkins)

by iterative numerical procedures [48, Section 10.2]. The methods for numerical minimization of $J(\theta)$ update the estimate $\hat{\theta}$ by the following equation:

$$\hat{\theta}^{(i+1)} = \hat{\theta}^{(i)} + \alpha f^{(i)}, \tag{D.42}$$

where $f^{(i)}$ is the search direction based on the information about $J(\theta)$ acquired on the previous iterations and α is a positive constant determined so that an appropriate decrease in value of $J(\theta)$ is obtained. Depending on the information supplied by the user to determine $f^{(i)}$, numerical minimization methods can be divided into three groups:

• methods using only values of $J(\theta)$,
• methods using values of $J(\theta)$ as well as its gradient,
• methods using values of $J(\theta)$, its gradient and its Hessian matrix.

The typical member of third group corresponds to Newton algorithms, where the search direction is obtained as

$$f^{(i)} = -[J''(\hat{\theta}^{(i)})]^{-1}J'(\hat{\theta}^{(i)}), \tag{D.43}$$

where $J'(\hat{\theta}^{(i)})$ and $J''(\hat{\theta}^{(i)})$ are the gradient and the Hessian matrix of loss function calculated with the estimates from previous iteration.

The important subclass of second group is *quasi-Newton methods* which are used to estimate the Hessian in order to form innovation term in accordance with (D.42). Algorithms of first group either form gradient estimates by difference approximation and then proceed as quasi-Newton methods or have other specific search patterns.

In MATLAB, the prediction error method is implemented in several functions of System Identification Toolbox such as `armax`, `bj`, and `pem`. The user can select search method for minimization of (D.12) between Gauss–Newton [48, Section 10.2], adaptive Gauss–Newton [204], `lasqnonlin` [205], Levenberg–Marquardt, and steepest gradient algorithm [48, Section 10.2]. By default search method option is `auto` which means that the descent direction is calculated using successively each of above mentioned algorithms at each iteration. The iterations continue until a sufficient reduction of error is achieved.

Analogously to (D.10), the general description of MIMO system is obtained as

$$\begin{aligned} y(k) &= G(q^{-1},\theta)u(k) + v(k), \\ v(k) &= H(q^{-1},\theta)e(k), \end{aligned} \tag{D.44}$$

where the input $u(k)$ is an m-dimensional vector, the output $y(k)$ is p-dimensional vector, the additive output disturbance $v(k)$ is p-dimensional vector, $G(q^{-1},\theta)$ and $H(q^{-1},\theta)$ are proper rational transfer matrices and the p-dimensional vector $e(k)$ is zero mean white noise with covariance matrix Λ and θ is vector with unknown model parameters, which must be estimated from experimental data. The estimate $\hat{\theta}_N$ is obtained by (D.13) where the cost function takes the form

$$J(\theta) = \frac{1}{N}\sum_{k=1}^{N} e^T(k,\theta)e(k,\theta), \tag{D.45}$$

The p-dimensional prediction error is calculated by the following equation:

$$e(k, \theta) = y(k) - \hat{y}(k/k - 1, \theta), \tag{D.46}$$

where $\hat{y}(k/k - 1, \theta)$ is a p-dimensional vector with one-step-ahead predicted outputs. The estimates are determined after parametrization of (D.44) and calculation of $\hat{y}(k/k - 1, \theta)$. The one-step-ahead predictor can be easily generalized for MIMO case after parametrization of (D.44). Most of the regarded SISO polynomial models can be generalized for MIMO systems. The simplest case is a generalization of ARX model (D.15):

$$\begin{aligned} y(k) + A_1 y(k - 1) + A_2 y(k - 2) + \cdots + A_{na} y(k - na) = B_1 u(k - 1) \\ + B_2 u(k - 2) + \cdots + B_{nb} u(k - nb) + e(k), \end{aligned} \tag{D.47}$$

where A_i are $p \times p$ matrices and B_i are $p \times m$ matrices. The element a_i^{js} of matrix A_i gives the relation between the current value of the jth output $y_j(k)$ and the delay by i-samples value of sth output $y_s(k - i)$. The corresponding element b_i^{js} of the matrix B_i gives the relation between the current value of the jth output $y_j(k)$ and the delay by i-samples value of sth input $u_s(k - i)$. After introducing the matrix polynomials

$$\begin{aligned} A(q^{-1}) &= I + A_1 q^{-1} + A_2 q^{-2} + \cdots + A_{na} q^{-na}, \\ B(q^{-1}) &= B_1 q^{-1} + B_2 q^{-2} + \cdots + B_{nb} q^{-nb}, \end{aligned} \tag{D.48}$$

the MIMO ARX model can be rewritten as

$$A(q^{-1}) y(k) = B(q^{-1}) u(k) + e(k), \tag{D.49}$$

which is a special case of the MIMO model (D.44), where $G(q^{-1}, \theta) = A^{-1}(q^{-1}) B(q^{-1})$ and $H(q^{-1}, \theta) = A^{-1}(q^{-1})$. The inverse $A^{-1}(q^{-1})$ of matrix polynomial is calculated in straightforward way, due to the fact that $G(q^{-1}, \theta)$ is a $p \times m$ matrix, whose ijth element is a rational transfer function from input u_j to output y_i. The very important issue in MIMO case is parametrization of (D.47). This means which elements of matrices in (D.47) should be included in parameters θ. If we include all elements of matrices and define the $[na.p + nb.m] \times p$ matrix with unknown parameters θ and $[na.p + nb.m]$ dimensional regression vector $\varphi(k)$ as

$$\theta = [A_1, A_2, \ldots, A_{na}, B_1, B_2, \ldots, B_{nb}]^T, \varphi(k) = \begin{bmatrix} -y(k - 1) \\ \vdots \\ -y(k - na) \\ u(k - 1) \\ \vdots \\ u(k - nb) \end{bmatrix}, \tag{D.50}$$

according to (D.14), the predicted output is obtained as

$$\hat{y}(k/k - 1, \theta) = \theta^T \varphi(k). \tag{D.51}$$

Then, the MIMO analogue of linear regression model is

$$y(k) = \theta^T \varphi(k) + e(k). \tag{D.52}$$

The cost function (D.45) becomes

$$J(\theta) = \frac{1}{N} \sum_{k=1}^{N} (y(k) - \theta^T \varphi(k))^T (y(k) - \theta^T \varphi(k)),$$

with the estimate

$$\hat{\theta}_N = \left[\frac{1}{N} \sum_{k=1}^{N} \varphi(k) \varphi^T(k) \right]^{-1} \frac{1}{N} \sum_{k=1}^{N} \varphi(k) y^T(k). \tag{D.53}$$

Equation (D.52) can be regarded as p different linear regressions written on top of each other. Each of them uses the same regression vector. When additional structure is imposed by parametrization we cannot use model (D.52), since the different outputs will not employ the identical regression vectors. Then, the d-dimensional vector θ' with unknown parameters and a $p \times d$ matrix $\varphi'^T(k)$ must be formed so to represent the MIMO ARX model as

$$y(k) = \varphi'^T(k)\theta' + e(k), \tag{D.54}$$

with

$$\theta' = col(\theta) = \begin{bmatrix} \theta^1 \\ \theta^2 \\ \vdots \\ \theta^p \end{bmatrix}, \varphi'(k) = \varphi(k) \otimes I_p,$$

where θ^i is the ith column of matrix θ, I_p is the p-dimensional identity matrix and symbol $\otimes$ means the Kronecker product. For a regression model (D.54) the cost function (D.45) becomes

$$J(\theta) = \frac{1}{N} \sum_{k=1}^{N} (y(k) - \varphi'^T(k)\theta')^T (y(k) - \varphi'^T(k)\theta'), \tag{D.55}$$

with estimate

$$\hat{\theta}'_N = \left[\frac{1}{N} \sum_{k=1}^{N} \varphi'(k) \varphi'^T(k) \right]^{-1} \frac{1}{N} \sum_{k=1}^{N} \varphi'(k) y(k). \tag{D.56}$$

As can be seen from expressions (D.53) and (D.56), if we use the parametrization (D.51), then to determine the estimates it is sufficient to invert a $[na.p + nb.m] \times [na.p + nb.m]$ dimensional matrix, since to determine the parameters of model (D.54) we should invert a $[na.p + nb.m].p \times [na.p + nb.m].p$ dimensional matrix.

The MIMO analogue of ARMAX model is

$$y(k) + A_1 y(k - 1) + A_2 y(k - 2) + \cdots + A_{na} y(k - na)$$
$$= B_1 u(k - 1) + B_2 u(k - 2) + \cdots + B_{nb} u(k - nb) + C_1 e(k - 1)$$
$$+ C_2 e(k - 2) + \cdots + C_{nc} e(k - nc), \tag{D.57}$$

which is a special case of model (D.44) for $G(q^{-1}, \theta) = A^{-1}(q^{-1})B(q^{-1})$ and

$$H(q^{-1}, \theta) = A^{-1}(q^{-1})C(q^{-1}).$$

If we choose $G(q^{-1}, \theta) = F^{-1}(q^{-1})B(q^{-1})$ and

$$H(q^{-1}, \theta) = D^{-1}(q^{-1})C(q^{-1}),$$

then the MIMO analogue of Box–Jenkins (BJ) model will be obtained. After parametrization of MIMO ARMAX and BJ models, the parameter's estimates can be calculated by prediction error method. In MATLAB, the prediction error method in case of MIMO black-box polynomial models is implemented in the same functions as in SISO case. This means that the user can use functions `arx`, `pem`, `armax`, `bj`, `oe` for estimation of MIMO models. As we mentioned above, the parametrization of MIMO models is important issue which is out of scope of this book and we refer the reader to the references given in the end of the appendix where this topic is described in sufficient details.

As an alternative to (D.10), the relationship between output, input, and noise signals can be written in form of discrete-time linear state-space model

$$\begin{aligned} x(k+1) &= A_d(\theta)x(k) + B_d(\theta)u(k) + \eta(k), \\ y(k) &= C(\theta)x(k) + \zeta(k), \end{aligned} \tag{D.58}$$

where $y(k)$ is a p-dimensional output vector, $u(k)$ is m-dimensional input vector, $x(k)$ is n-dimensional state vector, $\eta(k)$ is process noise acting on the states, $\zeta(k)$ is measurement noise. Significant advantage of state-space model in comparison with polynomial models is that we have to specify only the order n. In (D.58), $\eta(k)$ and $\zeta(k)$ are assumed to be independent random variables with zero mean values and covariances:

$$E\left\{\eta(k)\eta^T(k)\right\} = R_\eta(\theta), E\left\{\zeta(k)\zeta^T(k)\right\} = R_\zeta(\theta), E\left\{\eta(k)\zeta^T(k)\right\} = R_{\eta\zeta}(\theta).$$

θ is a vector that parameterizes the model, including unknown coefficients in system matrices and noise covariances. All elements of model (D.58) can be free or the model can have some specific structure with few parameters in vector θ if we have some a priori information about physical meaning of states and noises. The assumption that the noises are white is not always realistic. In cases of colored noises, the state-space model (D.58) can be used after extra modeling and extension of state vector with additional states. Analogously to the input–output model (D.10), we have to estimate parameters θ from input/output data in such manner that the model (D.58) predicts well the output (model produces small prediction error). The predicted output $\hat{y}(k, \theta)$ for state-space model with Gaussian noises is calculated by the Kalman filter:

$$\begin{aligned} \hat{x}(k+1, \theta) &= A_d(\theta)\hat{x}(k, \theta) + B_d(\theta)u(k) + K(\theta)(y(k) - C(\theta)\hat{x}(k, \theta)) \\ \hat{y}(k, \theta) &= C(\theta)\hat{x}(k, \theta) \end{aligned}, \tag{D.59}$$

where $\hat{x}(k, \theta)$ is the predicted value of the state vector at time instant k. The matrix gain $K(\theta)$ is obtained from system matrices and noise covariances by

$$K(\theta) = [A_d(\theta)\bar{P}(\theta)C^T(\theta) + R_{\eta\zeta}(\theta)][C(\theta)\bar{P}(\theta)C^T(\theta) + R_\zeta(\theta)]^{-1}, \tag{D.60}$$

where $\bar{P}(\theta)$ is evaluated as the positive semidefinite solution of the discrete-time Riccati equation:

$$
\begin{aligned}
\bar{P}(\theta) = {} & A_d(\theta)\bar{P}(\theta)C^T(\theta) + R_\eta(\theta) - [A_d(\theta)\bar{P}(\theta)C^T(\theta) + R_{\eta\varsigma}(\theta)] \\
& \times [C(\theta)\bar{P}(\theta)C^T(\theta) + R_\varsigma(\theta)]^{-1}[A_d(\theta)\bar{P}(\theta)C^T(\theta) + R_{\eta\varsigma}(\theta)]^T
\end{aligned} \quad \text{(D.61)}
$$

The predicted output thus can be formed by the filter

$$
\begin{aligned}
\hat{y}(k,\theta) = {} & C(\theta)[qI - A_d(\theta) + K(\theta)C(\theta)]^{-1}B_d(\theta)u(k) \\
& + C(\theta)[qI - A_d(\theta) + K(\theta)C(\theta)]^{-1}K(\theta)y(k).
\end{aligned} \quad \text{(D.62)}
$$

The positive semidefinite matrix $\bar{P}(\theta)$ is the covariance matrix of the state estimate error

$$
\bar{P}(\theta) = M\left\{(x(k) - \hat{x}(k,\theta))(x(k) - \hat{x}(k,\theta))^T\right\}. \quad \text{(D.63)}
$$

Denoting the output prediction error as

$$
e(k) = y(k) - \hat{y}(k,\theta), \quad \text{(D.64)}
$$

the state-space model (D.58) can be rewritten as

$$
\begin{aligned}
x(k+1) &= A_d(\theta)x(k) + B_d(\theta)u(k) + K(\theta)e(k) \\
y(k) &= C(\theta)x(k) + e(k).
\end{aligned} \quad \text{(D.65)}
$$

This simpler representation is the *innovation form* of the state-space model and has only one disturbance source $e(k)$. Note that both (D.65) and (D.58) lead to the same predictor model (D.59). Equation (D.65) can be regarded as special case of general model (D.10) if

$$
\begin{aligned}
W(q^{-1},\theta) &= C(\theta)[qI - A_d(\theta) + K(\theta)C(\theta)]^{-1}B_d(\theta), \\
H(q^{-1},\theta) &= C(\theta)[qI - A_d(\theta) + K(\theta)C(\theta)]^{-1}K(\theta).
\end{aligned} \quad \text{(D.66)}
$$

The parameters of model (D.65) are estimated by minimization of cost function (D.45) with prediction error (D.64) and predictor (D.59). This minimization is performed by the same iterative algorithms as ones used in case of polynomial models. In MATLAB, the estimation of state-space model in innovation form by the PEM method is performed by System Identification Toolbox function `ssest`. The parameters may be obtained after parametrization of state-space model, which can be performed in various manners depending on a priori information. When we do not have knowledge about internal structure of model, the free parametrization is appropriate to use. Then any elements of matrices A_d, B_d, C, K are adjustable by estimation procedure. The algorithm chooses the state variables basis automatically to obtain well-conditioned calculations and state variables do not have physical meanings. The second possibility is to use canonical parametrization. It represents the state-space model in form with reduced number of parameters since many elements of matrices A_d, B_d, C, K are fixed to zeros and ones. MATLAB function `ssest` supports three types of canonical parameterizations: companion form—the characteristic polynomial appears in right most column of matrix A_d [48, Section 4.3, Example 4.2]; modal decomposition form—the A_d matrix is block diagonal with each block corresponding to a cluster with nearby modes [173]; observable canonical form—the free parameters appear only in the selected rows of A_d matrix, in B_d and K matrices [48, Appendix 4.A]. In case of canonical parametrization, the state variables and parameters again do not have physical meanings. The third possibility is to use *structured parametrization*, in which we exclude specific parameters from estimation procedure by setting them to specific values. This parametrization is useful when we have knowledge for structure of the state-space model, for example, if it is obtained from physical laws.

Alternatively to the PEM method, the parameters of state-space model (D.65) can be estimated by noniterative methods such as the subspace methods. The subspace approach gives very useful algorithms for state-space model estimation which are particularly appropriate for MIMO systems identification. These algorithms allow numerically very reliable implementations and produce models with good quality. If desired, the model quality may be improved by using the obtained model as initial estimate for PEM methods. Detailed description of subspace approach can be found in [48,206,207]. The family of subspace methods for estimation of the state-space model presented in innovation form is implemented in MATLAB System Identification Toolbox function n4sid. It gives possibility to choose the weighting scheme for singular value decomposition among MOESP algorithm, canonical variable algorithm, and SSARX—a subspace identification method that uses an ARX estimation based algorithm to compute the weighting matrices. Using an SSARX weighting scheme allows obtaining unbiased estimates when using data from closed-loop experiment [49]. The maximum prediction horizon and the number of past inputs and past outputs that are used in prediction also can be defined by user or can be chosen by algorithm on the basis of the Akaike information criterion.

After presenting the basic linear models and commonly used methods for their parameters estimation, we may continue with the next issue in identification procedure—model structure selection. This task involves following steps:

- Choosing the type of model, which in generally involves selection between linear and nonlinear models, between black-box state-space or polynomial models and structure parameterized state-space model,
- Choosing a number of candidate models in model set, which involves selection of possible orders of state-space model and possible degrees of polynomials in black-box polynomial models,
- Estimating the model performance by some quality metrics and choosing the best model in set of candidate models.

Note that the best model from given candidate model set may not describe sufficiently "well" plant dynamics; then the designer must redefine the set of candidate model (this can be done by increasing model order or by using candidate models with more complex structure).

Choice of the model type depends on priori knowledge for identified system. If we have such information, it is reasonable to incorporate it in model structure (e.g., we may choose structure parameterized state-space model). In lack of any prior information, it is advisable to use different type black-box models. It is well known that to reduce the bias of estimates a model with more parameters should be used, but the variance of estimates is increased with increasing the parameter numbers. Thus, the best model structure is trade-off between flexibility and parsimony of the model. The general advice for choice of model types in model set is "first try simple model," which means that one should go into sophisticated model structure only if the simple ones do not pass the validation tests. Usually, identification procedure begins with estimation of impulse-response and frequency-response models from input/output data to gain insight into system dynamic. For example, from impulse response model, the assumption for value of time delay can be done, and from frequency response, the assumption for model order can be made. Then, the set of ARX or state-space models is formed. These models are sufficiently simple because they have few design parameters (for state-space model only the order n should be chosen and for ARX model the polynomial degrees na and nb should be chosen). Also, their parameters are estimated by fast noniterative methods, which provide easy estimation of large number of models of various orders. Selection of best model is performed on the basis of model quality metrics. Obvious quality metrics is the mean square error:

$$J(\hat{\theta}) = \frac{1}{N} \sum_{k=1}^{N} e^2(k, \hat{\theta}), \tag{D.67}$$

where $e(k, \hat{\theta}) = y(k) - \varphi^T(k)\hat{\theta}$ is the residual calculation for the estimated model. Thus, the best model in the given model set will be the one that produces smallest value of (D.67). The significant disadvantage of index (D.67) is that it takes into account only the fit between measured and model outputs and does not take into account model order. As a result, the loss function decreases with increasing model order. It continues to decrease after a correct model structure has been reached because the additional parameters adjust themselves to features of the particular realization of the noise. This is known as *overfit* in model, and this extra fit has no significance to us since the estimated model will apply to data with different noise realization. The effect of over fitting can be overcome by modifying quality metrics. The modification is done by introducing additional *penalty* term in (D.67). This term should increase the value of metrics when overfitting is occurred. The most common modified quality metrics are

- Akaike's Information Criterion (AIC):

$$
\begin{aligned}
J_{\text{aic}}(\hat{\theta}, N) &= \ln\left(J\left(1 + \frac{2d}{N}\right)\right), \\
&= \ln(J) + \frac{2d}{N}, N \gg d
\end{aligned} \tag{D.68}
$$

  where d is the number of estimated parameters.
- Rissanen's Minimum Description Length (MDL) criterion:

$$J_{\text{mdl}}(\hat{\theta}, N) = J\left(1 + \frac{d \ln(N)}{N}\right). \tag{D.69}$$

- Akaike's final prediction error

$$\text{FPE} = \det\left(\frac{1}{N} E^T E\right)\left(\frac{1 + \dim \theta/N}{1 - \dim \theta/N}\right). \tag{D.70}$$

The obtained model order by AIC is good compromise between model complexity and data fitting, while the leading aim of MDL criteria is to obtain the lower order model. In MATLAB, the best model among given ARX model set according to (D.67) or (D.68) or (D.69) may be determined by System Identification Toolbox function `selstruc`. Before using function `selstruc`, user forms the model set by function `struc` and evaluates loss function (D.67) by function `arxstruc`.

The best state-space model among given model set is commonly determined by the Hankel singular values of model (see Section 3.5), which measure the contribution of each state to the input/output behavior. Hankel singular values are to the model order what singular values are to matrix rank. In particular, small Hankel singular values determine states that can be discarded to simplify the model. In MATLAB, the best state-space model among given set may be determined by System Identification Toolbox function `n4sid` which plots the Hankel singular values of models.

After choosing the best model in given model set, we must not forget that this model is the best among the finite number of models with specific structure, and it may not explain well system dynamics. Due to that we should perform validation of obtained model.

D.1.3   Model validation

As a result of the parameter estimation procedure, the "best" model from given model set is obtained. The general question then is whether this "best" model is "good" enough. This question has several aspects such as

- Does the model agree sufficiently well with the measure data?
- Is the model good enough for our purpose?

The first question means whether there is a functional closeness between the model and identified system. The second one means whether the obtained model can be used for aim which induces the modeling process. The ultimate validation then is to test whether this aim may be achieved using obtained model. For example, when model is used for controller design, then it will be "valid" if the controller provides control system performance. However, it is impossible, dangerous, or costly to test all possible models in real world. Instead, various tests for model validation are used.

Note that for tests which compare model outputs to measured one and perform residual analysis, two types of dataset must be used: one for estimating the models (estimation data) and other for validating the models (validation data). If the same dataset is used for estimating and validating the model, there is a risk to overfitting data. When model validation process uses independent dataset, this process is called *crossvalidation*.

D.1.3.1   Test consistency of model input–output behavior

For black-box models, our interest is focused on their input–output properties. These are displayed in time domain by transient responses while in frequency domain—by frequency responses. Impulse and step responses plots help us to validate how well a model captures the dynamics. For example, we can estimate an impulse or step response from the data using correlation analysis (nonparametric model) and then compare the correlation analysis result with transient responses of the parametric models. Because nonparametric and parametric models are derived using different algorithms, agreement between these models increases confidence in the parametric model results. It is always a good practice to display transient response of estimated model with confidence interval, which is obtained from variances of estimated parameters. The confidence interval corresponds to the range of response values with a specific probability of being the actual response of the system. If we assume that the estimates have a Gaussian distribution, then the 99% confidence interval corresponds to 2.58 standard deviations.

The frequency response plots help us to validate how well a model captures the dynamics in wide frequency range, which is their main advantage against time response. Analogously, we can estimate a frequency response from the data using spectral analysis (nonparametric model) and then compare the spectral analysis result with frequency response of estimated parametric models. The frequency response agreement between these models increases confidence in the parametric model results. Again we can display frequency response of estimated model with confidence interval. If we choose the estimated model to be a nominal one and transform the confidence interval into model uncertainty, then a model with unstructured uncertainty is obtained. This uncertainty model plays very important role in framework of robust controller design.

Note that the comparison between frequency and time responses of nonparametric models and parametric ones should not be used in presence of feedback (closed-loop identification procedure) because feedback makes nonparametric models unreliable.

D.1.3.2 Test of output signals

This test checks the model's ability to reproduce measured data by comparison of predicted or simulated model output with measured one. This comparison is made by plotting together the measured and predicted or simulated signals and by calculating some numerical fit. The most used numerical fit is

$$FIT = \left(1 - \frac{\|y - \hat{y}\|_2}{\|y - \bar{y}\|_2}\right) \times 100, \tag{D.71}$$

where $\hat{y}$ is a model output and $\bar{y}$ is a mean value of y. Notice that 100% corresponds to a perfect fit, and 0% indicates that the fit is no better than guessing the output to be a constant ($\hat{y} = \bar{y}$). The negative fit is worse than 0% and can occur for the following reasons: the estimation algorithm failed to converge; the minimization of one-step-ahead prediction is performed during the estimation but in validation the simulated output is used; the validation dataset was not preprocessed in the same way as the estimation data.

The plot of measured output and predicted or simulated output shows what features the model is capable to reproduce and what feature is not capture. The discrepancies can be due to noise or model errors but we see the combined effects of these sources.

D.1.3.3   Test of parameters confidence interval

This test checks whether the estimated models have too many parameters. Checking is made by comparison of estimated parameters with their standard deviations. If confidence interval contains zero, this is indication for too large model order. The model order should be reconsidered and we can try to estimate reduced order model. If the estimated standard deviations are all large, this is again indication of too large model order.

D.1.3.4   Pole zero test

This test checks whether we may reduce the order of estimated model. The model poles and zeros with their confidence intervals are plotted. When confidence intervals of a pole–zero pair overlap, this overlap indicates a possible pole–zero cancelation. Then, we try to estimate lower order model. The obtained low-order model should be validated by test of output signal and test of residuals. If the plot indicates pole–zero cancelations, but reducing model order degrades the fit, then the extra poles probably describe noise. In this case, a different model structure that decouples system dynamics and noise should be chosen. For example, ARMAX, OE, or BJ model structures with an A or F polynomial of an order equal to that of the number of uncanceled poles can be tried.

D.1.3.5   Test of residuals

The difference between measured output and estimated model output is called *residual*:

$$e(k) = y(k) - \hat{y}(k, \hat{\theta}).\qquad\qquad(D.72)$$

It is clear that the residuals carry information about the quality of model. If we calculate some basic statistics as

$$e_{\max} = \max_k |e(k)|\,, \bar{e} = \frac{1}{N}\sum_{k=1}^{N} e^2(k),\qquad\qquad(D.73)$$

the intuitive interpretation of (D.73) would then be like this "The model never produce the larger residual than $e_{\max}$ and an average error larger than $\bar{e}$, for all the data we have seen." It is likely that such a bound will hold also for "future data." The value of (D.73) will be limited if the residuals do not depend on particular input signal used in input–output dataset, since the model should work for a range of possible inputs. To check this, it is reasonably to study the covariance between residuals and past inputs:

$$\hat{R}_{eu}(\tau) = \frac{1}{N}\sum_{k=1}^{N} e(k)u(k - \tau).\qquad\qquad(D.74)$$

If this number is small, we have some reasons to believe that (D.73) have relevance when the model is applied to other inputs. Another explanation of importance $\hat{R}_{eu}(\tau)$ to be small is as follows: if dependence between residuals and past inputs exists, then there is some part of output signal which originates from input signal, but it is not captured correctly by the model. This means that the model could be improved.

Similarly, if we find that the autocorrelation of residuals

$$\hat{R}_e(\tau) = \frac{1}{N}\sum_{k=1}^{N} e(k)e(k-\tau) \tag{D.75}$$

is small that is indication of good model estimate. Indeed, it is well known that to obtain unbiased model estimates the residuals should be zero mean white noise. Moreover, if values of (D.75) are not small means that some part of $e(k)$ can be predicted from past data. Thus, we have again some dynamics that is not predicted by output and model can be improved.

The simple way to check the correlation between residuals and past inputs is to investigate whether $\hat{R}_{eu}(\tau), k = 1, 2, \ldots, M$ is normally distributed with zero mean and variance

$$P_{eu} = \sum_{k=-M}^{M} R_e(k)R_u(k),$$

$$R_e(\tau) = \frac{1}{N}\sum_{k=1}^{N-\tau} e(k)e(k-\tau), \tau = 0, 1, 2, \ldots, M, \ 25 < M < N - \dim\theta, \tag{D.76}$$

$$R_u(\tau) = \frac{1}{N}\sum_{k=1}^{N-\tau} u(k)u(k-\tau), \tau = 0, 1, 2, \ldots, M, \ 25 < M < N - \dim\theta.$$

This is equivalent with probability of 99% to

$$\left|\frac{\hat{R}_{eu}(\tau)}{\sqrt{\hat{R}_e(0)\hat{R}_u(0)}}\right| \leq \frac{2.58}{N}, \tau = 1, 2, \ldots, M. \tag{D.77}$$

If (D.77) is not hold, then the hypothesis that $e(k)$ and $u(k-\tau)$ are independent should be rejected.

Similarly, the simple way to check the correlation of residuals with themselves is to investigate whether the normalized correlation function $\hat{R}_e(\tau)/\hat{R}_e(0)$ is normally distributed, which is equivalent with probability of 99% to

$$\left|\frac{\hat{R}_e(\tau)}{\hat{R}_e(0)}\right| \leq \frac{2.58}{N}, \tau = 1, 2, \ldots, M. \tag{D.78}$$

If (D.78) does not hold, then the hypothesis that residuals are "white" should be rejected.

The test checked the dependence between residuals and past inputs is called *independence test* and the test checked autocorrelation of residuals is called *whiteness test*. Passing the whiteness test means that the obtained parameters estimate is unbiased, model is functionally close to identified system and the estimated noise model is good. Passing the independence test means that the model captures whole significant input–output dynamics. In MATLAB, these tests are performed with function `resid`. It plots the autocorrelation functions of residuals and crosscorrelation function between residuals and inputs along with the 99% confidence limits which are the expressions in the right side of (D.77) and (D.78). Such plot gives better insight on the correctness of model structure. For example, if the identified system has time delay of one sample but we assume in model time delay of two samples, then correlation between $u(k - 1)$ and $e(k)$ can be seen.

Another version of independence test is to estimate dynamics between residuals and inputs which means to estimate model of model error. Then the model's frequency response plot along the confidence regions will give picture of what frequency range the model has not captured the system dynamics. Depending on the aims of identification, the model then could be accepted even when (D.77) is violated, because the errors are in frequency range which is out of interest. The MATLAB function `resid` may also estimate a high-order FIR model between residual and inputs:

$$e(k) = \theta^T \varphi(k), \ \varphi(k) = [u(k-1), u(k-2), \ldots, u(k-n)]^T. \tag{D.79}$$

The frequency response plot of (D.79) along with 99% confidence region is shown. The 99% confidence region marks statistically insignificant response. A response falling inside the confidence region in the frequency range of interest indicates a reliable model.

Generally, the good model should pass both the whiteness and the independence tests except in the following cases: For output-error (OE) models, the modeling focus is on dynamics model $G(q^{-1}, \theta)$ and not on the disturbance model $H(q^{-1}, \theta)$. Then, it is significant to have independence of residuals and input and we should pay less attention to the results of the whiteness of residuals; correlation between residuals and input for negative lags is not necessarily an indication of an inaccurate model. When current residuals at sample k affect future input values, there might be feedback in identified system. In the case of feedback, we should concentrate on the positive lags in the independence test.

D.2   Identification of linear gray-box model

The gray-box approach is appropriate when we can derive model structure from physical laws but do not know the values of model parameters and noise model. The main advantage of gray-box models in comparison to black-box ones is that the model structure reveals physical relations between variables and model parameters have physical means. After obtaining model structure, we estimate noise model and physical parameters by identification procedure using input–output dataset. Most often, by gray-box approach, continuous-time models are evaluated because most laws of physics are expressed in continuous time.

Examples for black-box and gray-box identifications are given in Chapter 2.

D.3   Notes and references

The identification of linear system models is considered in many books, among them Isermann and Münchhof [208], Landau and Zito [4], Ljung [48], Overschee and De Moor [206], Verhaegen and Verdult [207]. The methods, presented in Ljung [48], are implemented in the System Identification Toolbox of MATLAB, see [173].

Appendix E
Interfacing IMU with target microcontroller

Analog devices MEMS IMU *ADIS16405* integrates three gyroscopes and three accelerometers orthogonally oriented and also include three magnetometers. A functional diagram of the *ADIS16405* IMU is presented in Figure E.1.

Measurement process is controlled by finite state machines (digital control module in Figure E.1). Measurement results and sensor configuration are located at dedicated memory and can be accessed from external devices with SPI port. SPI stands for Serial Peripheral Interface (synchronous duplex master-slave serial communication). Therefore, angular velocity and acceleration data values are recorded at corresponding memory addresses and will be read by SPI. Interconnection between

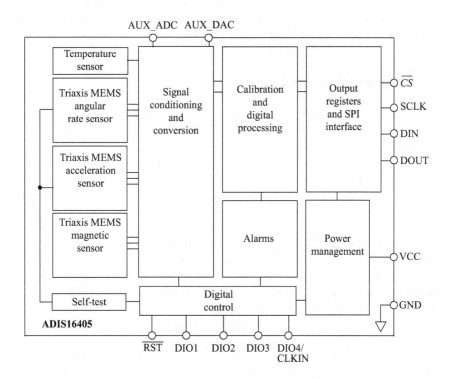

Figure E.1   ADIS16405 functional diagram

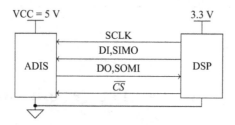

Figure E.2   Interconnection between ADIS16405 and TMS320F28335

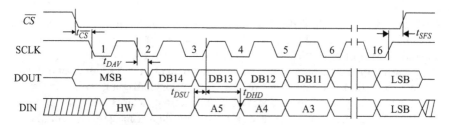

Figure E.3   Communications timings for ADIS16405 SPI

microcontroller and IMU with SPI is presented in Figure E.2. Detailed technical information about *ADIS16405* is available in [58].

E.1   Driving SPI communication

SPI communication is composed of four logical lines. Typical transients are illustrated on Figure E.3. Low logical level of the signal $\overline{CS}$ (Chip select) enables data transfer from the device. When there are several slave devices to a single SPI master, then each slave is on a separate $\overline{CS}$. SCLK signal is a clock signal generated by the master device (*TMS320F28335* in our case). SIMO (slave in master out) is output for the master and input for the slave devices and SOMI (slave out master in) is output for the slave and input for the master device.

DOUT signal on Figure E.3 carries a single 16-b message sent from the sensor to microcontroller (MSB, DB14, ..., DB1, LSB) where MSB stands for Most Significant Bit. The first bit is *valid* after interval $t_{\overline{CS}}$, and every next bit is *valid* after t_{DAV} referenced to falling edge of the clocking signal. To read a single bit of information, one has to store the level of DOUT line within limited time frame, while the data is *valid*. In opposite case with probability $p \geq 0.5$, there will be read erroneous information. On Figure E.3, validity time frame for a single bit is defined by parameters t_{DAV} and $t_{\overline{CS}}$.

The reader can observe analogical situation for the line DI in Figure E.2. It is used by master device to send data to the slave and the first bit signifies the type of operation (read or write), second bit is always zero. There is a requirement data to be set on

Listing E.1   Initialization of microcontroller SPI port

```
1   SpiaRegs.SPICCR.bit.SPISWRESET = 0; // SPI Reset
2   SpiaRegs.SPICCR.bit.SPICHAR = 15;  // Character length control
3   SpiaRegs.SPICCR.bit.SPILBK = 0;    // Configure SPI for normal mode
4
5   // 1(0) = Data is output on falling ( rising ) edge and input on rising ( falling ) edge.
6   SpiaRegs.SPICCR.bit.CLKPOLARITY = 1;
7   SpiaRegs.SPICTL.bit.TALK = 1;          // Transmit enable, bit 1,
8   SpiaRegs.SPICTL.bit.MASTER_SLAVE = 1; // SPI configured as a master.
9   // 1 = Data is output one half-cycle before the first  rising / falling  edge
10  SpiaRegs.SPICTL.bit.CLK_PHASE = 0;
11
12  SpiaRegs.SPIPRI.bit .FREE = 1; // Free run, continue SPI operation regardless of suspend
13  SpiaRegs.SPIBRR = 40;          // Baud rate, bit
14  SpiaRegs.SPIFFTX.bit.SPIFFENA = 1;   // SPI_A FIFO mode
15  SpiaRegs.SPIFFCT.all= 20;            // FIFO transmit delay
16  SpiaRegs.SPICCR.bit.SPISWRESET = 1; // Enable SPI
```

a line for time t_{DSU} before the rising edge of the clock signal and to stay valid on the line for time t_{DHD} after this edge. Obtained timing information concerning SPI sensor port is necessary for programing *SPI communication module* of the microcontroller. The table attached to Figure E.3 shows some parameter's values taken from *ADIS16405* datasheet document [58]. Microcontroller SPI is set up with the program from Listing E.1.

Peripheral devices of C28x microcontrollers are programed with the help of *read* or *write* operations executed on specialized addresses. Peripheral device registers form a separate address space. Each registers controls some aspect of device function. Each peripheral devices user guide can be found in a separate document—[209] for SPI [210] for SC, [211] for ADC and so on. In the file *DSP2833x_Device.h*, there is defined structure SpiaRegs of type struct which names control registers and bitfields [212]. This structure is mapped to peripheral address space during linking stage of code generation process.

ADIS16405 device starts its inertial measurements right after power is supplied. Sample time for data acquisition can be tuned. Its default value is $T_s = 0.0012$ s. At the end of current sample interval, measured data are stored into sensor memory. Listing E.2 defines reading sensor data over SPI. This function can be invoked from *timer interrupt* which occurs periodically or can be invoked when the sensor triggers an interrupt.

Write operation to SPITXBUF field of SpiaRegs structure fills input SPI queue of the microcontroller [209]. The code above writes addresses of the sensor registers which contain inertial measurements. While SPI input queue is not empty, SPI module shifts queued data over output line complying to timing setup. Concurrently to data sending the microcontroller receives messages from SPI_{MISO} line. Received messages are stored in SPIFFRX queue. This queue should be empty in initial state for correct interpretation of the received messages. Sensor measurements are stored as 12 b word [58] which requires application of the 0x3FFF mask on the received data.

Listing E.2   Reading sensor memory over SPI

```
1   #define SUPPLY_OUT 0x0200
2   #define XGYRO_OUT 0x0400
3   ...
4   #define ZACCL_OUT 0x0E00
5   #define STATUS_REG 0x3C00
6
7   void Measure() {
8       unsigned short buf1; int i;
9
10      while (SpiaRegs.SPIFFRX.bit.RXFFST > 0) buf1 = ~SpiaRegs.SPIRXBUF;
11
12      SpiaRegs.SPITXBUF = SUPPLY_OUT; SpiaRegs.SPITXBUF = XGYRO_OUT;
13      SpiaRegs.SPITXBUF = YGYRO_OUT; SpiaRegs.SPITXBUF = ZGYRO_OUT;
14      SpiaRegs.SPITXBUF = XACCL_OUT; SpiaRegs.SPITXBUF = YACCL_OUT;
15      SpiaRegs.SPITXBUF = ZACCL_OUT; SpiaRegs.SPITXBUF = STATUS_REG;
16
17      while (SpiaRegs.SPIFFRX.bit.RXFFST < 7) ; //Wait to read. Flush the TX fifo .
18
19      buf1 = SpiaRegs.SPIRXBUF; //Don't care word
20
21      for (i=0;i<7;i++) {
22          buf1 = (SpiaRegs.SPIRXBUF) & 0x3FFF; //Read and filter next word
23          Data_Host[i] = buf1; // Put in packet for the host (Simulink model)
24      }
25  }
```

E.2   Design of Simulink® interface block

In simulation mode, the inertial sensor block is only a placeholder (Figure E.4). It has defined output ports with appropriate types to allow integration of the sensor data with the rest of the model. The block is introduced as S-function ADIS_Driver which is kind of C program compiled as a MEX file [9]. S-function defines only initialization routine about setup of data type and port configuration and size (Listing E.3).

There is defined single vector output port with seven components which are 16-b unsigned numbers. Another function defines the block sample time as inherited. The S-function is compiled with command mex ADIS_Driver.c, which generates the ADIS_Driver.*mex* file. The interface between Simulink generated C code with sensor SPI driving code there is defined by a TLC file (Listing E.4).

TLC syntax is documented in [213]. ADIS_Driver TLC file defines following model functions:

- BlockTypeSetup—includes adis_lib.h which declares low level sensor driving functions and variables;
- Start—triggers initialization procedure of the microcontroller SPI device;
- Outputs—invoked once in every sample interval of model execution. The *Measure()* function is called and the received data from the sensor stored in the Data_Host structure are assigned to the block outputs.

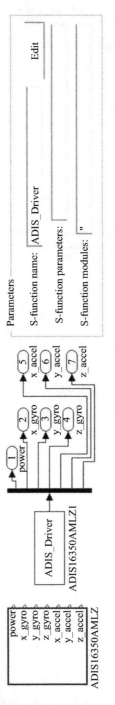

Figure E.4   Interface block for ADIS IMU in Simulink

Listing E.3   S-function definition of the IMU block

```
1   #define S_FUNCTION_NAME ADIS_Driver
2   #define S_FUNCTION_LEVEL 2
3
4   #include "simstruc.h"
5
6   static void mdlInitializeSizes (SimStruct *S)
7   {
8       ...
9       ssSetNumContStates(S, 0);
10      ssSetNumDiscStates(S, 0);
11
12      if (!ssSetNumOutputPorts(S, 1)) return;   // One output port − port 0
13      ssSetOutputPortWidth(S, 0, 7);            // Dimension of the signal on port 0
14      ssSetOutputPortDataType(S, 0, SS_UINT16); // Type of the signal at port 0
15
16      ssSetNumSampleTimes(S, 1);
17      ssSetNumRWork(S, 0); ssSetNumIWork(S, 0); ssSetNumPWork(S, 0);
18      ssSetNumModes(S, 0); ssSetNumNonsampledZCs(S, 0);
19          ...
20  }
21
22  static void mdlInitializeSampleTimes (SimStruct *S)
23  {
24      ssSetSampleTime(S, 0, INHERITED_SAMPLE_TIME);
25      ssSetOffsetTime (S, 0, 0.0);
26  }
```

Listing E.4   TLC interface between Simulink code and Driver code

```
1   %implements ADIS_Driver "C"
2
3   %function BlockTypeSetup(block, system) void
4       %<LibAddToCommonIncludes("adis_lib.h")>
5   %endfunction
6
7   %function Start (block,system) Output
8       init_board_spi_port ();
9   %endfunction
10
11  %function Outputs(block,system) Output
12  /* %<Type> Block: %<Name> (\%<ParamSettings.FunctionName>) */
13  {
14      Measure();
15      %assign y1 = LibBlockOutputSignal(0,"","",0)
16      %<y1> = Data_Host[0];
17      %assign y2 = LibBlockOutputSignal(0,"","",1)
18          ...
19      %assign y7 = LibBlockOutputSignal(0,"","",6)
20      %<y7> = Data_Host[6];
21  }
22  %endfunction
23  ...
```

Listing E.5   ADIS16405 generated code

```
1   /* Model step function */
2   void DSP_Embedded_LQRC2_kalman_psi_step(void)
3   {
4     ...
5     /* S−Function Block: <S1>/ADIS16405AMLZ (ADIS_Driver) */
6     {
7       Measure();
8       DSP_Embedded_LQRC2_kalman_psi_B.ADIS16405AMLZ[0] = Data_Host[0];
9       DSP_Embedded_LQRC2_kalman_psi_B.ADIS16405AMLZ[1] = Data_Host[1];
10      DSP_Embedded_LQRC2_kalman_psi_B.ADIS16405AMLZ[2] = Data_Host[2];
11      DSP_Embedded_LQRC2_kalman_psi_B.ADIS16405AMLZ[3] = Data_Host[3];
12      DSP_Embedded_LQRC2_kalman_psi_B.ADIS16405AMLZ[4] = Data_Host[4];
13      DSP_Embedded_LQRC2_kalman_psi_B.ADIS16405AMLZ[5] = Data_Host[5];
14      DSP_Embedded_LQRC2_kalman_psi_B.ADIS16405AMLZ[6] = Data_Host[6];
15    }
16    ...
17  }
```

Listing E.5 shows the result from code generation about the sensor which was governed by TLC file from Listing E.2. One can see that assigned TLC placeholders had got their actual values. For example DSP...B.ADIS16405AMLZ[1] replaces %<y2> placeholder.

Appendix F
Measuring angular velocity with hall encoder

The rotational Hall encoder converts the angular velocity to electrical signal. Hall elements are semiconductor devices which change its conductivity as a function of applied magnetic field. A magnetic disk is coupled to the shaft of an electric motor. Along the disk boundary, there are evenly distributed magnetic pole pairs (N or S). Close to the magnetic disk, there are mounted both Hall elements which generate a logical signal reflecting current polarity (N or S) generating two shifted square pulse waveforms (Figure F.1). Their frequency ($2/T_{\mathrm{imp}}$) is proportional to the angular velocity and phase shift τ depends on the direction of rotation. Figure F.2 shows the interconnection between rotational encoder and microcontroller *TMS320F28335*. The angular velocity is calculated as

$$\widehat{\omega} = (-1)^s \frac{360}{2T_{\mathrm{imp}} nR} \ (\mathrm{deg/s}), \tag{F.1}$$

where $s \in \{0, 1\}$ is a logical signal for rotation direction, T_{imp} is the pulse width in seconds, $n = \omega_{out}/\omega_{in}$ is the gear ratio, and R is the number of pole pairs of the magnetic disk (defining angular resolution).

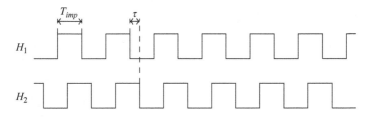

Figure F.1   Constant angular velocity encoder signals. $H_1(t) = H_2(t + \tau)$.

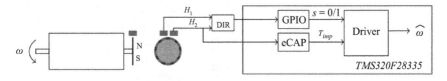

Figure F.2   Experimental setting for measuring angular velocity with Hall encoder

The signal s is generated from external logic. The pulse train H_1 is clock signal for a D trigger and H_2 drives the corresponding D input.

$$s(t) = D(t) \wedge (\text{CLK}(t) \wedge \text{CLK}(t - \delta)) \tag{F.2}$$

$$s(t) = H_2(t) \wedge (H_1(t) \wedge \overline{H}_1(t - \delta)) = H_1(t - \tau) \wedge (H_1(t) \wedge \overline{H}_1(t - \delta)) \tag{F.3}$$

The delay δ is much smaller than phase difference τ. The sign of τ depends on direction of rotation. The signal s feeds the general purpose input output (GPIO) port of the microcontroller.

The duration T_{imp} can be precisely measured with enhanced Capture (eCAP) module of the C28x microcontroller. The eCAP detects series of events in a logical signal and measures the time between them.

$$T_{\text{imp}} = \text{floor}((t_2 - t_1)/T_{clk})T_{clk} = N_{clk}T_{clk},$$

$$H_2(t_1 < t < t_2) = 1, \quad H_2(t_1 - \delta) = 0, \quad H_2(t_2 + \delta) = 0, \tag{F.4}$$

where T_{clk} is the CPU clock signal period ($T_{clk} = 1/150 \times 10^6$). Some other important parameters are—64 pole pairs for the encoder and 29:1 ratio for the gear box.

A simplified diagram of eCAP component from [214] is presented in Figure F.4. The input signal goes to edge detectors for registering the events. Event qualifier block is a 2-b counter which increases on each edge detected. When a new event is detected, then the value of 32-b dedicated counter is written to a corresponding event register. This 32-b counter is reset when detect a rising edge of the monitored signal. Timing of eCAP component is presented in Figure F.5.

Therefore, the Event qualifier bock restarts the counter when rising edge is detected from the encoder and writes the counter value in the register when falling edge is detected. In this way, the information about impulse duration goes into the

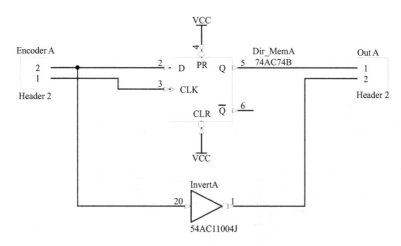

Figure F.3   Direction detection based on H_1 and H_2

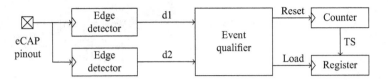

Figure F.4   General structure of the eCAP component

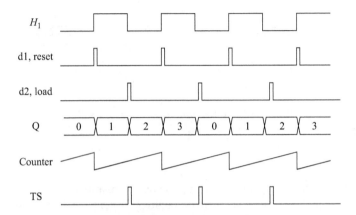

Figure F.5   Timing of eCAP event capture

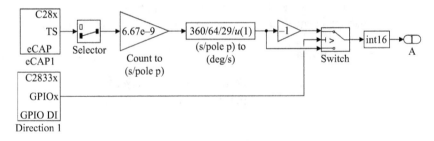

Figure F.6   Simulink® model for code generation

microcontroller. After calculation with (F.4) and (F.1), one can reach the value of angular velocity.

The angular velocity is calculated by *Driver* program according to (F.1) which has T_{imp} and s as input parameters. Simulink Data Flow model of the program is presented on Figure F.6. There are *eCAP* and *GPIO DI* blocks from the C2000 Simulink library.

All measurements $\widehat{\omega} = \omega + \Delta_\omega$ carry some numerical error Δ_ω. To analyze these measurement errors, we build a simulation model of the experimental setup shown in Figure F.7.

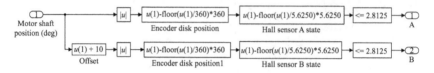

Figure F.7   Simulink model for angular velocity measurement

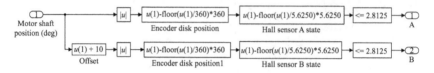

Figure F.8   Model of magnetic disk and Hall elements

The shaft rotation signal goes to the model of rotational encoder containing magnetic disk and Hall elements models (Figure F.8). As a result, H_1 and H_2 waveforms are modeled (A and B in the figure). They go to the cycle-accurate model of the C28x microcontroller (Figure F.9). It models eCAP hardware component and program execution.

Rotational encoder model (Figure F.8) has two data flow channels. The absolute value of the shaft angle equals the angle of magnetic disk rotation in [0 360] deg.

$$\alpha_{disk} = \alpha_{shaft} \textbf{ mod } 360 \tag{F.5}$$

The magnetic disk has 64 pole pairs hence each pair covers $360/64 = 5.625$ deg. Next block calculates the disk angle in the limits of one pole pair.

$$\alpha_{pole} = \alpha_{disk} \textbf{ mod } 5.625 \tag{F.6}$$

The N and S poles of the pair cover $5.625/2 = 2.8125$ deg, respectively. The Hall element generates a logical signal level depending on pole type. So its mathematical descriptions is

$$H(t) = \begin{cases} 1, & \alpha_{pole} \leq 2.8125 \\ 0, & \alpha_{pole} > 2.8125 \end{cases}. \tag{F.7}$$

Second encoder channel is modeled accounting for a time shift in its input signal caused by spatial shift of the Hall elements.

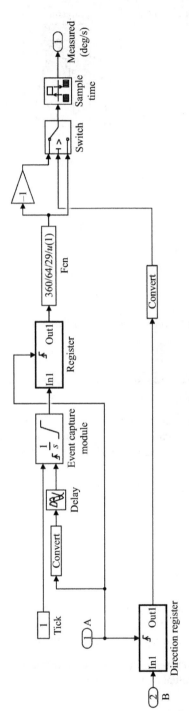

Figure F.9 Microcontroller model

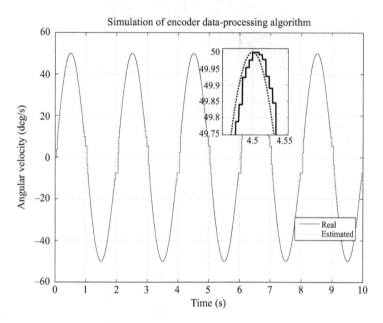

Figure F.10   Simulation of rotation encoder data acquisition by the microcontroller

Microcontroller model of encoder signal processing (Figure F.9) has the following subsystems:

- Model of the direction detector from Figure F.3. The trigger is modeled as a *Triggered Subsystem* block which is left empty (direct throughput). So the output becomes equal to the input on the rising edge of the trigger port.

$$y(t) = \begin{cases} u(t), & \overline{T}(t-\delta) \wedge T(t) \\ y(t-1), & T(t-\delta) \vee \overline{T}(t) \end{cases} \qquad (\text{F.8})$$

$$T(t) = u(t-\tau)$$

$$y(t) = \begin{cases} 1, & \tau < 0 \\ 0, & \tau > 0 \end{cases} \qquad (\text{F.9})$$

- A model of eCAP module organized around integrator with input 1 for modeling of the 32-b counter. The integrator is restarted after delay δ of the encoder signal rising edge. Same rising edge causes write of the integrator value into the register before its restart. The register is modeled as *Triggered Subsystem*.
- A model for program calculating the angular velocity (Figure F.6).

Figure F.10 shows a comparison of angular velocity measurement of the encoder model with the original sinusoidal signal (0.5 Hz).

References

[1] Kostrikin A.I. and Manin Yu I. *Linear Algebra and Geometry*. Overseas Publishers Association, Amsterdam, The Netherlands, 1997. ISBN 90 5699 049 7.

[2] Bishop R.H., editor. *Mechatronic Systems, Sensors and Actuators: Fundamentals and Modeling*. CRC Press, Boca Raton, FL, 2nd edition, 2008. ISBN 978-0-8493-9258-0.

[3] Wittenmark B., Åström K.J., and Årzén K.-E. Computer control: An overview. *IFAC Professional Briefs*, 2002. [Online]. Available at: https://www.ifac-control.org/publications/list-of -professional-briefs.

[4] Landau I.D. and Zito G. *Digital Control Systems: Design, Identification and Implementation*. Springer-Verlag, London, 2006. ISBN 1-84628-055-9.

[5] The MathWorks, Inc., Natick, MA. *Fixed-Point Designer User's Guide*, 2017. [Online]. Available at: http://www.mathworks.com/help/pdf_doc/ fixedpoint/FPTUG.pdf.

[6] IEEE Computer Society. *IEEE Std 754®-2008, IEEE Standard for Floating Point Arithmetic*. New York, NY, 2008. DOI 10.1109/IEEESTD.2008. 5976968.

[7] Higham N.J. *Accuracy and Stability of Numerical Algorithms*. Society for Industrial and Applied Mathematics, Philadelphia, PA, 2nd edition, 2002. ISBN 0-89871-521-0.

[8] Franklin G.F., Powell J.D., and Workman M.L. *Digital Control of Dynamic Systems*. Addison Wesley Longman, Inc., Menlo Park, CA, 3rd edition, 1998. ISBN 0-201-33153-5.

[9] The MathWorks, Inc., Natick, MA. *Simulink Coder User's Guide*, 2017. [Online]. Available at: http://www.mathworks.com/help/pdf_doc/ rtw/rtw_ug.pdf.

[10] Hristu-Varsakelis D. and Levine W.S., editors. *Handbook of Networked and Embedded Control Systems*. Birkhäuser, Boston, 2005. ISBN 978-0-8176-3239-7.

[11] Marwedel P. *Embedded System Design: Embedded Systems Foundations of Cyber-Physical Systems*. Springer, Dordrecht, The Netherlands, 2nd edition, 2011. ISBN 978-94-007-0256-1, DOI 10.1007/978-94-007-0257-8.

[12] Popovici K., Rousseau F., Jerraya A.A., and Wolf M. *Embedded Software Design and Programming of Multiprocessor System-on-Chip*. Springer, New York, NY, 2010. ISBN 978-1-4419-5566-1, DOI 10.1007/ 978-1-4419-5567-8.

[13] Westcot T. *Applied Control Theory for Embedded Systems*. Elsevier, Amsterdam, 2006. ISBN 978-0-7506-7839-1.

[14] Basten T., Hamberg R., Reckers F., and Verriet J., editors. *Model-Based Design of Adaptive Embedded Systems*. Springer, New York, NY, 2013. ISBN 978-1-4614-4820-4, DOI 10.1007/978-1-4614-4821-1.

[15] Ledin J. *Embedded Control Systems in C/C++: An Introduction for Software Developers Using MATLAB*. CRM Books, San Francisco, CA, 2004. ISBN 1578201276.

[16] Lozano R., editor. *Unmanned Aerial Vehicles Embedded Control*. Wiley-ISTE, Chichester, UK, 2010. ISBN 978-1-84821-127-8.

[17] Forrai A. *Embedded Control System Design: A Model Based Approach*. Springer, Berlin, 2013. ISBN 978-3-642-28594-3, DOI 10.1007/978-3-642-28595-0.

[18] Fraden J. *Handbook of Modern Sensors: Physics, Designs, and Applications*. Springer, New York, NY, 4th edition, 2010. ISBN 978-1-4419-6465-6, DOI 10.1007/978-1-4419-6466-3.

[19] Bräunl T. *Embedded Robotics: Mobile Robot Design and Applications with Embedded Systems*. Springer, Berlin, 3rd edition, 2008. ISBN 978-3-540-70533-8.

[20] Isermann R. *Mechatronic Systems: Fundamentals*. Springer-Verlag, London, 2005. ISBN 1-85233-930-6.

[21] Åström K. and Wittenmark B. *Computer-Controlled Systems: Theory and Design*. Dover Publications, Inc., Mineola, NY, 3rd edition, 2008. ISBN 9780486486130.

[22] Isermann R. *Digital Control Systems*, volume 1 of *Fundamentals, Deterministic Control*. Springer-Verlag, Berlin, 2nd edition, 1989. ISBN 978-3-642-86419-3, DOI 10.1007/978-3-642-86417-9.

[23] Isermann R. *Digital Control Systems*, volume 2 of *Stochastic Control, Multivariable Control, Adaptive Control, Applications*. Springer-Verlag, Berlin, 2nd edition, 1989. ISBN 978-3-642-86422-3, DOI 10.1007/978-3-642-86420-9.

[24] Fadali M.S. and Visioli A. *Digital Control Engineering: Analysis and Design*. Elsevier, Amsterdam, 2nd edition, 2013. ISBN 978-0-12-394391-0.

[25] Houpis C.H. and Lamont G.B. *Digital Control Systems: Theory, Hardware, Software*. McGraw-Hill, Inc., New York, NY, 2nd edition, 1992. ISBN 978-0070305007.

[26] Levine W.S., editor. *Control System Fundamentals*. The Control Handbook. CRC Press, Boca Raton, FL, 2nd edition, 2011. ISBN 978-1-4200-7363-8.

[27] Pelgrom M. *Analog-to-Digital Conversion*. Springer, Cham, Switzerland, 3rd edition, 2017. ISBN 978-3-319-44970-8, DOI 10.1007/978-3-319-44971-5.

[28] Moler C.B. *Numerical Computing with MATLAB®*. Society for Industrial and Applied Mathematics, Philadelphia, PA, 2004. ISBN 978-0-89871-660-3, DOI 10.1137/1.9780898717952.

[29] Muller J.-M., Brisebarre N., de Dinechin F., *et al.* *Handbook of Floating-Point Arithmetic*. Birkhäuser, Boston, 2010. ISBN 978-0-8176-4704-9, DOI 10.1007/978-0-8176-4705-6.

[30] Koren I. *Computer Arithmetic Algorithms*. A K Peters, Ltd., Natick, MA, 2nd edition, 2002. ISBN 1-56881-160-8.

[31] Laub A.J. *Computational Matrix Analysis*. Society for Industrial and Applied Mathematics, Philadelphia, PA, 2012. ISBN 978-1-611972-20-7.

[32] Overton M.L. *Numerical Computing with IEEE Floating Point Arithmetic*. Society for Industrial and Applied Mathematics, Philadelphia, PA, 2001. ISBN 0-89871-571-7.

[33] Isermann R. Mechatronic systems—Innovative products with embedded control. *Control Engineering Practice*, 16:14–29, 2008. DOI 10.1016/j.conengprac.2007.03.010.

[34] The MathWorks, Inc., Natick, MA. *Embedded Coder User's Guide*, 2017. [Online]. Available at: http://www.mathworks.com/help/pdf_doc/ecoder/ecoder_ug.pdf.

[35] Hardin D.S., editor. *Design and Verification of Microprocessor Systems for High-Assurance Applications*. Springer, 2010. ISBN 978-1-4419-1538-2, DOI 10.1007/978-1-4419-1539-9.

[36] Marwedel P. and Goossens G., editors. *Code Generation for Embedded Processors*. The Springer International Series in Engineering and Computer Science. Springer, New York, NY, 2002. ISBN 978-1-4613-5983-8, DOI 10.1007/978-1-4615-2323-9.

[37] Mogensen T. *Basics of Compiler Design*. University of Copenhagen, 3rd edition, 2010. [Online]. Available at: http://www.diku.dk/ torbenm/ Basics.

[38] Noergaard T. *Embedded Systems Architecture: A Comprehensive Guide for Engineers and Programmers*. Elsevier, Amsterdam, 2nd edition, 2013. ISBN 978-0-12-382196-6.

[39] Pedroni V.A. *Circuit Design with VHDL*. MIT Press, Cambridge, MA, 2004. ISBN 978-0262162241.

[40] Dubey R. *Introduction to Embedded System Design Using Field Programmable Gate Arrays*. Springer-Verlag, London, 2009. ISBN 978-1-84882-015-9, DOI 10.1007/978-1-84882-016-6.

[41] Kilts S. *Advanced FPGA Design: Architecture, Implementation, and Optimization*. John Wiley & Sons, Inc., Hoboken, NJ, 2007. 978-0470054376.

[42] Sass R. and Schmidt A.G., editors. *Embedded Systems Design with Platform FPGAs: Principles and Practices*. Elsevier, Amsterdam, 2010. ISBN 978-1-4614-4820-4, DOI 10.1007/978-1-4614-4821-1.

[43] Goodwin G.C., Graebe S.F., and Salgado M.E. *Control System Design*. Prentice-Hall International, Inc., Upper Saddle River, NJ, 2001. ISBN 0-13-958653-9.

[44] Lewis F.L., Xie L., and Popa D. *Optimal and Robust Estimation: With an Introduction to Stochastic Control Theory*. CRC Press, Boca Raton, FL, 2nd edition, 2008. ISBN 978-1-4200-0829-6.

[45] Hendricks E., Jannerup O., and Sørensen P.H. *Linear Systems Control: Deterministic and Stochastic Methods*. Springer-Verlag, Berlin, 2008. ISBN 978-3-540-78485-2.

[46]   Van Loan C.F. Computing integrals involving the matrix exponential. *IEEE Transactions on Automatic Control*, AC-23:395–404, 1978. DOI 10.1109/TAC.1978.1101743.

[47]   Kalman R.E. and Bucy R.S. New results in filtering and prediction theory. *Transactions of the ASME, Series D: Journal of Basic Engineering*, 83:95–108, 1961.

[48]   Ljung L. *System Identification: Theory for the User*. Prentice-Hall International, Inc., Englewood Cliffs, NJ, 2nd edition, 1999. ISBN 978-0136566953.

[49]   Jansson M. Subspace identification and ARX modelling. *IFAC Proceedings Volumes*, 36:1585–1590, 2003. https://doi.org/10.1016/S1474-6670(17)34986-8.

[50]   Balas G., Chiang R., Packard A., and Safonov M. *Robust Control Toolbox™ User's Guide*. The Mathworks, Inc., Natick, MA, 2016. [Online]. Available at: http://www.mathworks.com.

[51]   Tóth R., Lovera M., Heuberger P.S.C., Corno M., and Van den Hof P.M.J. On the discretization of linear fractional representations of LPV systems. *IEEE Transactions on Control Systems Technology*, 20:1473–1489, 2012. DOI 10.1109/TCST.2011.2164921.

[52]   Gugercin S., Antoulas A.C., and Zhang H.P. An approach to identification for robust control. *IEEE Transactions on Automatic Control*, 48:1109–1115, 2003. DOI 10.1109/TAC.2003.812821.

[53]   Van den Hof P. Identification of experimental models for control design. In *Proceedings of the 18th Instrumentation and Measurement Technology Conference, (IMTC 2001)*, volume 2, pages 1155–1162, Budapest, Hungary, May 2001. IEEE. DOI 10.1109/IMTC.2001.928260.

[54]   Venkatesh S. Identification of uncertain systems described by Linear Fractional Transformations. In *Proceedings of the 42nd IEEE Conference an Decision and Control*, volume 2, pages 5532–5537, Maui, Hawaii, USA, December 2001. IEEE. DOI 10.1109/CDC.2003.1272518.

[55]   Aggarwal P., Syed Z., Noureldin A., and El-Sheimy N. *MEMS-Based Integrated Navigation*. Artech House, Norwood, MA, 2010. ISBN 978-1-60807-043-5.

[56]   Allan D.W. Statistics of atomic frequency standards. *Proceedings of the IEEE*, 54:221–230, 1966. DOI 10.1109/PROC.1966.4634.

[57]   IEEE Aerospace and Electronic Systems Society. *IEEE Std 952-1997, IEEE Standard Specification Format Guide and Test Procedure for Single-Axis Interferometric Fiber Optic Gyros*. New York, NY, 1998. DOI 10.1109/IEEESTD.1998.86153.

[58]   Analog Devices, Inc., Norwood, MA. *ADIS16405: Triaxial Inertial Sensor with Magnetometer*, 2016. [Online]. Available at: http://www.analog.com.

[59]   de Silva C.W. *Modeling and Control of Engineering Systems*. CRC Press, Boca Raton, FL, 2009. ISBN 978-1-4200-7686-8.

[60]   Egeland O. and Gravdahl J.T. *Modeling and Simulation for Automatic Control*. Marine Cybernetics, Trondheim, Norway, 2002. ISBN 82-92356-01-0.

[61]   Fishwick P.A., editor. *Handbook of Dynamic System Modeling*. Chapman & Hall/CRC, Boca Raton, FL, 2007. ISBN 978-1-4200-1085-5.

[62]   Golnaraghi F. and Kuo B.C. *Automatic Control Systems*. John Wiley & Sons, Inc., Hoboken, NJ, 9th edition, 2009. ISBN 978-0470048962.

[63]   Karnopp D.C., Margolis D.L., and Rosenberg R.C. *System Dynamics: Modeling, Simulation, and Control of Mechatronic Systems*. John Wiley & Sons, Inc., Hoboken, NJ, 5th edition, 2012. ISBN 978-0-470-88908-4.

[64]   Kulakowski B.T., Gardner J.F., and Shearer J.L. *Dynamic Modeling and Control of Engineering Systems*. Cambridge University Press, Cambridge, UK, 3rd edition, 2007. ISBN 978-0-521-86435-0.

[65]   Ljung L. and Glad T. *Modeling of Dynamic Systems*. Prentice-Hall International, Inc., Englewood Cliffs, NJ, 1994. ISBN 0-13-597097-0.

[66]   Spong M.W., Hutchinson S., and Vidyasagar M., editors. *Robot Modeling and Control*. John Wiley & Sons, Inc., New York, NY, 2006. ISBN 978-0-471-64990-8.

[67]   Boubaker O. and Iriarte R., editors. *The Inverted Pendulum in Control Theory and Robotics: From Theory to New Innovations*. The Institution of Engineering and Technology, London, 2017. ISBN 978-1-78561-320-3.

[68]   Ogata K. *Modern Control Engineering*. Prentice-Hall International, Inc., Upper Saddle River, NJ, 5th edition, 2009. ISBN 978-0136156734.

[69]   Franklin G.F., Powell J.D., and Emami-Naeini A. *Feedback Control of Dynamic Systems*. Prentice-Hall International, Inc., Upper Saddle River, NJ, 7th edition, 2014. ISBN 978-0133496598.

[70]   Åström K.J. *Introduction to Stochastic Control Theory*. Academic Press, New York, NY, 1970. ISBN 0-486-445-31-3, Republished by Dover Publications, 2006.

[71]   Bryson Jr, A.E. and Ho Y.C. *Applied Optimal Control: Optimization, Estimation, and Control*. Taylor and Francis, New York, NY, 1975. ISBN 978-0-89116-228-5.

[72]   Maybeck P.S. *Stochastic Models, Estimation and Control*, volume 1. Academic Press, New York, NY, 1979. ISBN 0-12-480701-1.

[73]   Speyer J.L. and Chung W.H. *Stochastic Processes, Estimation, and Control*, volume 17 of *Advances in Design and Control*. Society for Industrial and Applied Mathematics, Philadelphia, PA, 2nd edition, 2008. ISBN 978-0-89871-655-9.

[74]   Kalman R.E. A new approach to linear filtering and prediction problems. *Transactions of the ASME, Series D: Journal of Basic Engineering*, 82:35–45, 1960.

[75]   Anderson B.D.O. and Moore J.B. *Optimal Filtering*. Prentice-Hall International, Inc., Englewood Cliffs, NJ, 1979. ISBN 0-13-638122-7.

[76]   Crassidis J.L. and Junkins J.L. *Optimal Estimation of Dynamic Systems*. CRC Press, Boca Raton, FL, 2nd edition, 2011. ISBN 978-1439839850.

[77]   Gibbs B.P. *Advanced Kalman Filtering, Least-Squares and Modeling: A Practical Handbook*. John Wiley & Sons, Inc., Hoboken, NJ, 2011. ISBN 978-0-470-52970-6.

[78]   Simon D. *Optimal State Estimation: Kalman, H_∞, and Nonlinear Approaches*. John Wiley & Sons, Inc., Hoboken, NJ, 2006. ISBN 978-0-471-70858-2.

[79]   Grewal M.S. and Andrews A.P. *Kalman Filtering: Theory and Practice with MATLAB®*. John Wiley & Sons, Inc., New York, NY, 4th edition, 2014. ISBN 978-1118851210.

[80]   Levine W.S., editor. *Control System Advanced Methods*. The Control Handbook. CRC Press, Boca Raton, FL, 2nd edition, 2011. ISBN 978-1-4200-7364-5.

[81]   Bittanti S. and Garatti S. System identification and control: A fruitful cooperation over half a century, and more. In *Proceedings of the 31st Chinese Control Conference*, pages 6–15, Hefei, China, July 2012. IEEE. ISBN 978-1-4673-2581-3.

[82]   Zhou K., Doyle J.C., and Glover K. *Robust and Optimal Control*. Prentice-Hall International, Inc., Upper Saddle River, NJ, 1996. ISBN 0-13-456567-3.

[83]   Groves P.D. *Principles of GNSS, Inertial, and Multisensor Integrated Navigation Systems*. Artech House, Boston, 2014. ISBN 978-1-58053-255-6.

[84]   Titterton D.H. and Weston J.L. *Strapdown Inertial Navigation Technology*. The Institution of Electrical Engineers, Stevenage, Hertfordshire, UK, 2nd edition, 2004. ISBN 0 86341 358 7.

[85]   Farrell J.A. *Aided Navigation: GPS with High Rate Sensors*. McGraw Hill, New York, NY, 2008. ISBN 978-0-07-164266-8.

[86]   Siouris G.M. *An Engineering Approach to Optimal Control and Estimation Theory*. John Wiley & Sons, Inc., New York, NY, 1996. ISBN 978-0471121268.

[87]   Skogestad S. and Postlethwaite I. *Multivariable Feedback Control: Analysis and Design*. John Wiley and Sons Ltd., Chichester, England, 2nd edition, 2005. ISBN 0-470-01167-X.

[88]   Freudenberg J.S. and Looze D.P. Right half plane poles and zeros and design tradeoffs in feedback systems. *IEEE Transactions on Automatic Control*, AC-30:555–565, 1985. DOI 10.1109/TAC.1985.1104004.

[89]   Åström K. and Murray R.M. *Feedback Systems: An Introduction for Scientists and Engineers*. Princeton University Press, Woodstock, Oxfordshire, 2008. ISBN 978-0-691-13576-2.

[90]   Dorf R.C. and Bishop R.H. *Feedback Control Theory*. Pearson Education, Inc., Upper Saddle River, NJ, 2010. ISBN 978-0136024583.

[91]   Doyle J.C., Francis B.A., and Tannenbaum A.R. *Feedback Control Theory*. Macmillan Publishing Company, New York, NY, 1992. ISBN 0-02-330011-6.

[92]   Helton J.W. and Merino O. *Classical Control Using H^∞ Methods: Theory, Optimization, and Design*. Society for Industrial and Applied Mathematics, Philadelphia, PA, 1998. ISBN 0-89871-419-2.

[93]   Freudenberg J.S. and Looze D.P. *Frequency Domain Properties of Scalar and Multivariable Feedback Systems*. Springer-Verlag, Berlin, 1988. ISBN 978-3-540-18869-8.

[94] Leong Y.P. and Doyle J.C. Effects of delays, poles, and zeros on time domain waterbed tradeoffs and oscillations. *IEEE Control Systems Letters*, 1:122–127, 2017. DOI 10.1109/LCSYS.2017.2710327.

[95] Lurie B.J. and Enright P.J. *Classical Feedback Control with MATLAB*. CRC Press, Boca Raton, FL, 2nd edition, 2012. ISBN 9781439860175.

[96] Åström K. Limitations on control system performance. *European Journal of Control*, 6:2–20, 2000. DOI 10.1016/S0947-3580(00)70906-X.

[97] Seron M.M., Braslavsky J.H., and Goodwin G.C. *Fundamental Limitations in Filtering and Control*. Springer-Verlag, London, 1997. ISBN 3-540-76126-8.

[98] Callier F.M. and Desoer C.A. *Linear System Theory*. Springer-Verlag, New York, NY, 1991. ISBN 0-387-97573-X.

[99] Green M. and Limebeer D.J.N. *Linear Robust Control*. Dover Publications, Mineola, NY, 2012. ISBN 978-0486488363.

[100] Maciejowski J.M. *Multivariable Feedback Design*. Addison-Wesley, Wokingham, England, 1989. ISBN 0-201-18243-2.

[101] Morari M. and Zafiriou E. *Robust Process Control*. Prentice-Hall International, Inc., Englewood Cliffs, NJ, 1989. ISBN 978-0137821532.

[102] Zhou K. and Doyle J.C. *Essentials of Robust Control*. Prentice-Hall International, Inc., Upper Saddle River, NJ, 1998. ISBN 0-13-790874-1.

[103] Sánchez-Peña R.S. and Sznaier M. *Robust Systems: Theory and Applications*. John Wiley & Sons, Inc., New York, NY, 1998. ISBN 0-471-17627-3.

[104] Dullerud G.E. and Paganini F. *A Course in Robust Control Theory: A Convex Approach*. Springer Science, New York, NY, 2000. ISBN 978-1-4419-3189-4.

[105] Braatz R.D., Young P.M., Doyle J.C., and Morari M. Computational complexity of μ calculations. *IEEE Transactions on Automatic Control*, 39:1000–1002, 1994. DOI 10.1109/9.284879.

[106] Fan M.K.H., Tits A.L., and Doyle J.C. Robustness in the presence of mixed parametric uncertainty and unmodeled dynamics. *IEEE Transactions on Automatic Control*, 36:25–38, 1991. DOI 10.1109/9.62265.

[107] Young P.M., Newlin M.P., and Doyle J.C. Computing bounds for the mixed μ problem. *International Journal of Robust and Nonlinear Control*, 5:573–590, 1995. DOI 10.1002/rnc.4590050604.

[108] Young P.M., Newlin M.P., and Doyle J.C. Let's get real. In Francis B.A. and Khargonekar P.P., editors, *Robust Control Theory*, volume 66 of *The IMA Volumes in Mathematics and its Applications*, pages 143–173. Springer-Verlag, New York, NY, 1995. ISBN 978-1-4613-8451-9.

[109] Roos C. and Beanic J.-M. A detailed comparative analysis of all practical algorithms to compute lower bounds on the structured singular value. *Control Engineering Practice*, 44:219–230, 2015. http://dx.doi.org/10.1016/j.conengprac.2015.06.006.

[110] Gu D.-W., Petkov P.Hr., and Konstantinov M.M. *Robust Control Design with MATLAB®*. Springer-Verlag, London, 2nd edition, 2013. ISBN 978-1-4471-4681-0.

494   *Design of embedded robust control systems using MATLAB® /Simulink®*

[111]   Anderson B.D.O. and Moore J.B. *Optimal Control: Linear Quadratic Methods*. Prentice-Hall International, Inc., Englewood Cliffs, NJ, 1989. ISBN 0-13-638651-2.

[112]   Gahinet P., Nemirovski A., Laub A.J., and Chilali M. *LMI Control Toolbox User's Guide*. The MathWorks, Inc., Natick, MA, 1995.

[113]   Glover K. and Doyle J.C. State-space formulae for all stabilizing controllers that satisfy an H_∞ norm bound and relations to risk sensitivity. *Systems & Control Letters*, 11:167–172, 1988. DOI 10.1016/0167-6911(88)90055-2.

[114]   Gahinet P. and Apkarian P. A linear matrix inequality approach to H_∞ control. *International Journal of Robust and Nonlinear Control*, 4:421–448, 1994. DOI 10.1002/rnc.4590040403.

[115]   Åström K. and Hägglund T. *PID Controllers*. Instrument Society of America, Research Triangle Park, NC, 2nd edition, 1995. ISBN 1-55617-516-7.

[116]   Johnson M.A. and Moradi M.H., editors. *PID Control: New Identification and Design Methods*. Springer-Verlag, London, 2005. ISBN 978-1-85233-702-5.

[117]   Visioli A. *Practical PID Control*. Springer-Verlag, London, 2006. ISBN 978-1-84628-585-1.

[118]   O'Dwyer A. *PI and PID Controller Tuning Rules*. Imperial College Press, London, 3rd edition, 2009. ISBN 978-1-84816-242-6.

[119]   Ang K.H., Chong G., and Li Y. PID control system analysis, design, and technology. *IEEE Transactions on Control Systems Technology*, 13:559–576, 2005. DOI 10.1109/TCST.2005.847331.

[120]   Wang Q.-G., Ye Z., Cai W.-J., and Hang C.-C. *PID Control for Multivariable Processes*. Springer-Verlag, Berlin, 2008. ISBN 978-3-540-78481-4, DOI 10.1007/978-3-540-78482-1.

[121]   Trimeche A., Sakly A., Mtibaa A., and Benrejeb M. PID controller using FPGA technology. In Yurkevich V.D., editor, *Advances in PID Control*. InTech, Rijeka, Croatia, 2011. ISBN 978-953-307-267-8.

[122]   Vilanova R. and Visioli A., editors. *PID Control in the Third Millennium: Lessons Learned and New Approaches*. Springer-Verlag, London, 2012. ISBN 978-1-4471-2424-5, DOI 10.1007/978-1-4471-2425-2.

[123]   Kalman R.E. Contributions to the theory of optimal control. *Boletin de la Sociedad Matemática Mexicana*, 5:102–119, 1960.

[124]   Kwakernaak H. and Sivan R. *Linear Optimal Control Systems*. John Wiley & Sons, Inc., New York, NY, 1972. ISBN 0-471-51110-2.

[125]   Doyle J.C. and Stein G. Multivariable feedback design: Concepts for a classical/modern synthesis. *IEEE Transactions on Automatic Control*, AC-26:4–16, 1981. DOI 10.1109/TAC.1981.1102555.

[126]   Stein G. and Athans M. The LQG/LTR procedure for multivariable feedback control design. *IEEE Transactions on Automatic Control*, AC-32:105–114, 1987. DOI 10.1109/TAC.1987.1104550.

[127]   McFarlane D. and Glover K. A loop shaping design procedure using $\mathcal{H}_\infty$ synthesis. *IEEE Transactions on Automatic Control*, AC-37:749–769, 1992. DOI 10.1109/9.256330.

[128] Boyd S., El Ghaoui L., Feron E., and Balakrishnan V. *Linear Matrix Inequality in Systems and Control Theory*, volume 15 of *Studies in Applied and Numerical Mathematics*. Society for Industrial and Applied Mathematics, Philadelphia, PA, 1994. DOI 10.1137/1.9781611970777.

[129] El Ghaoui L. and Niculescu S.-l., editors. *Advances in Linear Matrix Inequality Methods in Control*, volume 2 of *Advances in Design and Control*. Society for Industrial and Applied Mathematics, Philadelphia, PA, 2000. ISBN 0-89871-438-9.

[130] Nesterov Y. and Nemirovskii A. *Interior-Point Polynomial Algorithms in Convex Programming*, volume 13 of *Studies in Applied and Numerical Mathematics*. Society for Industrial and Applied Mathematics, Philadelphia, PA, 1994. ISBN 0-89871-319-6.

[131] Gahinet P. Reliable computation of H_∞ central controllers near the optimum. In *Proceedings of the 1992 American Control Conference*, pages 738–742, Chicago, Illinois, June 1992. ISBN 0-7803-0210-9.

[132] Gahinet P. and Laub A.J. Numerically reliable computation of optimal performance in singular H_∞ control. *SIAM Journal on Control and Optimization*, 35:1690–1710, 1997. DOI 10.1137/S0363012994269958.

[133] Stoorvogel A.A. Numerical problems in robust and H_∞ optimal control. Technical Report 1999-13, NICONET Report, 1999. Available at http://www.icm.tu-bs.de/NICONET/.

[134] Lucas Nülle GmbH, Kerpen-Sindorf, Germany. *Compact Level Control Kit Including Vessel, Tank, Pump and Sensors*, 2017. [Online]. Available at: https://www.lucas-nuelle.com/317/pid/6708/apg/3353/Compact-level-control-kit-including-vessel,-tank, -pump-and-sensors–.htm.

[135] Arduino. *Arduino MEGA 2560 & Genuino MEGA 2560*, 2017. [Online]. Available at: https://www.arduino.cc/en/Main/arduinoBoardMega2560/.

[136] The MathWorks, Inc., Natick, MA. *Simulink Support Package for Arduino Hardware*, 2017. [Online]. Available at: http://www.mathworks.com/hardware-support/ arduino-simulink.html.

[137] Arduino. *Arduino IDE*, 2017. [Online]. Available at: https://www.arduino.cc/en/Main/Software.

[138] Slavov Ts. and Puleva T. Multitank system water level control based on pole placement controller and adaptive disturbance rejection. In *Preprints of the IFAC Workshop Dynamics and Control in Agriculture and Food Processing*, pages 219–224, Plovdiv, Bulgaria, June 2012. ISBN 978-954-9641-54-7.

[139] Gavrilets V., Mettler B., and Feron E. Dynamic model for a miniature aerobatic helicopter. Technical Report MIT-LIDS Report LIDS-P-2580, Massachusetts Institute of Technology, Cambridge, MA, 2003.

[140] Mollov L., Kralev J., Slavov T., and Petkov P. μ-Synthesis and hardware-in-the-loop simulation of miniature helicopter control system. *Journal of Intelligent and Robotic Systems*, 76:315–351, 2014. DOI 10.1007/s10846-014-0033-x.

[141] Padfield G.D. *Helicopter Flight Dynamics: The Theory and Application of Flying Qualities and Simulation Modelling*. Blackwell Publishing Ltd., Oxford, 2nd edition, 2007. ISBN 978-1-4051-1817-0.

[142] Gavrilets V. *Autonomous Aerobatic Maneuvering of Miniature Helicopters*. PhD thesis, Department of Aeronautics and Astronautics, Massachusetts Institute of Technology, Cambridge, MA, 2003.

[143] Mettler B. *Identification Modeling and Characteristics of Miniature Rotorcraft*. Kluwer Academic Publishers, Boston, 2003. ISBN 978-1-4419-5311-7, DOI 10.1007/978-1-4757-3785-1.

[144] Dadkhah N. and Mettler B. Control system design and evaluation for robust autonomous rotorcraft guidance. *Control Engineering Practice*, 21:1488–1506, 2013. DOI 10.1016/j.conengprac.2013.04.011.

[145] Boubdallah S. and Siegwart R. Design and control of a miniature quadrotor. In Valavanis K.P., editor, *Advances in Unmanned Aerial Vehicles. State of the Art and the Road to Autonomy*, chapter 6, pages 171–210. Springer, Dordrecht, The Netherlands, 2007. ISBN 978-1-4020-6113-4.

[146] Moorhouse D.J. and Woodcock R.J. Background information and user guide for mil-f-8785c, military specification—flying qualities of piloted airplanes. Technical Report AFWAL-TR-81-3109, Wright-Patterson Air Force Base, Ohio, 1982.

[147] McFarland R.E. A standard kinematic model for flight simulation at NASA-AMES. Technical Report NASA CR-2497, Computer Sciences Corporation, Mountain View, CA, 1975.

[148] Barr N.M., Gangsaas D., and Schaeffer D.R. Wind models for flight simulator certification of landing and approach guidance and control systems. Technical Report FAA-RD-74-206, Boeing Commercial Airplane Company, Seattle, WA, 1974.

[149] Waslander S. and Wang C. Wind disturbance estimation and rejection for quadrotor position control. In *Proceedings of AIAA Infotech&Aerospace Conference*, pages 1–14, Seattle, Washington, April 2009. DOI 10.2514/6.2009-1983.

[150] Castillo P., Lozano R., and Dzul A.E. *Modelling and Control of Mini-Flying Machines*. Springer-Verlag, London, 2005. ISBN 1-85233-957-8.

[151] Castillo-Effen M., Castillo C., Moreno W., and Valavanis K.P. Control fundamentals of small miniature helicopters—a survey. In Valavanis K.P., editor, *Advances in Unmanned Aerial Vehicles. State of the Art and the Road to Autonomy*, chapter 4, pages 73–118. Springer, Dordrecht, The Netherlands, 2007. ISBN 978-1-4020-6113-4.

[152] Nonami K., Kendoul F., Suzuki S., Wang W., and Nakazawa D. *Autonomous Flying Robots. Unmanned Aerial Vehicles and Micro Aerial Vehicles*. Springer, Tokyo, 2010. ISBN 978-4-431-53855-4, DOI 10.1007/978-4-431-53856-1.

[153] Raptis I.A. and Valavanis K.P. *Linear and Nonlinear Control of Small-Scale Unmanned Helicopters*. Springer Science+Business Media, Dordrecht, The Netherlands, 2011. ISBN 978-94-007-0022-2, DOI 10.1007/978-94-007-0023-9.

[154] Sandino L.A., Bejar M., and Ollero A. A survey on methods for elaborated modeling of the mechanics of a small-size helicopter. Analysis and comparison. *Journal of Intelligent and Robotic Systems*, 72:219–238, 2013. DOI 10.1007/s10846-013-9821-y.

[155] Bramwell A.R.S., Done G., and Balmford D. *Bramwell's Helicopter Dynamics*. Butterworth-Heinemann, Oxford, 2nd edition, 2001. ISBN 978-0-7506-5075-5.

[156] Wang X. and Zhao X. A practical survey on the flight control system on small-scale unmanned helicopter. In *Proceedings of the 7th World Congress on Intelligent Control and Automation (WCICA 2008)*, pages 364–369, Chongqing, China, June 2008. DOI 10.1109/WCICA.2008. 4592952.

[157] La Civita M., Papageorgiou G., Messner W.C., and Kanade T. Design and flight testing of a high-bandwidth $\mathcal{H}_\infty$ loop shaping controller for a robotic helicopter. *Journal of Guidance, Control, and Dynamics*, 29:485–494, 2006.

[158] Boukhnifer M., Chaibet A., and Larouci C. H-infinity robust control of 3-DOF helicopter. In *9th International Multi-Conference on Systems, Signals and Devices (SSD-12)*, pages 1–6, Chemnitz, Germany, March 2012. DOI 10.1109/SSD.2012.6198011.

[159] Poslethwaite I., Prempain E., Turkoglu E., Turner M.C., Ellis K., and Gubbels A.W. Design and flight testing of various $\mathcal{H}_\infty$ controllers for the Bell 205 helicopter. *Control Engineering Practice*, 13:383–398, 2005. DOI 10.1016/j.conengprac.2003.12.008.

[160] Kureemun R., Walker D.J., Manimala B., and Voskuijl M. Helicopter flight control law design using H_∞ techniques. In *Proceedings of the 44th IEEE Conference on Decision and Control, and the European Control Conference*, pages 1307–1312, Seville, Spain, December 2005. DOI 10.1109/ CDC.2005.1582339.

[161] Cai G., Chen B.M., Dong X., and Lee T.H. Design and implementation of robust and nonlinear flight control system for an unmanned helicopter. *Mechatronics*, 21(5):803–820, 2011. DOI 10.1016/j.mechatronics. 2011.02.002.

[162] Shim H. *Hierarchical Flight Control System Synthesis for Rotorcraft-based Unmanned Aerial Vehicles*. PhD thesis, University of California, Berkeley, CA, 2000.

[163] Yuan W. and Katupitiya J. Design of a μ-synthesis controller to stabilize an unmanned helicopter. In *CD-ROM Proceedings of the 28th Congress of the International Council of the Aeronautical Sciences*, pages 1–9, Brisbane, Australia, September 2012. Paper number ICAS 2012-11.5.2.

[164] Leith D.J. and Leithead W.E. Survey of gain-scheduling analysis and design. *International Journal of Control*, 73:1001–1025, 2000. DOI 10.1080/002071700411304.

[165] Kralev J., Slavov Ts., and Petkov P. Design and experimental evaluation of robust controllers for a two-wheeled robot. *International Journal of Control*, 89:2201–2226, 2016. DOI 10.1080/00207179.2016.1151940.

[166]   Chan R.P.M., Stol K.A., and Halkyard C.R. Review of modelling and control of two-wheeled robots. *Annual Reviews in Control*, 37:89–103, 2013. DOI 10.1016/j.arcontrol.2013.03.004.

[167]   Hu J.-S. and Tsai M.-C. Design of robust stabilization and fault diagnosis for an auto-balancing two-wheeled cart. *Advanced Robotics*, 22:319–338, 2008. DOI 10.1163/156855308X292600.

[168]   Muhammad M., Buyamin S., Ahmad M.N., and Nawawi S.W. Dynamic modeling and analysis of a two-wheeled inverted pendulum robot. In *Third International Conference on Computational Intelligence, Modelling and Simulation CIMSiM 2011*, pages 159–164, Langkawi, Malaysia, September 2011. IEEE. DOI 10.1109/CIMSim.2011.36.

[169]   Vermeiren L., Dequidt A., Guerra T.M., Rago-Tirmant H., and Parent M. Modeling, control and experimental verification on a two-wheeled vehicle with free inclination: An urban transportation system. *Control Engineering Practice*, 19:744–756, 2011. DOI 10.1016/j.conengprac.2011.04.002.

[170]   Zhuang Y., Hu Z., and Yao Y. Two-wheeled self-balancing robot dynamic model and controller design. In *Proceeding of the 11th World Congress on Intelligent Control and Automation*, pages 1935–1939, Shenyang, China, 2014. IEEE. DOI 10.1109/WCICA.2014.7053016.

[171]   Jahaya J., Nawawi S.W., and Ibrahim Z. Multi input single output closed loop identification of two wheel inverted pendulum mobile robot. In *2011 IEEE Student Conference on Research and Development (SCOReD)*, pages 138–143, Cyberjaya, Malaysia, December 2011. IEEE. DOI 10.1109/SCOReD.2011.6148723.

[172]   Alarfaj M. and Kantor G. Centrifugal force compensation of a two-wheeled balancing robot. In *11th International Conference on Control, Automation, Robotic and Vision (ICARCV)*, pages 2333–2338, Singapore, December 2010. IEEE. DOI 10.1109/ICARCV.2010.5707337.

[173]   Ljung L. *System Identification Toolbox™ User's Guide*. The MathWorks Inc., Natick, MA, 2016. [Online]. Available at: http://www.mathworks.com.

[174]   Spectrum Digital, Inc., Stafford, TX. *eZdspTMF28335 Technical Reference*, 2007. [Online]. Available from: http://c2000.spectrumdigital.com/ezf28335/docs/ezdspf28335c_techref.pdf.

[175]   Inc. Segway. *Segway Personal Transporters*. Bedford, NH, 2012. http://www.segway.com.

[176]   Yamamoto Y. *NXTway-GS (Self-Balancing Two-Wheeled Robot) controller design)*, 2009. [Online]. Available at: http://www.mathworks.com/matlabcentral/fileexchange/19147-nxtway-gs-self-balancing-two-wheeled-robot-control ler- design.

[177]   Double Robotics. *Double*. Sunnyvale, CA, 2014. http://www.double robotics.com.

[178]   Lee H. and Jung S. Balancing and navigation control of a mobile inverted pendulum robot using sensor fusion of low cost sensors. *Mechatronics*, 22:95–105, 2000. DOI 10.1016/j.mechatronics.2011.11.011.

[179] Ren T.-J., Chen T.-C., and Chen C.-J. Motion control for a two-wheeled vehicle using a self-tuning PID controller. *Control Engineering Practice*, 16:365–375, 2008. DOI 10.1016/j.conengprac.2007.05.007.

[180] Qiu C. and Huang Y. The design of fuzzy adaptive PID controller of two-wheeled self-balancing robot. *International Journal of Information and Electronics Engineering*, 5:193–197, 2015. DOI 10.7763/IJIEE.2015.V5.529.

[181] Fang J. The LQR controller design of two-wheeled self-balancing robot based on the particle swarm optimization algorithm. *Mathematical Problems in Engineering*, 2014: p. 6, 2014. Article ID 729095, DOI 10.1155/2014/729095.

[182] Lupián L.F. and Avila R. Stabilization of a wheeled inverted pendulum by a continuous-time infinite-horizon LQG optimal controller. In *IEEE Latin American Robotic Symposium LARS'2008*, pages 65–70, Natal, Rio Grande do Norte, Brazil, October 2008. IEEE. DOI 10.1109/LARS.2008.33.

[183] Sun C., Lu T., and Yuan K. Balance control of two-wheeled self-balancing robot based on linear quadratic regulator and neural network. In *Fourth International Conference on Intelligent Control and Information Processing (ICICIP)*, pages 862–867, Beijing, China, June 2013. IEEE. DOI 10.1109/ICICIP.2013.6568193.

[184] Muralidharan V. and Mahindrakar A.D. Position stabilization and waypoint tracking control of mobile inverted pendulum robot. *IEEE Transactions on Control Systems Technology*, 22:2360–2367, 1993. DOI 10.1109/TCST.2014.2300171.

[185] Nawawi S.W., Ahmad M.N., and Osman J.H.S. Real-time control system for a two-wheeled inverted pendulum mobile robot. In Fürstner I., editor, *Advanced Knowledge Application in Practice*, chapter 16, pages 299–312. Sciyo, Rijeka, Croatia, 2010. ISBN 978-953-307-141-1.

[186] Azimi M.M. and Koofigar H.R. Model predictive control for a two wheeled self balancing robot. In *Proceeding of the 2013 RSI/ISM International Conference on Robotics and Mechatronics*, pages 152–157, Tehran, Iran, February 2013. IEEE. DOI 10.1109/ICRoM.2013.6510097.

[187] Sayidmarie O.K., Tokhi M.O., Almeshal A.M., and Agouri S.A. Design and real-time implementation of a fuzzy logic control system for a two-wheeled robot. In *17th International Conference on Methods and Models in Automation and Robotics (MMAR)*, pages 569–572, Miedzyzdroje, Poland, August 2012. IEEE. DOI 10.1109/MMAR.2012.6347823.

[188] Xu J.-X., Guo Z.-Q., and Lee T.H. Design and implementation of a Takagi–Sugeno-type fuzzy logic controller on a two-wheeled mobile robot. *IEEE Transactions on Industrial Electronics*, 60:5717–5728, 2013. DOI 10.1109/TIE.2012.2230600.

[189] Li Z., Zhang Y., and Yang Y. Support vector machine optimal control for mobile wheeled inverted pendulums with unmodelled dynamics. *Neurocomputing*, 73:2773–2782, 2010. DOI 10.1016/j.neucom.2010.04.009.

[190] Raffo G.V., Ortega M.G., Madero V., and Rubio F.R. Two-wheeled self-balanced pendulum workspace improvement via underactuated robust

nonlinear control. *Control Engineering Practice*, 44:231–242, 2015. DOI 10.1016/j.conengprac.2015.07.009.

[191]   Ruan X. and Chen J. H_∞ robust control of self-balancing two-wheeled robot. In *Proceedings of the 8th World Congress on Intelligent Control and Automation*, pages 6524–6527, Jinan, China, July 2010. IEEE. DOI 10.1109/WCICA.2010.5554171.

[192]   Shimada A. and Hatakeyama N. Movement control of two-wheeled inverted pendulum robots considering robustness. In *The Society of Instrument and Control Engineers (SICE) Annual Conference 2008*, pages 3361–3366, Tokyo, Japan, August 2008. IEEE. DOI 10.1109/SICE.2008.4655245.

[193]   Dinale A., Hirata K., Zoppi M., and Murakami T. Parameter design of disturbance observer for a robust control of two-wheeled wheelchair system. *Journal of Intelligent and Robotic Systems*, 77:135–148, 2015. DOI 10.1007/s10846-014-0142-6.

[194]   Wu J., Liang Y., and Wang Z. A robust control method of two-wheeled self-balancing robot. In *6th International Forum on Strategic Technology (IFOST)*, pages 1031–1035, Harbin, Heilongjiang, China, August 2011. IEEE. DOI 10.1109/IFOST.2011.6021196.

[195]   Yau H.-T., Wang C.-C., Pai N.-S., and Jang M.-J. Robust control method applied in self-balancing two-wheeled robot. In *Second International Symposium on Knowledge Acquisition and Modeling, KAM '09*, pages 268–271, Wuhan, China, 2009. IEEE. DOI 10.1109/KAM.2009.234.

[196]   Horn R.A. and Johnson C.R. *Matrix Analysis*. Cambridge University Press, New York, NY, 2nd edition, 2013. ISBN 978-0-521-83940-2.

[197]   Bernstein D.S., editor. *Matrix Mathematics: Theory, Facts and Formulas*. Princeton University Press, Princeton, NJ, 2009. ISBN 978-0-691-13287-7.

[198]   Laub A.J. *Matrix Analysis for Scientists & Engineers*. Society for Industrial and Applied Mathematics, Philadelphia, PA, 2005. ISBN 0-89871-576-8.

[199]   Antsaklis P.J. and Michel A.N. *Linear Systems*. Birkhäuser, Boston, 2006. ISBN 978-0-8176-4434-5.

[200]   Fairman F.D. *Linear Control Theory: The State Space Approach*. John Wiley & Sons Ltd., Chichester, England, 1998. ISBN 0-471-97489-7.

[201]   Kailath T., editor. *Linear Systems*. Prentice-Hall International, Inc., Englewood Cliffs, NJ, 1980. ISBN 978-0135369616.

[202]   Wonham W.M. *Linear Multivariable Control: A Geometric Approach*. Springer-Verlag, New York, NY, 3rd edition, 1985. ISBN 0-387-96071-6.

[203]   Papoulis A. *Probability, Random Variables and Stochastic Processes*. McGraw Hill, Inc., New York, NY, 3rd edition, 1991. ISBN 0-07-048477-5.

[204]   Wills A., Ninness B., and Gibson S. On gradient-based search for multivariable system estimates. *IFAC Proceedings Volumes*, 38:832–837, 2005. DOI 10.3182/20050703-6-CZ-1902.00140.

[205]   The MathWorks, Inc., Natick, MA. *Optimization Toolbox User's Guide*, 2017. [Online]. Available at: http://www.mathworks.com/help/pdf_doc/optim/optim_tb.pdf.

[206] Overschee P. and De Moor B. *Subspace Identification of Linear Systems: Theory, Implementation, Applications.* Kluwer Academic Publishers, Boston, MA, 1996. ISBN 978-1-4613-8061-0, DOI 10.1007/978-1-4613-0465-4.

[207] Verhaegen M. and Verdult V., editors. *Filtering and System Identification: A Least Squares Approach.* Cambridge University Press, New York, NY, 2007. ISBN 978-0-521-87512-7.

[208] Isermann R. and Münchhof M. *Identification of Dynamic Systems: An Introduction with Applications.* Springer, Heidelberg, 2011. ISBN 978-3-540-78878-2, DOI 10.1007/978-3-540-78879-9.

[209] Texas Instruments. *TMS320x2833x, 2823x DSC Serial Peripheral Interface (SPI) Reference Guide, Rev. A,* June 2009. [Online]. Available at: http://www.ti.com/product/TMS320F28335/ technicaldocuments.

[210] Texas Instruments. *TMS320x2833x, 2823x Serial Communications Interface (SCI) Reference Guide, Rev. A,* July 2009. [Online]. Available at: http://www.ti.com/product/TMS320F28335/ technicaldocuments.

[211] Texas Instruments. *TMS320x2833x, 2823x Analog-to-Digital Converter (ADC) Module Reference Guide, Rev. A,* October 2007. [Online]. Available at: http://www.ti.com/product/TMS320F28335/ technicaldocuments.

[212] Texas Instruments. *Programming TMS320x28xx and 28xxx Peripherals in C/C++ Application Report, Rev. D,* January 2013. [Online]. Available at: http://www.ti.com/product/TMS320F28335/ technicaldocuments.

[213] The MathWorks, Inc., Natick, MA. *Simulink Coder Target Language Compiler,* 2017. [Online]. Available at: http://www.mathworks.com/products/help/pdf_doc/rtw/rtw_tlc.pdf.

[214] Texas Instruments. *TMS320x280x, 2801x, 2804x Enhanced Capture (eCAP) Module,* September 2007. [Online]. Available at: http://www.ti.com/lit/ug/spru807b/spru807b.pdf.

Index